Commercializing Biobased Products
Opportunities, Challenges, Benefits, and Risks

RSC Green Chemistry

Titles in the Series:

1: The Future of Glycerol: New Uses of a Versatile Raw Material
2: Alternative Solvents for Green Chemistry
3: Eco-Friendly Synthesis of Fine Chemicals
4: Sustainable Solutions for Modern Economies
5: Chemical Reactions and Processes under Flow Conditions
6: Radical Reactions in Aqueous Media
7: Aqueous Microwave Chemistry
8: The Future of Glycerol: 2nd Edition
9: Transportation Biofuels: Novel Pathways for the Production of Ethanol, Biogas and Biodiesel
10: Alternatives to Conventional Food Processing
11: Green Trends in Insect Control
12: A Handbook of Applied Biopolymer Technology: Synthesis, Degradation and Applications
13: Challenges in Green Analytical Chemistry
14: Advanced Oil Crop Biorefineries
15: Enantioselective Homogeneous Supported Catalysis
16: Natural Polymers Volume 1: Composites
17: Natural Polymers Volume 2: Nanocomposites
18: Integrated Forest Biorefineries
19: Sustainable Preparation of Metal Nanoparticles: Methods and Applications
20: Alternative Solvents for Green Chemistry: 2nd Edition
21: Natural Product Extraction: Principles and Applications
22: Element Recovery and Sustainability
23: Green Materials for Sustainable Water Remediation and Treatment
24: The Economic Utilisation of Food Co-Products
25: Biomass for Sustainable Applications: Pollution Remediation and Energy
26: From C–H to C–C Bonds: Cross-Dehydrogenative-Coupling
27: Renewable Resources for Biorefineries
28: Transition Metal Catalysis in Aerobic Alcohol Oxidation
29: Green Materials from Plant Oils
30: Polyhydroxyalkanoates (PHAs) Based Blends, Composites and Nanocomposites

31: Ball Milling Towards Green Synthesis: Applications, Projects, Challenges
32: Porous Carbon Materials from Sustainable Precursors
33: Heterogeneous Catalysis for Today's Challenges: Synthesis, Characterization and Applications
34: Chemical Biotechnology and Bioengineering
35: Microwave-Assisted Polymerization
36: Ionic Liquids in the Biorefinery Concept: Challenges and Perspectives
37: Starch-based Blends, Composites and Nanocomposites
38: Sustainable Catalysis: With Non-endangered Metals, Part 1
39: Sustainable Catalysis: With Non-endangered Metals, Part 2
40: Sustainable Catalysis: Without Metals or Other Endangered Elements, Part 1
41: Sustainable Catalysis: Without Metals or Other Endangered Elements, Part 2
42: Green Photo-active Nanomaterials: Sustainable Energy and Environmental Remediation
43: Commercializing Biobased Products: Opportunities, Challenges, Benefits, and Risks

How to obtain future titles on publication:
A standing order plan is available for this series. A standing order will bring delivery of each new volume immediately on publication.

For further information please contact:
Book Sales Department, Royal Society of Chemistry, Thomas Graham House, Science Park, Milton Road, Cambridge, CB4 0WF, UK
Telephone: +44 (0)1223 420066, Fax: +44 (0)1223 420247
Email: booksales@rsc.org
Visit our website at www.rsc.org/books

Commercializing Biobased Products

Opportunities, Challenges, Benefits, and Risks

Edited by

Seth W. Snyder
McCormick School of Engineering, Northwestern University, Evanston IL, USA
Email: seth.snyder@northwestern.edu

THE QUEEN'S AWARDS
FOR ENTERPRISE:
INTERNATIONAL TRADE
2013

RSC Green Chemistry No. 43

Print ISBN: 978-1-78262-039-6
PDF eISBN: 978-1-78262-244-4
ISSN: 1757-7039

A catalogue record for this book is available from the British Library

Published by The Royal Society of Chemistry,
Thomas Graham House, Science Park, Milton Road,
Cambridge CB4 0WF, UK

Registered Charity Number 207890

For further information see our web site at www.rsc.org

Printed in the United Kingdom by CPI Group (UK) Ltd, Croydon, CR0 4YY, UK

Preface

Biobased products have been the next "big thing" for about two decades. Products have higher profit margins than biofuels. The market volumes of products are more amenable to biobased feedstocks and biobased products. *Commercializing Biobased Products: Opportunities, Challenges, Benefits, and Risks* is a comprehensive review of the state of the industry. The book covers the process from feedstock sustainability to conversions and separations. In addition, the book highlights several emerging biobased products. Going beyond other volumes on the subject, *Commercializing Biobased Products* evaluates sustainability factors, supply chain factors, policy, and economics. Beyond a review of the technology, this volume exposes the reader to the full spectrum of opportunities, challenges, benefits, and risks of a robust biobased product market. The chapter contributors come from the US, Europe, and Asia and represent academia, government, national laboratories, and industry. The editor has a joint appointment in academia and a national laboratory and has worked extensively with government and industry.

Seth W. Snyder
Department of Chemical and Biological Engineering
McCormick School of Engineering
Northwestern University

RSC Green Chemistry No. 43
Commercializing Biobased Products: Opportunities, Challenges, Benefits, and Risks
Edited by Seth W. Snyder

Published by the Royal Society of Chemistry, www.rsc.org

Contents

Chapter 1 An Introduction to Commercializing Biobased Products: Opportunities, Challenges, Benefits, and Risks 1
Seth W. Snyder

Chapter 2 The Changing Landscape: A History and Evolution of Bio-based Products 8
Gene R. Petersen and Nichole D. Fitzgerald

2.1 Introduction and Background 8
2.2 Early Stages: Chemurgy (1900–1940) 12
2.3 A Changing World (1941–1979) 12
2.4 Resource Concerns (1973–2000) 13
2.5 Balancing Acts (2000–2020) 15
2.6 Examples of the Bio-based Product Landscape 19
2.7 Conclusions 23
Acknowledgements 23
References 23

Chapter 3 Bioenergy Crops: Delivering More Than Energy 25
M. Cristina Negri and Herbert Ssegane

3.1 The Context 25
3.2 Bioenergy Crops as Providers of Energy, Environmental and Ecosystem Services 26
3.2.1 Soil Physical Properties 28
3.2.2 Soil Carbon 29

RSC Green Chemistry No. 43
Commercializing Biobased Products: Opportunities, Challenges, Benefits, and Risks
Edited by Seth W. Snyder

Published by the Royal Society of Chemistry, www.rsc.org

3.2.3 Rooting Habit, Water Capture and Drought Avoidance 29
3.2.4 Water Quality 31
3.2.5 Nutrient Management 31
3.2.6 Water Use 32
3.2.7 Adaptation to Flooding and Drought 33
3.2.8 Biodiversity 34
3.2.9 Pest Suppression 36
3.3 Designing for Yields and Ecosystem Services 36
3.4 The Broader Picture on Implementation 38
References 40

Chapter 4 Butanol Production by Fermentation: Efficient Bioreactors 48
Adriano P. Mariano, Thaddeus C. Ezeji and Nasib Qureshi

4.1 Introduction 48
4.2 Traditional Technology: Batch Fermentation 50
4.3 Continuous Systems 51
4.3.1 Free Cell Continuous Bioreactors 51
4.3.2 Immobilized Cell Continuous Reactors 53
4.3.3 Cell Recycle Continuous Reactors 59
4.3.4 Continuous Bioreactors and Simultaneous Product Recovery 61
4.4 Perspectives and Concluding Remarks 65
References 65

Chapter 5 Catalysis's Role in Bioproducts Update 71
Kim Magrini-Bair, Derek R. Vardon and Gregg T. Beckham

5.1 Introduction 71
5.2 Catalyst Considerations 72
5.3 DOE's Top Value Added Chemicals from Biomass Revisited 73
5.4 Process Option for Biomass Conversion to Bioproducts 76
5.5 Biomass to Products through Deconstructed Molecules 77
5.5.1 Biomass-derived Syngas Upgrading 77
5.5.2 Biomass Pyrolysis Products Upgrading 77
5.6 Biomass to Products through Platform Chemicals 79
5.6.1 Succinic Acid 79
5.6.2 Ethanol 80
5.6.3 2,5-FurandicarBoxylic Acid 81

5.6.4 3-Hydroxypropanoic Acid 81
5.6.5 Glycerol 82
5.6.6 Sorbitol 83
5.6.7 Levulinic Acid 84
5.6.8 Itaconic Acid 84
5.6.9 3-Hydroxybutyrolactone 85
5.6.10 Glutamic Acid 85
5.6.11 Glucaric Acid 85
5.6.12 Aspartic Acid 85
5.7 Biomass to Products through One-pot Reactions 86
5.8 Conclusions 86
References 87

Chapter 6 Separations Technologies for Biobased Product Formation—Opportunities and Challenges 92
Bhanendra Singh, Anju Kumari and Saurav Datta

6.1 Introduction 92
6.2 Fundamentals of Separations Technologies 94
6.2.1 Adsorption 94
6.2.2 Crystallization 98
6.2.3 Liquid–Liquid (L–L) Extraction 100
6.2.4 Membrane Separations Technologies 103
6.3 Application of Separations Technologies for Biobased Product Recovery 108
6.3.1 Algal Biomass-derived Valuable Products 110
6.3.2 Organic Acids 116
6.3.3 Furan Derivatives 121
6.3.4 Sugar Alcohols 122
6.4 Summary 126
Acknowledgements 127
References 127

Chapter 7 Lignin as Feedstock for Fibers and Chemicals 132
Steven W. Peretti, Ryan Barton and Regis Teixeira Mendonca

7.1 Introduction 132
7.2 Lignin Fundamentals 134
7.2.1 Lignin Structure and Chemistry 134
7.2.2 Lignin from Industrial Processes 138
7.3 Lignin Depolymerization 141
7.3.1 Catalytic Depolymerization 143
7.3.2 Biological Depolymerization 145

7.4 Functional Materials from Lignin 148
7.4.1 Adhesives 148
7.4.2 Copolymers or Polymer Additives 151
7.4.3 Foams and Gels 152
7.4.4 Carbon Fibers 153
7.5 Final Thoughts 155
References 155

Chapter 8 Update on Research and Development of Microbial Oils 166
Yanna Liang

8.1 Introduction 166
8.2 Identifying Low- or Zero-value Carbon Sources 167
8.2.1 Carbon Sources from Industry Waste Streams 168
8.2.2 Carbon Sources from Agricultural Waste Streams 172
8.3 Maximizing Oil Productivity through Biochemical Approaches 175
8.4 Maximizing Oil Productivity through Molecular Biology Techniques 178
8.4.1 Increasing Production of Fatty Acids through a Systems Biology Approach 178
8.4.2 Increasing Production of Microbial Oils through Engineering Oleaginous Species 180
8.5 Concluding Remarks 183
References 184

Chapter 9 Bioprocessing of Cost-competitive Biobased Organic Acids 190
Yupo J. Lin, Jamie A. Hestekin, Michael P. Henry and Norman Sather

9.1 Introduction 190
9.2 Current Technology for Organic Acid Production with Bioconversion Processes 191
9.3 Separative Bioreactors—Designs for Integrated Bioprocessing 192
9.3.1 The Separative Bioreactor 193
9.3.2 Innovative Electrodeionization Technology to Capture Organic Acids 195
9.4 Demonstration of Separative Bioreactor Performance for Organic Acid Production 197
9.4.1 Enzyme Separative Bioreactor for the Production of Gluconic Acid and Sorbitol 197

9.4.2 Integrated Fermentation Separative Bioreactor 201
9.4.3 Economics of Organic Acid Separations by EDI-SB 210
9.5 Summary 212
References 213

Chapter 10 CO_2 Conversion to Chemicals with Emphasis on using Renewable Energy/Resources to Drive the Conversion 215
Rich Masel, Zengcai Liu, Di Zhao, Qingmei Chen, Dale Lutz and Laura Nereng

10.1 Introduction 215
10.2 Reacting CO_2 with Fossil Fuels 216
10.2.1 Chemical Production from CO_2 216
10.3 Reducing CO_2 using Renewable Energy/Resources to Drive the Conversion 221
10.3.1 Producing H_2 using Renewable Energy, Reacting H_2 with CO_2 221
10.3.2 Electrochemically Reducing CO_2 to Chemicals and Fuels 222
10.4 Concentrated Solar for CO_2 Conversion 232
10.5 Summary 233
Conflict of Interest Statement 233
Acknowledgements 233
References 233

Chapter 11 Methodological Considerations, Drivers and Trends in the Life Cycle Analysis of Bioproducts 258
Jennifer B. Dunn, Felix K. Adom, Norman F. Sather and Jeongwoo Han

11.1 Introduction 258
11.1.1 Life Cycle Analysis of Bioproducts 259
11.1.2 Bioproduct LCA Results in the Literature 261
11.1.3 Feedstock Choice 263
11.1.4 Study Background and Motivation 264
11.2 Methodology 267
11.3 Results 274
11.3.1 Fossil Fuel Consumption and GHG Emissions of Conventional and Biomass-derived Compounds 274

11.3.2 Influence of End-of-life Assumptions on GHG Emissions of Biomass-derived Polyethylene 276
11.3.3 Cradle-to-gate Water Consumption of Bioproducts 278
11.4 Conclusions 279
References 280

Chapter 12 Design and Planning of Sustainable Supply Chains for Biobased Products **285**
JaeSuk Park, Dajun Yue and Fengqi You

12.1 Introduction 285
12.2 Biomass-to-chemical Pathways 286
12.3 Problem Statement 286
12.4 Model Formulation 288
12.4.1 Constraints 289
12.4.2 Costs, Revenues, and Economic Objective 291
12.4.3 GHG Emissions and Environmental Objective 293
12.5 Solution Algorithm 294
12.6 Case Study 294
12.6.1 Input Data 294
12.6.2 Results and Discussion 297
12.7 Conclusion 299
Nomenclature 300
References 302

Chapter 13 US Government Bioproducts Policy "Watch What We Do, Not What We Say" **304**
Robert E. Kozak

13.1 US Bioproducts Policy: Words, but No Deeds 304
13.2 Why Is There a Lack of US Bioproduct Policy? 307
13.2.1 In Washington, Size Matters 307
13.2.2 US Biochemical Production: A New, Small Industry 308
13.2.3 Accidental Policy in the United States 309
13.3 Could a Similar Approach Be Used for Bioproducts? 311
13.3.1 Cost Effectiveness of Using Bioproducts to Reduce GHG Impacts 311

13.3.2 A Competing GHG Reduction Approach: Using E30 (30% Ethanol) to Meet US Fuel Economy Standards 312
13.4 Possible Strategies to Develop a US Bioproducts Policy Framework 313
13.5 Conclusion: "Watching What They Do, Not What They Say" 313
References 314

Chapter 14 Study on Investment Climate in Bio-based Industries in the Netherlands 315
Lara Dammer and Michael Carus

14.1 Introduction and Executive Summary 315
14.2 Objective and Methodology 316
14.3 Investment Climate and Barriers for Investment 317
14.3.1 Definitions and Standards 319
14.3.2 Knowledge and Education 320
14.3.3 Infrastructure 321
14.3.4 Public Procurement 321
14.3.5 Public Funding Structures 322
14.3.6 Tax Policy 323
14.3.7 Other 324
14.4 Strengths and Weaknesses of the Netherlands as a Location for Bio-based Economy 324
14.5 Conclusions and Recommendations 327
14.5.1 Level Playing Field—the Competition Triangle 329
14.5.2 Sustainability and Incentives—Two Sides of One Coin 329
14.5.3 Recommended Measures 331
References 334

Chapter 15 A Monte Carlo-based Methodology for Valuing Refineries Producing Aviation Biofuel 336
Damian Blazy, Matthew N. Pearlson, Bruno Miller and Rebekah E. Bartlett

15.1 Introduction 336
15.2 Background and Review of Past Analyses 337
15.3 Limitations of DCFROR Models in Incorporating Uncertainty 338

15.4 Key Methodology, Data and Assumptions Used in the Monte Carlo Model 341
15.4.1 Monte Carlo Methodology Overview 341
15.4.2 Step 1: Constructing Uncertainty Profiles 341
15.4.3 Step 2: Constructing the Correlation Matrix 343
15.4.4 Step 3: Developing Assumptions 343
15.5 Steps 4–7: Monte Carlo Simulation—Results 346
15.6 Implications for Government Policy 347
References 351

Chapter 16 A Path Forward: Investment Cooperation between the United States and China in a Bioeconomy 352
Seth W. Snyder

16.1 Introduction 352
16.2 Why Invest in the Bioeconomy? 353
16.3 A Local Supply Chain 353
16.3.1 Feedstocks for the Bioeconomy 354
16.3.2 Biobased Feedstocks *vs.* Fossil Feedstocks 354
16.3.3 Processing 355
16.3.4 The Advantages of Small 356
16.4 Public Policy and the Bioeconomy 356
16.4.1 Aviation Biofuels as an Example of the Challenge 357
16.4.2 Biogas is a Bright Spot 357
16.5 Factoring in Risk 358
16.6 The Largest Challenge is Project Finance 358
16.6.1 Project Finance So Far 359
16.7 A Path Forward 360
16.7.1 Capital from Chinese Investors 360
16.8 The Bioeconomy in China 361
16.8.1 The Needs in China Are Even Greater 361
16.8.2 Progress in China 361
16.9 Deployment in the US and China 362
16.9.1 Benefits of Cooperation in the Bioeconomy 362
16.9.2 Impacts on the Environment and Greenhouse Gas Emissions 362
16.9.3 Agriculture 363

16.9.4 Challenges to Cooperation in the Bioeconomy 363
16.9.5 Making the Case for Cooperation on the Bioeconomy 363
16.9.6 Programs to Foster Cooperation Must Address Intellectual Property Rights 364
16.10 Potential Size of the Bioeconomy 364
16.11 Conclusions 365
Acknowledgements 365

Subject Index **366**

CHAPTER 1

An Introduction to Commercializing Biobased Products: Opportunities, Challenges, Benefits, and Risks

SETH W. SNYDER

Northwestern University, McCormick School of Engineering,
2145 Sheridan Road, Evanston, IL 60208, USA
Email: seth.snyder@northwestern.edu

Biobased products have been the next "big thing" for about two decades. Speak to a scientist or engineer in the field and you will hear a hundred reasons why investments should focus on products rather than fuels. Products have higher profit margins. The market volumes of products are more amenable to biobased feedstocks and biobased products. While biobased feedstocks are close to 50% oxygen, fuels contain very little oxygen. The oxygen content of biobased products such as alcohols, carboxylic acids and esters much more closely match the feedstock, and therefore, are more atom efficient. The editor has engaged in the field since 1999. While we have seen dramatic progress in commercialization since about 2013, production of biobased products are still orders of magnitude less than biofuels. While biofuels have received significant public support including tax credits, blending mandates, and state incentives, biobased products do not have mandates or economic incentives in the US.

RSC Green Chemistry No. 43
Commercializing Biobased Products: Opportunities, Challenges, Benefits, and Risks
Edited by Seth W. Snyder

Published by the Royal Society of Chemistry, www.rsc.org

Gasoline stations prominently display fuel prices at every major intersection reminding consumers (and voters) of the volatility of the energy markets. Ask your neighbor and they could probably tell you the price of fuel to three significant figures. Ask them what they pay for natural gas or electricity and they probably will be able to guess within a factor of three. Ask your neighbor about the embedded costs for the chemicals, materials, plastics, and solvents that are derived from the same fossil energy sources and they probably won't even understand your question. The simplest reason that we developed incentives for biofuels is that consumers are constantly reminded of the price of fuel. The original driver for biofuels incentives was to create a market for agriculture products, but public acceptance grew from constant reminders of fuel prices. Biobased products do not provide immediate and continuous price feedback so they haven't generated public support for policy incentives.

In this book we cover the opportunities, challenges, benefits, and risks of commercializing biobased products. As expected in a science and engineering series, we include chapters that consider feedstocks, conversions, and processing technology, as well as several potential products. There have been several books and journal issues dedicated to biobased products. Most of these publications have used experts to describe integrated processes to produce either specific biobased products or classes of products. Our intention is to expand the breadth of the reader's knowledge by considering supply chains, environmental impact, policy status, economic analysis, and a conjecture for a path forward. Therefore the aim of this book is to provide a comprehensive view of the state of the industry rather than the state of the technology.

In Chapter 2, "The Changing Landscape: a History and Evolution of Biobased Products", Petersen and Fitzgerald highlight the development and history of biobased products. The authors cover the growth of the industry and the range of end markets. The authors describe how changes in priorities along with advances in technology have created cycles of opportunity and deployment. They highlight the rationale and use of these materials and their expectations of growth. Petersen and Fitzgerald have both served as program managers at the Bioenergy Technologies Office (BETO) of the US Department of Energy (DOE). In that role they have been exposed to the emergence of the field and reviewed and monitored a major fraction of the cutting edge ideas and concepts in the bioeconomy. Petersen also led biobased products and bioenergy research programs at the National Renewable Energy Laboratory (NREL). In that role, Petersen was the co-author of the 2004 report "Top Chemicals from Biomass", the most highly cited and impactful study of the field.

In Chapter 3, "Bioenergy Crops: Delivering More than Energy", Negri and Ssegane indicate that much of the impacts of bioenergy and biomaterial cropping depend on how large scale deployment will occur. Designing production systems that purposefully incorporate sustainability objectives or ecosystem services along with the biomass feedstock is possible. The authors report the benefits of perennials over traditional row crops. Perennials such as switchgrass, miscanthus, other perennial grasses, and short rotation woody crops share a deeper root system, a general better ability to thrive on poorer soils, a lower dependence on fertilizer inputs, and at least for some, management options that can be friendly to wildlife. The authors indicate that they can design bioenergy landscapes that balance productivity and environmental performance, are socially acceptable and deliver much more than bioenergy and biobased products. Negri is a leader in sustainable bioenergy landscape design and has implemented phytotechnologies to improve environmental performance across the Midwest and Europe. She works closely with the DOE bioenergy technologies office on achieving sustainable landscape designs.

In Chapter 4, "Butanol Production by Fermentation: Efficient Bioreactors", Mariano, Ezeji, and Qureshi use their work as an example of butanol production to describe microbiology and fermentation technology. Butanol is a valuable solvent and has potential as a biofuel. They describe some of the classic limitations to fermentation where product inhibition limits product yields and concentrations. Low fermentation concentrations increase energy use for product recovery. The authors describe novel bioreactors for butanol fermentation using different advanced fermentation systems such as free cell continuous, immobilized cell continuous and cell recycle continuous membrane reactors, and integrated continuous processes where product can be simultaneously recovered using energy efficient product recovery techniques. Ezeji and Qureshi have collaborated since Ezeji was a graduate student in the laboratory of Dr Hans Blashek of the University of Illinois Urbana-Champaign.

In Chapter 5, "Catalysis's Role in Bioproducts Update", Magrini-Bair, Vardon, and Beckham define the major classes of reactions including dehydration, decarboxylation, decarbonylation, hydrogenolysis, esterification, and ketonization. The goal of most biomass catalysis is to create market value from oxygenated feedstocks and they present a crosscutting review of the progress over the past decade. Magrini-Bair leads a catalysis program at NREL and has collaborated with a range of academic, national laboratory, and industrial researchers.

In Chapter 6, "Separations Technologies for Biobased Product Formation—Opportunities and Challenges", Singh, Kumari, and Datta indicate that separations technologies can contribute 50% to overall production costs. The authors identify and define key separations technologies platforms for producing biobased products. They focus on broad classes of oxygenated species. Within each separations platform they define the driving forces, the target species, the specific benefits and limitations. Before becoming a Professor in India, Datta developed his expertise in separations in the US where he received his PhD under Professor DB Bhattacharya (University of Kentucky) and as served as a postdoctoral fellow with the editor at Argonne National Laboratory.

In Chapter 7, "Lignin as Feedstock for Fibers and Chemicals", Peretti, Barton, and Teixeira Mendonca indicate that because of its resistance to degradation, lignin has primarily been used as an energy resource in biorefineries. Lignin has about 60% excess energy and is available as a valuable carbon source. The authors focus on lignin structures that result from pretreatment technologies, and efforts to valorize that lignin through material property modification or depolymerization. The authors indicate that only 2% of available lignin is currently used in commercial products. Successful treatment technologies and broader classes of products are required for beneficial use of this resource. Professor Peretti is a leader in North Carolina State University's forest products research program. Forest products are a primary source of lignin.

In Chapter 8, "Update on Research and Development of Microbial Oils", Liang discusses the opportunity for large-scale production of microbial oils. The author indicates that there are a broad range of oleaginous microorganisms including yeasts, microalgae, fungi and bacteria capable of accumulating intracellular oils or lipids. Two high-value products—arachidonic acid and docosahexanoic acid—have reached commercial scale. The author highlights key advances made in finding low cost and renewable materials for microbial cultivation, identifying ways to improve oil/lipid productivity either from a biochemical engineering perspective or through systems biology approaches. The author suggests applications beyond low-value biodiesel. Professor Liang, a microbiologist by training, has developed a research program in biobased products and biofuels that spans the pathway from strain development through efficient product recovery.

In Chapter 9, "Bioprocessing of Cost-competitive Biobased Organic Acids", Lin, Hestekin, Henry, and Sather indicate that the high cost of product recovery is a major challenge to the production of organic acids. The authors highlight a bioprocessing method that integrates upstream bioconversion and downstream product separation into a continuous process. The bioprocess uses a membrane technology, resin wafer electrodeionization (RW-EDI), which results in a simpler bioprocess train with fewer unit operations, better pH control, reduced product inhibition from the organic acids, higher organic acid product concentrations and enhanced bioconversion rates and yields. They discuss the design and operation of the system and performance at the bench scale and pilot scale. The authors review the technical and economic viability for commercial production using the platform. The authors are long-term collaborators of the editor. Lin is an electrochemist with a long track record developing novel approaches for electrochemically-driven processes. Professor Hestekin, the editor's first postdoctoral fellow, has developed a crosscutting research program from strain selection through fermentation and product recovery.

In Chapter 10, "CO_2 Conversion to Chemicals with Emphasis on Using Renewable Energy/Resources to Drive the Conversion" Masel, Liu, Zhao, Chen, Lutz, and Nereng discuss the growing field of CO_2 utilization. The authors summarize two approaches: first, one in which hydrocarbons react with CO_2 to produce useful products; and second, one in which electricity from renewable energy is used to convert CO_2 into useful products. There is still more fundamental and practical work to be done, but it looks likely that both approaches will produce viable commercial processes within the next few years. While chemicals from CO_2 appear to outlie the core subject of this book, it is important to recognize that use of CO_2 achieves the primary goals of biobased products, decreases dependence on fossil carbon sources and decreases life-cycle greenhouse gas emissions. However, CO_2 utilization doesn't enable growth of markets from agricultural feedstocks, the other goal of the biobased supply chain. Masel is a well-established expert in CO_2 utilization. Before founding Dioxide Materials, he was a professor at the University of Illinois Urbana-Champaign, where some of his former PhD students went on to become members of the National Academy of Engineering.

In Chapter 11, "Methodological Considerations, Drivers and Trends in the Life Cycle Analysis of Bioproducts", Dunn, Adom, Sather, and Han assess the relative environmental performance of bioproducts compared to their conventional (fossil-based) counterparts. The authors identify opportunities to improve bioproducts' environmental impacts. The authors examine the application of life cycle analysis (LCA) including treatment of bioproduct feedstocks, conversion process analysis, and end-of-life assumptions. They highlight the significance of bioproduct end-of-life treatment. A critical issue in bioproduct LCA is treatment of co-products. The authors present results for life-cycle greenhouse gas (GHG) emissions and fossil energy consumption (FEC) for eight biobased products. With all biobased–fossil product comparisons, the biobased products exhibited lower life-cycle GHG emissions and FEC. The authors also consider life-cycle water consumption because of concerns about water consumption used to grow biomass and compare terrestrial feedstocks to algae. Dunn is a prime leader of bioenergy life-cycle analysis for the DOE Bioenergy Technologies Office. The team's work is a primary driver for assessing the impact of new feedstocks, conversions, and products in the bioeconomy.

In Chapter 12, "Design and Planning of Sustainable Supply Chains for Biobased Products", Park, Yue and You present a mathematical framework to model and optimize a supply chain for industrial chemical products derived from biomass. The authors present a multi-objective, multi-period mixed-integer linear programming (MILP) model that takes various aspects of the supply chain into account. They highlight the significant decisions required for creating a successful supply chain and the impacts of location and seasonality. The model is presented to co-optimize economic and environmental objectives and includes a case study to verify its viable functionality. Professor You is a highly cited academic who has achieved an h-index of 31 within his first four years as a faculty member.

In Chapter 13, "US Government Bioproducts Policy 'Watch What We Do, Not What We Say'", Kozak compares past speeches, commissioned reports and press releases to applicable legislation. The author highlights the lack of legislative, regulatory, or spending frameworks necessary to implement the stated bioproduct policy. The author indicates that legislation allows government agencies to avoid purchasing bioproducts as part of their normal course of business in contrast to the goal of becoming a first adopter. The author also examines how the divestment of downstream refining by integrated petroleum majors prevented the US biobased products industry from creating an effective bioproducts policy on their own. Kozak is the co-founder of Advanced Biofuels USA, a 501(c)(3) non-profit organization focused on increasing the professional and layman's knowledge of the state and potential impact of the bioeconomy.

In Chapter 14, "Study on Investment Climate in Bio-based Industries in the Netherlands", Dammer and Carus present a study of the barriers faced by small companies active in biobased economy when they want to acquire investment for their businesses. The study uses interviews and literature reviews and is focused exclusively on biobased chemicals and materials, not on food, feed or energy produced from biomass. The objective was to assess the investment climate for biobased industries in the Netherlands in comparison to other countries. In comparison to Chapter 13, the climate in Europe is quite distinct from the US and is partially a driver for the essay in Chapter 16. Carus is the Director of nova-Institute for Ecology and Innovation, the primary research institute highlighting the state of the European bioeconomy.

In Chapter 15, "A Monte Carlo-based Methodology for Valuing Refineries Producing Aviation Biofuel", Blazy, Pearlson, Miller and Bartlett present an analysis relating uncertainty in market and policy conditions to their impact on the long-term economics of biorefineries. While some readers may not consider aviation biofuels as biobased products, the competition between land-based transportation fuels and aviation biofuels for feedstock and biorefinery capacity is very complementary to supply chain challenges in biobased products. The authors describe a methodology for analysing capital budgeting decisions and valuation under uncertainty for such investments and indicate that the methodology can be used as a decision making tool for investment decision timing. The authors employ a commercially available technology for producing aviation-grade biofuel and renewable diesel for the assessment with well-understood capital and operating costs. The authors evaluate how the distribution of net present values (NPVs) using a discounted cash flow model was used to determine the profitability of such a project over its economic life. They include price uncertainty and government mandates and report that price support policies are necessary to reduce the uncertainty of profitability to commercially acceptable levels. Blazy initiated this analysis at Massachusetts Institute of Technology based on analytical tools developed by Pearlson. The analysis was completed as part of a study by the Midwest Aviation Sustainable Biofuels Initiative (MASBI) in 2013. Blazy collaborated closely with the editor on this project and it served as a primary driver for Chapter 16.

In Chapter 16, "A Path Forward: Investment Cooperation between the United States and China in a Bioeconomy", Snyder proposes that the US collaborates with China to build a binational biobased products industry.

As highlighted throughout this book, the US has substantial technical expertise in the field but lacks the economic and policy drivers to foster the industry. Society needs in China change the risk profile for investment and could become the tool to promote partnering on joint projects in which the countries develop parallel supply chains without directly competing for product markets. Snyder served as Argonne National Laboratory's bioenergy technology leader for a decade where he was engaged in production technologies across the bioeconomy supply chain. He developed this idea by synthesizing concepts developed by the Henry Paulson Institute on cooperation with China on environmentally sustainable projects with analysis from MASBI on risk barriers to commercialization in the bioeconomy.

In Commercializing Biobased Products we present a comprehensive view of the history, feedstocks, conversion technologies, products, impacts, policy, and economics of the industry. We believe that this broad view will provide a useful tool for the reader to consider the entire process to realize this industry. Many researchers have dedicated their careers to growing the industry and have experienced repeated frustrations as promising technologies and products do not cross the "valley of death". Success requires strong technology. Success also requires coherent policy that provides incentive to commercialize technologies and products that offer environmental and economic benefits to society.

CHAPTER 2

The Changing Landscape: A History and Evolution of Bio-based Products

GENE R. PETERSEN[a] AND NICHOLE D. FITZGERALD*[b]

[a] Las Vegas, Nevada, USA; [b] US Department of Energy, Bioenergy Technologies Office, 1000 Independence Ave., Washington, DC 20005, USA
*Email: Nichole.Fitzgerald@ee.doe.gov

2.1 Introduction and Background

The production of chemical products from biomass is experiencing a renaissance. End uses such as automobile manufacturing, housing, and consumer goods are driving the production of chemical intermediates and commodity chemicals from renewable resources. In recent history, bio-derived products are often seen as replacements for petroleum-derived products and a surge of interest in bio-derived products has correlated with oil supply concerns. Current trends suggest that for bio-based chemicals to be successful, they must perform as well or better than existing petroleum-derived chemicals at the same or lower prices. Producing building block chemicals from biomass is not a new concept and interest has waxed and waned over the past century. This chapter will illustrate how the pursuit of bio-derived products has fluctuated in popularity over the past century as a result of technological developments, global politics, and economic factors. A brief history of the use of these materials and what the future holds is provided. Considerations regarding the environmental impact of renewable and regenerable resources cannot be dismissed in light of our more

RSC Green Chemistry No. 43
Commercializing Biobased Products: Opportunities, Challenges, Benefits, and Risks
Edited by Seth W. Snyder

Published by the Royal Society of Chemistry, www.rsc.org

heightened ecological awareness and this aspect of bio-based products is discussed briefly.

In a previous publication, a thorough discussion of various rationales for bio-based chemical targets was presented.[1] Drivers for target selection included where to focus research and development funds, industrial product development, how to allocate government funding and environmental concerns. That report established that determining the outcome for a bio-based chemical project (product development, company portfolio needs, mitigating negative environmental impacts, funding priorities) was necessary and this provided the research, development, and deployment agendas for various institutions and funding agencies. A set of axioms for working with bio-based products is presented in Figure 2.1.

The United States Secretary of Agriculture defines bio-based products as "commercial or industrial products (other than food or feed) composed wholly or in significant part of biological products including renewable agricultural materials (plant, animal, and marine materials) or forestry materials."[2] Bio-based products can take many forms, but for the purpose of this chapter, bio-based products refer to products derived from small organic molecules, such as plastics, additives, and other chemical building blocks that can be obtained from the chemical or biological conversion of biomass. Although wood chips and subsequent products based on the material properties of wood fall into the USDA definition of a bioproduct, they will not be discussed here. This chapter focuses on chemicals that are synthesized from the molecules that are derived from the breakdown of

Type of bio-based product	Upside	Downside
Drop in replacement chemicals (Polyethylene glycol, acetic acid, acrylic, 1,4-butanediol, adipic acid)	• Markets exist (fungible) • Cost structures and growth potential understood • Market risk reduced	• Harsh cost competition (competes on price alone) • Competes against depreciated capital • Limited market differentiation
Functionally equivalent replacement (Phenolic resins, unsaturated polyesters, isosorbide, 1,3-propane diol (Sorona™), poly-hydroxybutyrates, high fructose corn syrup)	• New market opportunities may exist • Competitive petrochemical routes undeveloped • Cost tolerance more likely • Inherent properties of biomass utilized	• Cost competition remains real • Market performance potential is unclear • High capital risks • Long commercialization times likely
New functionality or properties (Engineered wood products, 5-hydroxymethyl furans, polylactic acid, levulinic acid and derivatives, dehydromucic acid)	• New market opportunities may exist • Competitive petrochemical analogs non-existent • Cost often not an issue initially • Inherent properties of biomass utilized	• Capital risks often high • Long commercialization times likely • Market development almost always required and not always clearly defined

Figure 2.1 Axioms for bio-based products (modified reproduction of a similar table published in ref. 2 by Wiley).

cellulose, hemicellulose, or lignin, either through biological or chemical processing to produce sugars or a bio-oil, or from the upgrading of biologically-derived syngas.

The most abundant starting materials for making bio-based chemicals are carbohydrates. As a result, most of the research and development challenges over the past few decades have focused on developing processes to liberate usable carbohydrate building blocks from crude biomass. There are a number of drivers to selecting potential bio-based products that have been outlined by Petersen, Bozell and White.[1] In addition, several papers have been written on potential bio-based products that could be made by conversion of carbohydrates and lignin.[3–6] As technology advances are made in liberating useful chemical building blocks from biomass, new bio-based products can be targeted.

Bio-based products can be roughly organized into three categories that describe how they relate to petroleum-derived products. First, a bio-based product that has identical structure and properties to a petroleum-based product is referred to as a direct replacement. There are many benefits of direct replacements, particularly that their properties have already been characterized and well-accepted by manufacturers and consumers. On the other hand, because petroleum feedstocks and biomass feedstocks have such different chemical compositions, particularly with respect to hydrogen and oxygen content, it does not always make sense to use a bio-based product for direct replacement of a petroleum-based product. In many cases, to resemble existing petroleum-derived chemicals, oxygen must be removed during the production of bio-based chemicals, resulting in lower mass efficiencies and loss of initial feedstock value. Often, a functional replacement is the best use of biomass to create a bio-based product. A functional replacement has very similar properties to a petroleum-based product, but the structure is different. The structure usually reflects the highly oxygenated nature of biomass starting materials. An example of a functional replacement is isosorbide as a replacement for bisphenol A (BPA). Although it is unclear what the properties of isosorbide are that make it a good replacement for BPA, several companies are pursuing corn- and cellulosic-derived isosorbide.[7] The third class of bio-based products is novel materials: these compounds do not resemble petroleum-based products in either structure or function. These novel materials represent an exciting future for the materials industry as new properties and capabilities are introduced. Polylactic acid (PLA) is one well-known example of a novel material made from biomass. PLA is now part of a large product line that includes substitutions for styrene-based food packaging and unique wicking fibers. PLA products are often biodegradable, a significant selling point for end uses where disposal is an important criteria for materials. As can be seen with PLA, bioproducts can do much more than replace traditional materials, they can also fill product gaps.[8]

Although it might seem like bio-based products are relatively new, in reality they have been used by man throughout history. Highly complex purified organic molecules, such as medicines such as taxol, polymers derived from plant materials such as cellophane, and mixtures of simple molecules, *e.g.* fuels, can all be classified as bio-based products. Chemicals extracted from a living organism, known as natural products, have served man as therapeutics, commodities and even weapons. Natural products have been used as traditional medicines and have long been a source of inspiration for the design of pharmaceuticals—almost half of FDA-approved drugs have structures similar or identical to natural products.[9] Ancient and modern civilizations used indigo, a natural dye isolated from a variety of plant sources, to color textiles. Although the technology for cultivation and extraction of indigo was known for thousands of years, the advent of modern synthetic organic chemistry allowed for synthetic indigo to be produced from petroleum more cheaply than ever before. In 1897, 19 000 tons of indigo were isolated from plant sources; by 2002, 17 000 tons of indigo were synthesized from petroleum derived chemicals.[10] As the understanding of organic chemistry improved, the process of isolating naturally occurring products was often replaced by chemical synthesis. Today, chemical synthesis is essential to almost every aspect of human life. Over time, the starting materials for synthesizing products have alternated between biomass sources and petroleum or coal sources, depending on economic, technical, and geopolitical factors.

Figure 2.2 illustrates the evolution of bio-based products since the advent of synthetic organic chemistry. Four phases of this evolution are identified: Chemurgy (1900–1940), the Changing World (1941–1979), Resource Concerns (1973–2000), and Balancing Acts (2000–2020). The historical, political, social, and environmental aspects of these phases are presented as well as

General Characteristic of Phases			
Chemurgy	**Changing World**	**Resource Concerns**	**Balancing Acts**
1900–1940	*1941–1979*	*1973–2000*	*2000–2020*
• Push to use agricultural products • Coal was a major source of chemicals • Petrochemical industry still in growth phase • Material needs met by wood, metals, stone and agricultural fibers	• Shortages due to world conflicts • Emerging petrochemical industry meets many material needs • Agricultural products more focused on food • Significant increases in crop yields • Coal use largely directed to production of electrical power	• Questions arise about indeterminate supplies of resources • Environmental movements more prominent • Desire to be green and return to natural materials • Increased markets for agricultural products pushed • Significant advances in biotech for agriculture • New product development opportunities arise	• Food *versus* fuel debates • Continued push for expanded roles of agriculture and forestry in more commodity markets • New materials from indigenous sources *(nanocrystalline cellulose, acetylated wood, carbon fibers)* • Price of oil fluctuates, sources are from unstable regions • Assessment of environmental impacts stressed

Figure 2.2 The changing landscape—evolution of bio-based products.

how bio-based products relate to other organic and inorganic material resources.

2.2 Early Stages: Chemurgy (1900–1940)

A very good perspective on the early stages of the bio-based product industry is discussed in a paper by Finlay.[11] In the 1920s, a small group of agriculturists, scientists, and industrial leaders pioneered the concept that "chemicals found in farm products could provide industry with needed raw materials". They termed this concept "chemurgy" to reflect the root words chemistry (chemi) and work (ergon). According to Finlay, the chemurgists had three major goals: "to develop new, nonfood uses of existing crops; to develop new farm commodities useful to industry to grow in lieu of surplus commodities; and to find profitable use for various agricultural wastes and residues." Anyone familiar with the modern biofuel and bioproducts industry will recognize these goals as also being current high priorities.

There were many early successes of the chemurgy movement. Early adopters included Henry Ford and Thomas Edison, who, at the Edison Institute in Dearborn, Michigan, researched new industrial uses for farm crops.[11] Their focus on the soybean led to the development of soy-based enamel, lubricants, and plastic car parts. In fact, Ford claimed that every car his company manufactured used a bushel of soybeans.[11] At the same time, alcohol fuels derived from corn and other crops became more widely adopted in many industrialized nations. In 1933, Iowa legislators passed a law that required fuels to include a 10% blend of grain alcohol, foreshadowing modern fuel blending.[11] The use of biomass for everyday products was becoming mainstream.

The chemurgy movement lost favor when crude oil started to be employed for chemicals and fuels.[6] Because crude can be distilled into many fractions, the versatility of this feedstock source initiated a large development in petrochemicals. During this time period, distillation, catalytic processing, and combinations of both were being developed for crude oil with major successes coming in the early 1940s with fluidized bed catalytic cracking. With the advances in chemical processing and chemical engineering, as well as the seemingly endless supply of crude oil, the attention to chemurgy waned. Little effort was undertaken except for isolated cases such as the production of citric acid, specialty grades of ethanol, and various uses of lignin for fuel or lignosulfonates for binder materials.

2.3 A Changing World (1941–1979)

World War II marked the beginning of many significant global changes, even for the fledgling bioproducts industry. At this point, sustainability and conservation of resources was not a primary concern of the chemical and petrochemical industry. Indeed, forests could be replanted, oil discovery was a regular occurrence, and a new industry was being developed in the area of plastics. Coal was generally employed for electrical power generation and only a small percentage of commodity chemicals were derived from that source. The agricultural world was also emerging in its own way. Better plant breeding and increased use of pesticides, often derived from crude oil fractions, put commodity crops like corn on a steep slope of increased productivity. This allowed increased high protein meat production to feed many growing nations. This was also the period of time when Nobel Laureate Norman Borlaug was touting pursuit of technology to improve agricultural yields, known as the "Green Revolution," to deal with potential shortfalls in food due to population growth.[12] Hence, more attention was given to increasing basic food supplies such as wheat and corn and other potential cereal feedstocks, while little attention was paid to deriving significant quantities of chemicals from agricultural feedstocks.

In addition to a new focus on improving agriculture for food production, new challenges arose when commodities were limited during and around World War II. For example, during WWII, sources of isoprenoids, largely for *hevea*-based rubber production, were not very accessible. This led to investigations of both bio-based and petroleum-based alternatives. Plants such as guayule (*Parthenium argentatum*) grown in the American southwest could be used to replace sources of *hevea* from the Far East. Interest in guayule has remained to this day since latex rubber from guayule is hypoallergenic.[13,14] Interest in alternative sources to rubber led to the development of the vulcanization process, allowing durable rubber products to be produced from petrochemical sources.[15] When liquid petroleum-derived fuel became scarce in Germany during this time period, scientists developed the Fisher–Tropsch process for converting gasified coal and natural gas into liquid fuels. Major advances in synthetic chemistry were made to answer the call for new routes to indispensable commodities from a variety of alternative feedstocks. While the focus was not on utilizing biomass as a primary feedstock, the need for alternative strategies for chemical synthesis paved the way for many later developments in bio-based products. The horticulture and agronomy advances of that time were also significant in advancing food production throughout the world as noted in Borlaug's Nobel Prize address.[12]

2.4 Resource Concerns (1973–2000)

From the 1970s to the turn of the century, several disparate factors contributed to a growing interest in bioproducts. Political crises, environmental concerns, biotechnology advances, and conversion technology improvements all led to a renewed interest in bioproducts, with objectives not unlike the goals of the chemurgists in the early 20th century.

In the 70s, several oil supply crises caused prices to skyrocket and many western countries grappled with a highly constrained fuel supply. Environmental fears were realized when the Exxon Valdez oil spill in Alaska occurred in 1989. More scrutiny was given to supply chains that could lead to oil spills or other negative impacts on the environment. These events made the public question the reliability of their source of oil, and the corresponding fuels and chemicals. The public and elected officials wanted a domestic source of energy that was not subject to the same political complications found with imported energy. This sentiment resulted in the Energy Security Act of 1980, which established a biomass energy program. The Energy Policy Act of 1992 instituted tax credits for employing renewable fuels such as ethanol as an oxygenate. These Acts paved the way for significant federal investment in research and development for biofuels and bio-based products. Additionally, due to the Energy Policy Act of 2005, the US Department of Agriculture started to classify renewable material content in products that led to affirmative action plans to procure bio-based materials in the 2000–2010 decade.[16] These early government policies encouraging alternative energy sources inspired many scientists and engineers to come together to discuss technical barriers to renewable fuels. Although professional meetings to address technical barriers to renewable fuels are commonplace today, these meetings established a growing interest in engaging our universities, government agencies and industry in addressing the potential of using home-grown resources rather than importing such resources. For instance, the Symposium for Biotechnology for Fuels and Chemicals and other professional meetings were initiated with the support of industry, academia, and government, and continue to this day. The reemergence of chemurgy-type efforts began again in earnest during this period of resource concerns and has continued into succeeding decades even though as noted earlier, the emphasis has fluctuated in intensity.

At the same time, advances were made in biotechnology beyond typical plant breeding techniques, resulting in significant agricultural yield improvements. For example, corn and soybean varieties resistant to herbicides such as Roundup entered the market (Roundup Ready® crops are a Monsanto product). These technical advances resulted in unprecedented high yields of plant biomass, which in turn became appealing as a potential source for carbon-based liquid fuels and products. Concurrently, significant technical advances were made in converting biomass into useful bio-based chemical feedstocks. The Department of Energy and the United States Department of Agriculture began to support efforts in better conversion technologies. In 1993, the DOE published an initial report on the techno-economic potential of producing bio-based products from agricultural and forest carbohydrates and phenolics (largely lignin).[6]

One early indicator of this progress was the development of lactic acid initially by Battelle Memorial Institute, which was further developed by small start-up companies (NatureWorks) and then by large agricultural commodity companies (Cargill). Using the 1993 report by Bozell and Landucci as the guide, the Department of Energy even had a short-lived Alternative Feedstocks Program that largely focused on developing technologies for conversion of sugars and lignin to commercial products.[6] Five DOE laboratories formed a virtual laboratory with the goal of working with industry in developing the industry of non-fuel bio-based products. Notably, this virtual laboratory consortium developed seminal work regarding production of succinic acid from carbohydrate sugars. Although the work did not become commercially viable at that time, there are now multiple commercial plants in the US, Canada and Europe producing succinic acid at modest commercial scales (see Table 2.3). Organic diacids were highlighted as a near term opportunity by the 1993 report[6] and in fact have become one of the bio-based product successes of the 21st century. The long development times from conception to commercial-scale production highlights the axioms of Figure 2.1.

2.5 Balancing Acts (2000–2020)

The 21st century has seen a convergence of many drivers, including fluctuating oil prices and secure domestic supply chains that have led to broader interest in bio-based products. The large number of publications addressing this area and the growing number of businesses investing in technology to produce bioproducts are evidence of this interest. For example, the trade journal *Biofuels Digest* (www.biofuelsdigest.com) publishes annual lists of up and coming companies involved in this space. Further, authors from the Department of Energy have produced a series of publications that have served as guides for the development of bio-based products. In 2004, the Department of Energy's Pacific Northwest National Laboratory and the National Renewable Energy Laboratory published a report attempting to impute value chain considerations to bio-based products from sugars.[3] This was followed by a similar report related to lignin.[5] This 2004 Top Ten report was succeeded by a follow-up paper in *Green Chemistry* in 2010 evaluating the potential based on advances in technology, highlighting the importance of bio-based products for realizing a biorefinery's strategic energy and economic goals.[4] In 2013, Imhof and Kees van der Waal published a book examining the potential for employing advanced catalytic processes in the production of bio-based products.[17]

As the popularity of bio-based products grows, the public has increasingly scrutinized the ethical, environmental, and economic implications of these products. As in previous decades, there is a continued interest in pursuing "green" and "natural" solutions to consumer product needs. However, the public now has a more refined understanding of environmental impacts and is less likely to support a bio-based product if the resource demands for producing the product are more severe than for a traditional petroleum-based product. Many companies acknowledge that there is no "green" premium for their goods and that they must compete on performance, including environmental metrics. Life cycle analysis is key to addressing these concerns. Currently, the US Department of Energy sponsors life cycle assessments of many products that could be most easily derived from biomass resources (see Chapter 11). Furthermore, the lessons from misguided attempts to replace problematic chemicals with equally bad alternatives, such as the use of methyl *tert*-butyl ether as a replacement for lead in gasoline, has heightened the awareness of regulatory agencies such as the Environmental Protection Agency to carefully evaluate new routes to chemicals from biomass.

The following figure depicts the cognizance one must give to environmental issues. Overend provides a sober perspective on impacts of bio-based products as shown in Figure 2.3. Such analyses must be performed as the bio-based product industry continues to grow.

In addition to the environmental impacts of pursuing bio-based impacts, the public is concerned about the implications of redirecting food resources to fuels and other products. Commonly known as the "food *vs.* fuel" debate, this ethical concern has implications for bio-based products, although the issue is considerably less pronounced because the volumes of resources required for products are substantially lower than what is required for fuels. Nonetheless, it is necessary to keep this balance in mind when developing bio-based products.

Another balancing act is simultaneously supporting the welfare of petrochemical industries and the agricultural and forestry economies. Typically, the agricultural and forestry sectors are viewed only as sources of food and materials. However, the production of chemicals and fuels from these sources is becoming increasingly substantial, and eventually could be competitive with some portions of the petrochemical industry. Both the petrochemical and agricultural industries contribute significantly to the US economy and there is a significant fraction of our labor force involved in these two sectors. For comparison, using United States Bureau of Labor statistics, the number of individuals involved in the chemical manufacturing and petroleum/coal manufacturing is about 900 000. The number of individuals in the agriculture and forestry industries reported by the Department

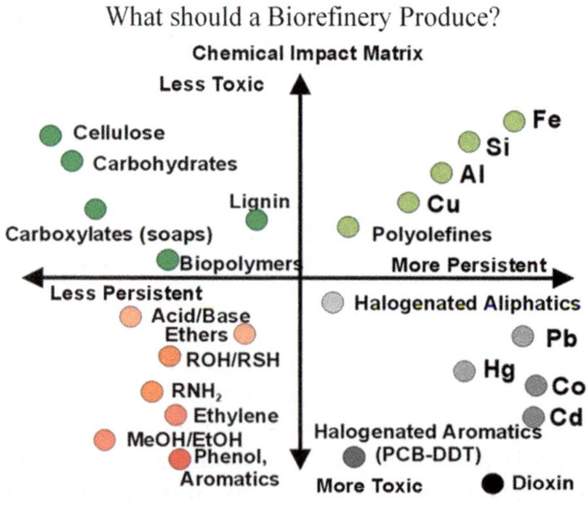

Figure 2.3 Environmental aspects of biorefinery products (reproduced by permission of the creator, Ralph Overend, personal communication).

of Labor is about 300 000.[18] Both of these manufacturing sectors seek economic growth. In some respects, the petrochemical industry has limited growth potential because their product line is relatively fixed and many industrialized nations are encouraging less petroleum consumption. In contrast, the agriculture and forestry industries may see employment increases in their sector, particularly with the emergence of biorefineries. These new biorefineries, several of which came online in 2014, employ between 40–60 individuals. If growth in such facilities continues, which appears to be the trend, the potential to increase this labor force grows significantly since it would take hundreds of these biorefineries to meet Energy Policy Act requirements. Hence, balancing support of the bioindustry and maintaining the viability of the petrochemical industry becomes a challenge for government policy makers.

Much of the world's reserves of oil are found in countries with unstable economies and policies. The United States Navy has underscored the problems with these uncertain sources by noting that a consistent mortality is associated with fuel shipments to the front lines. As a result, the Navy has made public pronouncements that their fleet is to be fueled by significant amounts of alternative fuel in the next two decades, including 50% less petroleum by 2014.[19] Specifically, they have promoted the "Great Green Fleet", which is the nickname of the group of ships that the US Navy has used to demonstrate the viability of biofuels for naval applications (see Chapter 13).

Currently, bioproducts are often discussed in the context of a biorefinery. With a petroleum refinery, fuels such as gasoline and diesel are the main products produced by volume, but they are supplemented by many value-added products, such as ethylene and other high value organic molecules. Many envision a biorefinery that produces both renewable biofuels and bioproducts as being the best way to ensure that biofuels production is profitable. According to Bozell and Petersen, "industrial adoption of renewable carbon requires a financial incentive to justify the use of unfamiliar building blocks, the development of processes to convert these building blocks to final products, and the capital investment necessary to take the technology to commercial scale. High value, lower volume bio-based chemicals provide this incentive".[4] As such, another clear driver for the interest in bio-based products is the realization that while fuel comprises a significant portion of a barrel of oil, there is a large fraction of crude oil that is used for other products. If we can replace a significant portion of a barrel of oil with bio-based products, more oil is available for fuel uses. However, there are significant challenges with this sentiment to overcome. In Figures 2.4–2.6, some of the barriers to pursuing bio-based products are illustrated by excerpts from documents presented in public meetings by Paul Bryan and Brian Duff, formerly of the US Department of Energy. These comments are grounded in the experience and observations by these two individuals who worked with large government portfolios of projects both at the research and demonstration level.

REALITIES

- "Biorefinery" will remain a misnomer for lignocellulosics
- Scale will be too small to produce a broad range of products economically
- Most Biorefineries will produce one or two primary products *at a time*
- Most high-value products have small markets
- Only a small number of plants can produce any given product
- Growth will be driven by flexibility of conversion technologies
- Existing producers will not yield markets willingly
- They will often have depreciated equipment and excellent market access
- To compete, bio-products must efficiently integrate feedstock, conversion technology, and market – every new plant will be a battle! *(value chain issues)*
- Fuels will often remain a major volume product
- In most cases, only one component of the feedstock will be convertible to a high-value product
- Biorefineries must derive significant income from fuels, heat and power

Figure 2.4 Realities of commercializing bio-based fuels and chemicals (courtesy of P. Bryan).

Pace is as important as direction

Cannot replace a multi-trillion dollar infrastructure overnight (*while this relates to transportation fuels – is it analogous to products?)*

Must balance pace and cost of disruption

Economics will drive transitions . . . at their own pace!

Figure 2.5 Integrating with the existing petrochemical industry (courtesy of P. Bryan).

Potential near term opportunities.

Breweries

Pulp & Paper plants – fuels, chemicals, power

Oilseed crushers/processors

Corn dry mills – fuels, chemicals

Petrochemical Refineries – fuels, chemicals, power

MRFs & Landfills – fuels, chemicals, power, materials

WWTP – fuels, chemicals, power

Pharma and contract fermentation operations

Figure 2.6 Opportunities exist for integration and expansion (courtesy of B. Duff).

2.6 Examples of the Bio-based Product Landscape

The best indicator of success for bio-based products is actual production and examples of commercial production continue to appear steadily around the world. Many countries with regenerable biomass resources seem to be interested in raising their standard of living by use of their indigenous resources, particularly if they don't have access to oil within their borders. Economies are complex and the attempts at including bio-based products within micro, meso, and macroeconomic structures possess their own set of circumstances and requirements that are far beyond the scope of this chapter. The historical perspective on the development of this industry typically shows that there is not a lot that is new under the sun. What is new is improved technologies and other drivers such as reduction of environmental impacts, the promise of "green" products, and an impact on our local economies. One interesting historical effort has been the interest in converting carbon dioxide into chemicals and fuels. While this is a thermodynamically challenging problem, the possibility of using this "waste" gas as a carbon source has intrigued scientists for many years with little real commercial progress.[20] However, to highlight the value of continued research and development, two companies, Ineos and Lanzatech, have actually implemented pilot and commercial scale operations using hybrid chemical

Table 2.1 Examples of existing bio-based products.

Chemical or intermediate	Manufacturer or developer	Purpose or use (if published, announced or expressed[a])
Fibers from existing wood products industry	Numerous	Construction, engineered wood products, vanillin, lignin products (sulfonates, binders, *etc.*)
Fibers from existing agricultural materials	Numerous including USDA's Forest Products Laboratory	Reinforcing additive (engineered plastics), paper, molded fiber products and composite fiber/plastic panels
RTP liquids (pyrolysis oils)	Ensyn	Food flavorings and ingredients; renewable liquid fuels
Citric acid	Cargil, Gadot, Jungbunzlauer, Citrique Belge, Archer Daniels Midland	Accidulent, food additive
Lignin-based phenols	LouisianaPacific, Weyerhauser, Georgia Pacific	Partial replacement for phenol in resins—*technical success, but not yet commercially viable.* Other lignin-based chemicals
Algal oils	Several	Speciality chemicals, biofuels, fluids
Cellulose fibers	Rayonier, Tembec, Eastman Chemicals, *etc.*	Rayon, adsorbent fibers (personal care products), filter tow, *etc.*

[a]As of 2014—typically found on company websites.

and biological approaches (*versus* typical Fischer–Tropsch technology) to the use of synthesis gas and carbon dioxide in producing fuels and chemicals as listed in Table 2.4.

What follows in this section are examples of successful, emerging and unsuccessful bio-based products as described in Tables 2.1–2.4. They illustrate the breadth and range of interests by industry, academics, and government. While not comprehensive, they provide the reader with a set of examples of ingenuity, perseverance, and industry in the pursuit of making our lives better through the production and use of bio-based products. Information was obtained typically from company and government web sites. Also, trade journals such as *Biofuels Digest* and specific sections of the online encyclopedia *Wikipedia* present lists and examples of bio-based product

Table 2.2 Examples of newly developed bio-based products.

Chemical or intermediate	Manufacturer or developer	Purpose or use (if expressed or announced[a])
Lactic acid and polylactic acid	Cargill (United States); NatureWorks LLC (United States); Synbra Technology B.V. (Netherlands); Purac (Netherlands)	Polylactides for polymer uses such as styrene replacements, wicking fibers, composites—Altus (Arkema) using 25–50% PLA in plexiglass formulations (Biostrength®). Market is expected to reach about $4 billion by 2020 (Digital Journal, 2014)
1,3-Propanediol	DuPont Tate & Lyle BioProducts	Sorona fiber—carpets, apparel, *etc.*; Zemea® alternative to glycols and glycerin for food and flavor products and a preservative-boosting humecant; and Susterra® speciality diol as a solvent or building block for fluids and resins
Propylene glycol	Archer Daniels Midland (Evolution Chemicals)	Industrial: used as a neat solution for various fluids and also as an intermediate for resins. USP grade used for fluids, cosmetics, food, flavoring, animal feed, *etc.*
Succinic acid	Reverida, France (joint venture of Royal DSM and Roquette Freres); Myriant – Lake Providence, LA; BioAmber, (Pomacle, France and eventually Ontario, Canada)	Intermediate for chemicals such as 1,4-butanediol, solvents and deicers. Intermediate for polyols, acrylics, and other polymers

[a]As of 2014 and generally described on company websites.

Table 2.3 Sampling of products on a commercial development track.

Chemical or intermediate	Manufacturer or developer	Purpose or use (if published, announced or expressed[a])
1,4-Dicarboxylic acids	Michigan Biotechnology Institute, Myriant, DSM, Roquette, BioAmber	Intermediate for chemicals such as 1,4-butanediol, solvents and deicers. Intermediate for polyols, acrylics, and other polymers
Algal oils	Solazyme, Sapphire Energy, Martek (now part of DSM), oil companies and many others worldwide	Fuels, fine and speciality chemicals
Biomass oils	KIOR, UOP-Honeywell (Ecofining™), Gas Technology Institute (TIGAS technology) with Carbona and Haldor Topsoe	Fuels and commodity chemicals
Biomass gasification	Gas Technology Institute with Carbona and Haldor Topsoe	TIGAS technology—DME intermediate to gasoline
Furanic derivatives	Avantium YXY technology	Polyester fiber for bottles—replace PET
1,4-Butanediol and butadiene	Genomatica (with BASF and Versalis)	Fiber feedstock, chemical intermediates
C2–C6 chain esters, acids, alcohols and alkenes	ZeaChem	Fuels, polyester feedstock source
Farnesene	Amyris (Biofene®)	Fuels, speciality chemicals, lubricants, polymer additives
Butanol and C4 alcohols	Butamax, Gevo, other small companies	Commodity chemicals and fuels
Ethylene	DOW, Braskem	Polymer intermediate—huge commodity chemical market
Clean sugars	Old Town Fuel and Fiber, BlueFire Renewbles, Motti & Ghisolfi (Chemtex), Renmatix, ICM and others	New commodity with potential large future, but competes with dextrose from conventional sources
C2–C3 intermediates	Enerkem	Fuels, commodity chemicals

[a]Primarily on company websites.

endeavors. Readers can refer to these sources for more up-to-date information as well as examining the portfolios of government and independent agencies.[21–23] The rationale for providing these tables is to illustrate the significant activity worldwide in these kinds of endeavors as multinational companies, large and small chemical companies, and small businesses are involved.

Table 2.4 Examples of new pilots and demonstration facilities.

Company	Location	Actual or proposed multiproduct outputs[a]	Stage of development (if stated or known[a])
Myriant	Lake Providence, LA	Succinic acid and other dicarboxylic acids	Commissioned in 2013 and operational in 2014
Amercian Process, Inc	Alpena, MI	Fiberboard, ethanol co-product, others AVAP process – Amercian Value Added Pulping	Commissioning and operation of a pilot plant completed in 2014—value-added uses of waste streams
Iogen	Ottawa, Canada	Cellulosic ethanol and other by-products	Been operating since 2004, a 10–20 tpd pilot plant
Gevo	Luverne, MN	Isobutanol and ethanol (side by side)	Commissioning and operation of a small scale commercial demonstration facility
GTI with CRI Catalysts	Illinois	Biofuels and other possible hydrocarbon products	50 kg d^{-1} pilot runs with very good economics of a thermochemical process—IH$^{2®}$ Process. Being licensed for commercial development
LanzaTech	Illinois (HQ) and Georgia (Pilot)	Biological fermentation of syngas to fuels and chemicals	Pilot scale facilities (Georgia and China) working towards demonstration scale commercial facilities in 2015. Announced partnerships with Invista for chemicals and a consortia for aviation fuels [Boeing, a bank (HSBC), General Electric, Virgin Air][b]
Coskata	Illinois	Biological fermentation of syngas to fuels and chemicals	Pilot scale facilities working towards demonstration scale facilities
Ineos – New Planet Bioenergy LLC	Florida	BRI process for fermentation of syngas to ethanol (8M gallons) with biopower (6 MWe) co product	Commercial demonstration initiated in 2013

[a]As of 2014—information usually from company web pages.
[b]Press releases from LanzaTech published in 2014.

2.7 Conclusions

The efforts to develop bio-based products (and fuels) arise from multiple drivers, such as political, social, economic, and environmental concerns. True commercial viability will be driven by the market and successful strategies for producing bioproducts are still being sought. Recent successes, such as the dramatic increase in corn to ethanol production from a few million gallons to several billion gallons in less than a decade, suggest that rapid improvements in the biomass to products industry are possible. The promise of engaging local regenerable resources for the long-term sustainable production of our basic chemical and material feedstocks has always been present. The improvements in science, engineering and technology are now making the promise a greater reality.

Acknowledgements

The authors thank Katy Christiansen, Glenn Doyle, and Liz Moore for providing their knowledge, edits, and constructive criticism to this chapter.

References

1. G. R. Petersen, J. Bozell and J. White, *Catalytic Process Development for Renewable Materials*, ed. P. Imhof and J. Kees van der Waal, Wiley Press, May, 2013, ch. 2 – Selecting Targets, http://www.wiley.com/WileyCDA/WileyTitle/productCd-3527331697.html.
2. Farm Security and Rural Investment Act of 2002, Public Law – 107-171-May 13, 2002, Title IX – ENERGY, Sec 9001 Definitions (2).
3. T. Werpy and G. Petersen, *Top Value Added Chemicals from Biomass. Volume I – Results of Screening for Potential Candidates from Sugars and Synthesis Gas*, U. S. D. o. Energy, DOE/GO-102004-1992, 2004.
4. J. J. Bozell and G. R. Petersen, Technology development for the production of biobased products from biorefinery carbohydrates – the US Department of Energy "Top 10" revisited, *Green Chem.*, 2010, **12**, 539.
5. J. J. Bozell, J. E. Holladay, J. F. White, and D. Johnson, *Top Value-Added Chemicals from Biomass: Volume II – Results of Screening for Potential Candidates from Biorefinery Lignin*, PNNL 16983, October, 2007.
6. J. J. Bozell and R. Landucci, Alternative Feedstocks Program Technical and Economic Assessment: Thermal/Chemical and Bioprocessing Components, 1993, p. 227, NREL Report No. TP-20646.
7. One example of many: ADM Adds Isosorbide to its Industrial Chemical Product Line http://www.adm.com/news/_layouts/PressReleaseDetail.aspx?ID = 259 August 11, 2010.
8. B. Vijayendran, Commentary: Bio products from biorefineries – trends, challenges and opportunities, *J. Bus. Chem.*, September, 2010.

9. K. Gransalke, Mother Nature's Drug Cabinet, *Lab Times*, February, 2011, **11**(1), 16–19, Drug Discovery – Is Mother Nature still the number one source for promising new drugs?.
10. *Plant Resources of Tropical Africa 3. Dyes and Tannins*, ed. P. C. M. Jansen and D. Cardon, PROTA Foundation, Wageningen, Netherlands/Backhuys Publishers, Leiden, Netherlands/CTA, Wageningen, Netherlands, 2005.
11. M. R. Finlay, Old Efforts at New Uses: A Brief History of Chemurgy and the American Search for Biobased Materials, *J. Ind. Ecol.*, 2004, 7(3–4), 33–46.
12. Nobel Lecture by Norman Borlaug, December 11, 1970. http://www.nobelprize.org/nobel_prizes/peace/laureates/1970/borlaug-lecture.html.
13. S. Macrae, M. G. Gilliland and J. Van Staden, Rubber Production in Guayule: Determination of Rubber Producing Potential, *Plant Physiol.*, August 1986, **81**(4), 1027–1032, DOI: 10.1104/pp.81.4.1027.
14. M. N. Dastoor, W. W. Schubert and G. R. Petersen, Preliminary results of *in vitro* propagation of guayule, *J. Agric. Food Chem.*, 1981, **29**, 686.
15. A. A. Morton, E. E. Magat and R. L. Letsinger, Polymerization VI. The alfin catalysts, *J. Am. Chem. Soc.*, 1947, **69**(4), 950–961.
16. Department of Agriculture Biobased Affirmative Procurement Program, May 8, 2006, http://www.dm.usda.gov/procurement/biobased/APP.pdf.
17. *Catalytic Process Development for Renewable Materials*, ed. P. Imhof and J. Kees van der Waal, Wiley Press, May, 2013.
18. US Department of Labor, Bureau of Labor Statistics: Chemical Industry – http://www.bls.gov/iag/tgs/iag325.htm#workforce – 808,000; Petroleum – http://www.bls.gov/iag/tgs/iag324.htm#workforce – 113,400; Agriculture – http://www.bls.gov/iag/tgs/iag11.htm#workforce – 267,390; and, Forestry and Logging – http://www.bls.gov/iag/tgs/iag113.htm#workforce – 36,530.
19. S. Closson, The military and energy: Moving the United States beyond oil, *Energy Policy*, October 2013, **61**, 306–316, DOI: 10.1016/j.enpol.2013.05.102.
20. G. Petersen, D. Viviani, K. Magrini-Bair, S. Kelley, L. Moens, P. Shepherd and D. DuBois, Non-Governmental Valorization of Carbon Dioxide, *Sci. Total Environ.*, 2005, **338**, 159–182.
21. Bioenergy Technologies Office, Energy Efficiency and Renewable Energy Office of the US Department of Energy, http://energy.gov/eere/bioenergy/research-development-demonstration-and-deployment.
22. United States Department of Agriculture (USDA) and its Agriculture Research Service (ARS) laboratories are active participants in biobased product R&D (some examples): http://www.ars.usda.gov/is/AR/archive/apr02/form0402.pdf; https://www.ars.usda.gov/is/AR/archive/apr02/indus0402.pdf; http://www.ars.usda.gov/IS/AR/archive/apr07/research0407.htm?pf=1.
23. E. de Jong, A. Higson, P. Walsh, M. Wellisch, IEA Bioenergy Task 42 Biorefinery Report: Bio-based Chemicals, Value Added Products from Biorefineries, 2012, http://www.ieabioenergy.com/publications/bio-based-chemicals-value-added-products-from-biorefineries/.

CHAPTER 3

Bioenergy Crops: Delivering More Than Energy

M. CRISTINA NEGRI* AND HERBERT SSEGANE

Argonne National Laboratory, 9700 S Cass Ave., Argonne, IL 60439, USA
*Email: negri@anl.gov

3.1 Thc Context

As world population is projected to increase by 2 billion humans between 2011 and 2050 and reach over 9 billion people globally,[1] and as dietary aspirations improve worldwide, the challenge of sustainably providing food, feed, shelter, fiber and ecosystem services for all is one of the greatest of this century. The World Resources Institute, jointly with the United Nations,[2] predicts that to satisfy the caloric needs of this growing population, food production will have to grow by 70% globally by 2050. Yet, current yield increase trends alone are not sufficient to meet this challenge.[3] Agriculture already uses 37% of the world's land,[2] and contributes 24% of greenhouse gases, mostly from soil agricultural practices and fertilization in particular. It is also the largest consumptive user of freshwater,[2] which in the US in 2010 stood at about 38% of withdrawals,[4] and globally at 70%.[2] Climate change is expected to magnify these challenges and compound them with others, such as shifting pest threats, increased CO_2 concentrations, rising temperatures and increased water stress.[5] The changing pattern of precipitation alternating intense storms with periods of lesser precipitations will also challenge agricultural production to seek drought and flood tolerant crops and a more diverse crop palette to mitigate and distribute risk.[6]

RSC Green Chemistry No. 43
Commercializing Biobased Products: Opportunities, Challenges, Benefits, and Risks
Edited by Seth W. Snyder

Published by the Royal Society of Chemistry, www.rsc.org

Conservation of soil, water, biodiversity and wildlife is essential to the provisioning of ecosystem services, both at the local scale and globally. There is a need to protect grasslands, wetlands, forests and other conservation-relevant areas from threats from urbanization, agricultural pressure and pollution. Converting uncultivated or forested land to crops, whether for food or bioenergy, has the potential for a carbon debt that may take decades to repay.[7–9] To meet productivity and conservation targets, a number of strategies have been proposed, including boosting crop yields, changing dietary targets, controlling population growth, increasing aquaculture production, adopting conservation agriculture techniques such as cover crops, precision agriculture, double cropping, *etc.* None of these strategies alone, however, are likely to deliver the desired results.

The development of bioenergy as a viable alternative to fossil fuels for transportation, for the production of many renewable bioproducts, and as the staple of household energy supply in developing countries,[10] enters this scene with promises and drawbacks: competing for land with food cropping, the potential for biodiversity loss and increased water quality and quantity impacts are arguments against bioenergy,[11] while a better performance in reducing greenhouse gas emissions against petroleum-based transportation fuels is its major advantage.[12] Yet, the extent of the potentially negative impacts is not universally fixed, but rather depends on the way these feedstock crops will be deployed on the land. Impacts on water quality and quantity, on soil quality and on biodiversity are also concerns in addition to increasing arable land to grow both bioenergy and food crops; expanding the current profit-maximizing production model for either food or bioenergy is bound to expand negative impacts as well.[11]

A possible key to a sustainable increase of the overall production of food, feed, bioenergy and bioproducts on a limited budget of land and water resources has been identified by proposing proactive designs of sustainable crop land intensification.[13–15] In this chapter, the authors provide a synopsis of the ecosystem services provisioning opportunities that can be offered by bioenergy crops, and provide examples of how proactively designing bioenergy deployments with the objective of building sustainability outcomes into the production model can contribute to an overall improved environmental performance and contribute to address the land use dilemma. While many different strategies will be necessary to deliver all the goods needed, it will be important that the development of a biobased economy be guided by sustainability criteria and tactics to contribute to the achievement of the immense societal goal of providing food, feed, energy, bioproducts, shelter, fiber, and ecosystem services to a growing world population.

3.2 Bioenergy Crops as Providers of Energy, Environmental and Ecosystem Services

Perennial bioenergy crops have distinctive traits that differentiate them from annual commodity crops such as soybeans and corn grown on agricultural land in the US. Compared to these, perennial crops such as switchgrass, *Miscanthus*, other perennial grasses, and short rotation woody crops share a deeper root system,[16] a general better ability to thrive on poorer soils, an assumed lower dependence on fertilizer inputs, and at least for some, management options friendlier to wildlife.[17] A deeper root system carries many potentially strategic benefits such as the ability to survive more extreme conditions of drought and flooding after establishment,[18] and the potential to sequester carbon at depth.[7,16,19–21] Poplars (*Populus* spp.) and willows (*Salix* spp.), have also shown the ability to capture shallow groundwater plumes, hydraulically contain them, and remove entrained pollutants by phytoremediation.[22] Lower fertilizer needs may lessen water quality concerns, lower the energy content of crops, and lower greenhouse gas (N_2O) emissions. Overall, a recent and thorough literature search has determined that the conversion of arable land to second generation bioenergy crops returned more positive than negative effects on ecosystem services, with climate and water quality regulation, erosion and flooding control and disease regulation the most represented positive impacts. Other transitions (such as conversion of marginal land or forest) showed a broader balance between negative, positive and neutral effects.[23]

There has been an increased interest in landscape design as a tool to maximize total output from the land in sustainable ways (http://energy.gov/eere/bioenergy/downloads/incorporating-bioenergy-sustainable-landscape-designs-workshop-two-summary). Landscape design, in essence a plan for resources allocation, is aimed at matching crop placement with land's productivity potential and environmental function in a comprehensive, holistic design that is optimized to deliver a suite of economic, societal and environmental outcomes. In the context of landscape design, the distinctive traits of lignocellulosic bioenergy crops discussed earlier are valuable tools to provide functional environmental services at the same time as the crops are grown for bioenergy. Placement of these crops on the landscape so that they can provide water filtering, erosion protection, and wildlife shelter where it is most needed is a way to lessen the environmental impact of bioenergy crops and ultimately of the commodity crops that share the landscape with them.[24] The placement of these bioenergy crops in subproductive or environmentally critical or risky land has the potential to provide environmental benefits with minimal disturbance of the overall production of commodity crops.[23,25] The resulting landscape could be a diverse, functionally sustainable environment, resilient to threats of climate change and economic downturns, and valued by the communities that call it home. The development of a biobased economy can indeed deliver, in addition to their intended final product, tangible goods that are valued by society even before crops are harvested.

Providing ecosystem services has economic value. Their potential synergism with trading markets or other systems to compensate farmers for cleaning water, reducing erosion, providing wildlife corridors and sequestering carbon is an increasingly attractive way to progress a conservation agenda (http://water.epa.gov/type/watersheds/trading/twg_index.cfm)

and is prime for development with the sustainable deployment of bioenergy crops.

This chapter illustrates some of the traits that make bioenergy crops capable of addressing some of the most important ecosystem functions and their indicators, when deployed strategically. While ecosystem services are systematically described and catalogued in the Millennium Ecosystem Assessment Reports (http://www.millenniumassessment.org/en/Index-2.html), a framework of relevant sustainability indicators for bioenergy crops is described by McBride *et al.*[26] As patch size, landscape position and other local environmental conditions strongly influence the expression of these ecosystem services, and these traits can provide wanted or unwanted effects depending on the context, it is important to note that strong, science-driven landscape design criteria need to fit each crop to the position on the landscape where it can provide the most benefit, to fully enable the realization of these potential benefits.

3.2.1 Soil Physical Properties

Plant growth and a number of environmental and ecosystem functions are dependent on soil physical properties such as soil bulk density, resistance to penetration, water infiltration capacity, and water holding capacity. These interrelated properties determine plant establishment, root penetration, runoff amounts, soil aeration, erosion potential, the ability to store water available for plant growth, and ultimately dramatically impact soil productivity. They have also been shown to impact soil carbon dynamics.[27,28] Management practices such as rotation, tilling and fertilizing, which are different among crops, and crop types also affect soil properties.[29] Literature reports that the deep and vigorous root system of perennial bioenergy crops can help reduce soil density and improve water infiltration rates compared to an annual crop such as corn.[30] Bonin *et al.* compared soil properties in corn, switchgrass and short rotation willow grown in experimental plots in Ohio and found that switchgrass plots had, after seven years, the lowest penetration resistance, lowest surface bulk density, and highest field moisture content of the three crops. Water infiltration rates appeared to be not statistically different among crops, however, they were higher in switchgrass. Soil under corn, and particularly no-till corn, had the highest penetration resistance, highest bulk density, and lowest moisture content.[31] Improvements in water infiltration rates are more likely to be measured after a longer timeframe, over 10 years, from the establishment of perennials[32,33] and may be aided by the establishment of both large root systems and a richer diversity of soil fauna brokered by roots, creating macropores in perennial plantings.[34] With considerable variations based on soil properties, small overall changes in soil bulk density were found by an on-farm, multi-state research effort in North Dakota (ND), South Dakota (SD), and Nebraska (NE) after five years of switchgrass cultivation, with the most relevant changes occurring at the

0.05–0.1 m depth, most decreases found in ND and SD, and small increases in NE.[35]

3.2.2 Soil Carbon

Perennial grasses in general are thought to increase soil carbon stock over time compared to row crops. A thorough review and analysis of soil carbon dynamics under different land use changes involving bioenergy crops can be found in ref. 36. This review suggests that cropland conversion to perennial bioenergy crops overall provided soil organic carbon (SOC) increases of 6–14% relative to the initial SOC levels, and that SOC increases may accrue on a deeper than previously considered soil depth. In another study, grassland was found to be able to potentially sequester about 45% of the carbon released each year by agriculture in the US.[37] Root development deeper in the soil profile, reduced tillage and other management requirements that minimally disturb the soil are credited for the increased accrual of SOC by perennials. In general, estimates of soil carbon increases from the literature when perennial grasses are introduced to cropland range from 1.1–1.2 Mg C ha^{-1} yr,[38,39] to 0.78–0.53 Mg C ha^{-1} yr carbon accrual,[40] or 0.4–0.7 Mg ha^{-1} yr.[7] Among perennial grasses, switchgrass appears to be quite a typical perennial grass as far as C sequestration potential is concerned.[41] In dedicated studies, conversion of cropland to switchgrass contributed to an average increase of 2.4 Mg C ha^{-1} in equivalent mass SOC, with changes varying by location from −5.6 to 7.1 Mg C ha^{-1}.[35] In a Texas study, switchgrass increased SOC, soil microbial biomass carbon, mineralizable carbon and particulate organic matter carbon compared to non-bioenergy row crops such as sorghum, wheat, soy, and peanuts.[42] Compared to corn, switchgrass increased soil C levels by 3 Mg ha^{-1} per year, with significantly higher root C levels below 30 cm compared to corn and willow.[43] *Miscanthus* was also shown to contribute to the accrual of C in the soils in which it was grown, with higher accumulation in longer-established stands.[44] Richter *et al.*[45] found that *Miscanthus* genotype is a relevant factor in SOC distribution over depth, and that while *Miscanthus* may cause a definite decline in original SOC concentration when introduced in non-cultivated land, the crop eventually more than compensated for this loss with its own SOC input. The NRCS reports a 9 Mg ha^{-1} in long-term storage under *Miscanthus*.[46] Willows are fast growing and very productive both above and below ground. In one study, soil C under willow was calculated to increase by 9 Mg ha^{-1} year, the highest compared to corn and switchgrass, reaching 130 Mg ha^{-1}, with fine roots representing about 60% of total root carbon.[43]

3.2.3 Rooting Habit, Water Capture and Drought Avoidance

Root distribution influences physical and chemical soil properties.[47,48] Root development through the soil profile influences aeration (bulk density), water capture, soil structure and carbon sequestration as root growth contributes to the accrual of organic carbon into soils and to respiration. Switchgrass, like other perennial grasses, exhibits a rapid root growth immediately after establishment, with a larger root than aboveground biomass growth compared to annuals until the plant reaches maturity. After harvest, shoot–root ratios are also in favor of roots at the onset of the growth season and then leveling off as the season proceeds. Roots of switchgrass have been shown to extend their period of growth into autumn.[49] This active presence of root systems from early spring to late in the autumn is an important trait to achieve water quality benefits as roots are ready to take up water and entrained solutes which are most available in subsurface and runoff flow early in the season when most annual crops are still too small to properly explore the soil in which they are grown. In switchgrass, rooting depths can reach more than 3 m, with the bulk of the roots being found in the first metre.[20] Garten and Wullschleger[50] found that 94% of coarse roots were in the upper 0.4 m of soil. Soil types and cultivars affect this distribution.[41] In addition to exposing the plant to the larger pools of nutrients in the subsoil, this distributed root system allows switchgrass to tap into moisture present in the upper soil layers until the late summer and then access water from the lower soil layers when the upper layers are drier. A soil's capacity for water will determine how much water is available to the growing switchgrass and contribute to mitigate drought impacts.[51] The rooting depth of short rotation woody crops has been reported as deep as 3 m[52] and more. While willows are considered phreatophytes, or plants that can access water from a saturated subsurface, root depth is influenced by soil saturation conditions as plants will not expend unnecessary energy in rooting at depth if water resources are available on the surface.[53,54]

Root density patterns in *Miscanthus* showing high root density at deeper soil layers (to 150–270 cm) was found by several authors,[55,56] and by Riche and Christian (2001) cited in ref 55. However, others found low root density below 90 cm, and the highest concentration above 35 cm,[57] probably due to differences in soil and water conditions. Some authors[45] found that roots of *Miscanthus* were up to an order of magnitude higher in the A horizon than in the B horizon with great variations with phenotypes. Tufted varieties showed higher root length density very near the plant center location than non-tufting varieties, while non-tufting varieties had higher densities than tufting ones in the off-row positions. Switchgrass, on the other hand, was more unanimously found to be able to grow deeper root systems, particularly under rainfed conditions, in which roots were found deeper than 270 cm.[57]

Water capture from deep soil layers was found to be generally greater under perennials than annuals, also because of their longer growth season. Ferchaud *et al.*[55] found a curvilinear asymptotic relationship between water capture and root density. Deep rooting, recorded for many perennials including short rotation woody crops, *Miscanthus* and switchgrass, is considered a strategy to maintain productivity during drought. It can also have the effect of causing higher water deficits (luxury consumption) thereby calling for a rigorous assessment of the tradeoffs between the need to conserve water and providing tolerance to droughty conditions (or, in case of excess precipitation, to provide flood or saturation management).

3.2.4 Water Quality

Rhizomatous warm season grasses such as switchgrass have been used extensively for conservation purposes in unharvested strips and hedges as they provide a physical means to constrict runoff flow and let solids settle.[58] Because perennial crops are on the ground for more than one year, their growth cycle minimizes soil disturbance and allows for long-term ground cover that is beneficial to mitigating overland flow. Switchgrass in a hedge or buffer strip was able to reduce nitrogen losses by runoff by 41% and sediment by 61%.[59] The effect seemed to be similar for dormant or active plants[60] pointing to the physical nature of the entrapment. The main crop traits useful for this function include upright stiff stems and rhizomatous growth habit which both switchgrass and *Miscanthus* possess. Switchgrass was also effective in reducing nutrient runoff from manure and fertilizer applications,[61] including dissolved and particulate phosphorus (48–38% respectively) and ammonia-N (60%). Besides overland flow, switchgrass has also been found to be able to improve water infiltration, measured as field-saturated hydraulic conductivity, and with this to improve the quality of water percolating to the subsurface compared to row crops in different landscape positions.[62] Willows are a common tool for riparian stabilization because of their rapid growth, stiff aboveground vegetative form, strong root system and coppicing growth.[63] *Miscanthus* has shown to be able, when mature, to reduce dramatically the nitrate-N concentration in drainage water, while during the establishment period, the small size of the crop may be responsible for an increase in the nitrate-N concentrations.[46]

3.2.5 Nutrient Management

Nitrogen accumulation in aboveground biomass of switchgrass and *Miscanthus* was shown to be highly seasonal[19] with summer concentrations being orders of magnitude higher than in winter for both crops because of N translocation to rhizomes in winter, which protects nitrogen from nitrification, denitrification and consequent loss as elemental nitrogen and nitrous oxide *via* air, or nitrate in water. This translocation has the potential of making both crops more efficient nitrogen users as they can in practice "recycle" nitrogen from year to year if harvest is scheduled in winter when most of the nitrogen is safely stored in the rhizomes. As much as 40–100 kg N ha^{-1} were estimated to be moved from shoots to roots and *vice versa* in switchgrass.[64] Other nutrients such as phosphorus and potassium are also recycled by these crops through litter deposited on the soil and overwinter decay, thus decreasing needs for fertilization. Willows have shown N exports of 18–86 kg N ha yr^{-1}, a large variation with clones, and a positive correlation between water use efficiency, productivity, and nitrogen export.[65] Research in the US North East has shown similar removals depending on yields,[66] which makes willows especially suited to remove large amounts of nitrogen from soils and water. Willows have been shown to deplete nitrate in their rhizosphere soil compared to adjacent bulk soil during the vegetative season in contrast with comparable bulk soil (Negri and Gopalakrishnan, unpublished).

Fertilization rates of switchgrass vary broadly in the literature. Generally, reported values are between 0 and 100 kg ha^{-1} N or 10–12 kg N per Mg biomass produced (Vogel (2004) cited in ref. 58). Recommendations for fertilization of shrub willows in the US North East are 100 lb ac^{-1} (112 kg ha^{-1}) in the second vegetative season.[67] It should be noted that willows appear to be accepting fertilizers well from various waste sources such as manure, biosolids, and wastewater.[66] *Miscanthus* fertilization rates vary dramatically, with recommendations for the US Midwest of applications of 5–9 lb N per ton of dry matter removed (Heaton *et al.* in ref. 46), and the expectation for *Miscanthus* to be able to thrive on existing N pools, which are recycled from year to year through rhizome storage,[46] at least for the first few years of growth. Similar to switchgrass, nitrogen use efficiency in *Miscanthus* was inversely correlated with nitrogen availability and a reduction of efficiency was found with increased water supply.[68]

The nitrogen use efficiency of switchgrass has not been well quantified, but is believed to be higher than row crops, for example at 66% compared to 50% of corn, because of the ability to store nitrogen in the rhizome as mentioned earlier.[69] However, others[64] reported that nitrogen use efficiency decreased at increasing fertilization rates with a yield advantage at approximately 9 kg biomass per kg applied N per year. It also appears that switchgrass is able to access N from sources unavailable to other crops, perhaps because of particular microbial associations, and the heavy reliance on mycorrhizae. Stout and Jung[70] indicated that switchgrass can take up 1.49–2.63 kg N ha^{-1} per day depending on soil N levels. Others[71] found that up to 22% of N could be taken up by roots deeper than 1.4 m.

3.2.6 Water Use

In general, studies suggest that perennials consume more water because of their longer growing period, deeper root system and higher productivity compared to annuals, however some studies show that perennial C4 grasses such as *Miscanthus* and switchgrass provided soil water deficits similar to annuals in addition to higher productivity and deeper root systems.[55] While potentially beneficial in locations that require better drainage, this consumption needs to be accounted for when planning for in areas with limited water resources. McIsaac found that *Miscanthus* had higher water use than switchgrass and a corn/soybean rotation. As water movement to streams was reduced from this increased consumption, inorganic N leaching was also found to be dramatically decreased with potentially useful changes in nutrient loadings leaving the watershed.[72] Models predicted a dramatic increase of seasonal evapotranspiration, by 58 and 36% respectively, for *Miscanthus* and switchgrass in the US Midwest.[73] McIsaac *et al.*[72] found that *Miscanthus* resulted in lower minimal soil moisture than switchgrass during four growing seasons in central Illinois. Both benefits (flood reduction) and costs (prolonged low flows) were predicted for the large scale deployment of *Miscanthus* and switchgrass in that particular study in Illinois, pointing again at the need to assess tradeoffs between possibly competing ecosystem services. Similar results pointing at the relationship between reduced water yields and reduced nutrient exports compared to corn were found elsewhere (Schilling *et al.* cited in ref. 58). For a comprehensive database of switchgrass water use efficiency, refer to Wullshleger *et al.*[74] Values were found to be higher for lowland than upland cultivars and averaged 21.6 with a range of 2.3–103.1 kg ha mm, and 68% of observations fell in the range of 10–30 kg ha mm interval.[75]

Willow is generally considered a substantial water user. Water use efficiency of willow was estimated to be 4.1–5.5 g kg^{-1} for a fertilized and irrigated stand in Sweden.[76] Transpiration rates measured on a three-year-old stand in June–July in the UK *via* a stem heat balance method were in the order of 6 mm day^{-1}, higher than any reported rate for agricultural crops in the same region. Importantly, willows appeared to be poor water savers, as they showed high stomatal conductance even under elevated soil and atmospheric humidity deficits.[77]

3.2.7 Adaptation to Flooding and Drought

Flooding and droughts are weather extremes that are expected to be more frequent under climate change scenarios. To mitigate damage and obtain viable yields, it has been proposed that floodplains would be ideal locations to grow flood tolerant bioenergy crops such as switchgrass, *Miscanthus* and willow.[78]

Miscanthus was found to be sensitive to reduced water availability throughout the main growing season[79] but especially at the end of the season.[68] *Miscanthus* in UK is water limited in many areas (Finch *et al.* (2004), cited in ref. 79). Any drought tolerance appears to be correlated with lower productivity: Zub and Brancourt-Hulmel[68] reported that there are differences between *Miscanthus* spp. in tolerance to drought, whereby *M. x sinensis* was found to be less productive but more drought tolerant than *M. x giganteus* as it is capable of reducing leaf water conductance under even mild water stress. This trait was not found in *M. x giganteus*, which was conversely found, in glasshouse trials, to be better able to tolerate flooding conditions.[80] Overall, *Miscanthus* was considered best suited where precipitation totaled 30″ (762 mm) per year.[46] The strong, upright stems of *Miscanthus* are also believed to be more resistant to the physical impacts of flooding than other grasses. Irrigation seems to help the establishment of newly planted *Miscanthus* rhizomes under drier conditions such as those found in southern Europe.[68] Available water capacity appears to be an important soil property in controlling water stress than precipitation, probably because of *Miscanthus*' high precipitation interception and evaporation rates which in practice limit effective rainfall (Finch *et al.* (2004) cited in ref. 79).

Switchgrass appears to be adaptable, given its broad native range, to both drought and flooding conditions, with latitude of origin of a cultivar having dramatic effects on production at different latitudes.[81,82] The lowland ecotype has been shown to be even more productive in flooded than normal conditions (Porter (1966), cited in ref 83) and has been proposed as a facultative wetland species. Conversely, the upland ecotype is generally more tolerant to dry conditions. In greenhouse stress tests, upland species did not, however, outperform lowland species and did not maintain higher photosynthetic rates under drought as hypothesized. Both suffered severe reductions in yields at −4 MPa water stress. Lowland ecotypes performed better than upland ecotypes, although with delayed germination, under flooding conditions. While germination under drought was maintained at varying levels, at 10% moisture only 55% of emerging plants survived.[83] Likewise, other authors state that switchgrass requires moist soil for at least one month after planting for effective germination and survival.[82] Once established, it appears that switchgrass tolerance to drought may be higher, albeit at a price of reduced productivity. Others confirm these results.[84]

Willows are commonly found in riparian environments because of their preference for water-rich environments. While maximum photosynthetic activity was found when ample soil moisture was available without soil waterlogging,[85] willows have shown to be quite tolerant to flooding conditions[86] and have adaptive mechanisms to survive waterlogged conditions.[63] Willow seedlings have also shown different tolerance to receding water tables,[86] with gradual declines being more favorable to seedling survival than abrupt changes, and important implications for plant establishment. Once established, specific willow species have been shown to be more tolerant of droughty conditions[87] as their prolific root system can adaptively develop into deeper soil layers and use moisture found there as discussed earlier.

3.2.8 Biodiversity

Biodiversity at the microbial scale was studied[88] in short rotation willow coppice compared to an agricultural system and to forest and grassland soils. After three years, short rotation coppice (SRC) in this study seemed to promote most of the soils' biological properties. Soil biological parameters proposed as indicators of soil biological health such as fungal abundance and earthworm diversity in the willow system were higher than the agro-system, but lower than the forest and grassland systems' soil. After three years from planting, the SRC system showed evidence of a transition towards a bacterial community structure close to grassland, and the highest earthworm density. These changes were attributed to the lack of tillage and pesticide use over these three years and to the direct effect of the willows' contribution to soil organic matter through root and leaf litter and exudates. Probably reflecting the woody litter chemical composition,[89] or the decreased nitrogen fertilization which favors bacteria,[90,91] the fungi–bacteria ratio was also higher in short rotation coppices than agrosystems.

The deployment of perennial bioenergy crops has profound implications for maintaining and increasing quality habitat for migratory birds during breeding, post-breeding, *en route* during migration and wintering. Patch size, heterogeneity and surrounding landscape (such as the absence of forested patches) are important to habitat creation for grassland birds, one of the most imperiled components of avian fauna. Switchgrass and mixed prairie grass–forbes have been shown to be able to provide significantly important habitat benefits for avifauna during post breeding and migratory periods, especially when they replace row crops.[92] Particularly, both switchgrass and mixed grasses–forbes combinations appeared to be equally useful to attract a diversity of abundant migratory birds and have the potential to act as spring and fall migratory stopover habitats for both generalists and grassland specialist birds.[93] However, these authors and others[17] recommended that adjustments to intensive cropping methods be considered to realize these benefits, such as adopting alternative cropping schedules and strategies to create within-crop structural heterogeneity beneficial to creating attractive habitats, and limiting the amount of chemicals used in crop production. Regarding *Miscanthus*, it appears that its impacts on biodiversity vary with its age, with younger and less dense stands reported to not be detrimental to wildlife, while older stands show lower biodiversity of insects and wildlife, probably because of their tall and dense growth.[46] In another study conducted in the UK, while *Miscanthus* hosted a greater abundance and diversity of birds than wheat in winter and summer, changes in the crop structure during breeding influenced the birds, showing that careful management of crop production for yields and wildlife is needed to facilitate bird habitat conservation throughout the birds' lifecycle.

Arthropods are another critical population for biodiversity, as they mediate ecosystem services such as pollination and pest suppression.[94] Among arthropods, insects alone have been shown to provide ecosystem services such as dung burial, pest control and pollination and recreation and fisheries worth almost $60 billion a year.[95] A comparison study of corn, switchgrass and mixed prairie grasses and forbes in Michigan showed that bees were more diverse and three to four times more abundant in switchgrass and prairie mixes than in corn, and that beneficial insects generally responded well to the increased vegetational diversity of prairie and switchgrass.[96] The presence of potential pathogenic organisms or vectors in the bioenergy crops was not studied. Compared to corn, switchgrass and mixed prairie were associated with a 230% and 324% increase in arthropod family diversity and 750% and 2700% arthropod biomass, respectively.[97]

3.2.9 Pest Suppression

Biological pest control by naturally present predatory arthropods is an ecosystem service that, when replacing direct costs to farmers, could be valued at more than $4 billion per year nationally.[98] Adding the value of the avoided potentially negative impacts on environmental and human health could increase this conservative estimate. Perennial cropping has been shown to host a more abundant, diverse and active predatory arthropod community compared to annual crops. Future expansion of perennial bioenergy crops on marginal lands in the US Midwest had a predicted increase in biocontrol index averaging 20% while the same land conversion to annual bioenergy crops was predicted to provide a decrease of 10% in the same index.[98] Moreover, the specific landscape location of perennial bioenergy crops relative to other habitats appeared to contribute to the extent of the biocontrol services produced by these crops, thus implying that the impact of increased diversity and abundance of predators found in bioenergy cropped areas may be felt at a larger than patch scale such as the landscape, with the bioenergy plantings acting as reservoirs of predators benefiting row crops nearby.[99,100] In a comparison study between willow, barley and wheat, willow in short rotation had the highest abundance of ground dwelling arthropod predators of the three crops, although predation rates did not vary detectably in the study, possibly because of the difference in preferred preys used in the experiment.[101] The presence of pests and their predatory species may, however, be altered in imported crops such as *Miscanthus* and therefore present a threat to both the bioenergy crop and neighboring row crops.[46,102] In fact, the most common potential concerns regarding *Miscanthus* have regarded its potential for invasiveness, given its non-native status in the US, and its potential for hosting insect pests of commercial crops.[46]

3.3 Designing for Yields and Ecosystem Services

Figure 3.1 shows several strategies to exploit the bioenergy crop traits discussed to design more sustainable landscapes to produce both biomass and ecosystem services. For example, buffers of deep rooted perennials within row crop fields can be designed to intercept both overland flow[103] and subsurface nutrient losses.[25]

In the latter example, subfield portions of land were identified that were more susceptible to nitrate losses and at the same time less suitable for the cost-effective production of corn (Figure 3.2). A buffer of deep rooted willows was planted in this subfield buffer strip to intercept the 120-cm-deep pool of dissolved nitrate and use it as a passively accessed fertilizer source (Figure 3.3). Modeled impacts of the buffer include total yields comparable to baseline, and nitrate exports reduced in comparison to the no-buffer case by 61 and 59% respectively for a switchgrass or willow buffer. Nitrous oxide emissions from the field were also reduced in the model by 5.5 and 10.8% respectively for switchgrass and willow.[25]

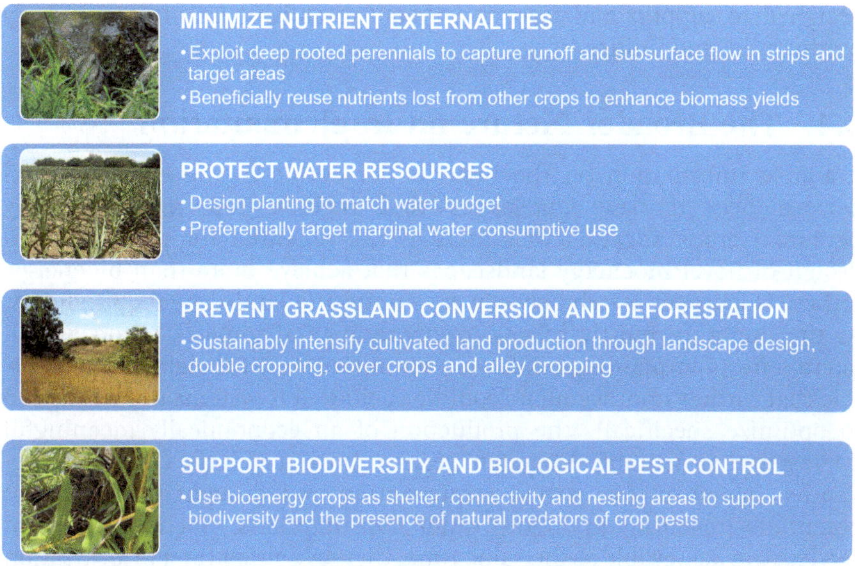

Figure 3.1 Strategies to leverage bioenergy crops' potential to maximize ecosystem services on the landscape (photo credit, third photo from top: USDA-NRCS).

Figure 3.2 Site analysis for the design of an effective bioenergy contour buffer to protect water quality: yield mapping and soil water mapping converge to identify a willow buffer location in a corn field.

In another example, researchers studied different designs of contour strips placed within row crops at the Neal Smith National Wildlife Refuge in Iowa. Results have shown that as little as 10% of field acreage converted to strips of prairie grasses could provide substantial and affordable improvements in sediment control, nutrient management and biodiversity[103] as well as provide bioenergy feedstock.[104] Likewise, landscapes can be designed to maximize biodiversity or pest suppression functions[100,105] and can be designed to support and favor the overall establishment of economically productive communities.[106]

3.4 The Broader Picture on Implementation

Available information on the benefits achievable in the environmental performance of new landscapes that include row and bioenergy perennial crops suggests that there is the potential for the successful design of novel bioenergy landscapes that achieve more than bioenergy, if there is a proper match between crop traits and landscape requirements. As the literature mostly relates to environmental impacts found as collateral effects in plantings that were optimized only for yields, it is plausible that further improvements could be achieved if systems were designed to optimize specifically the production of an economically meaningful harvest and measurable ecosystem service. As seen with the case of rooting depths, beneficial for carbon sequestration and drought tolerance but potentially resulting in higher water consumptions, not all ecosystem services are achievable at the same time, thereby calling for a prioritization of specific objectives as the basis, with specific local constraints, for the evaluation of necessary tradeoffs. An example of process for the prioritization, design and assessment of a sustainable bioenergy landscape to protect water quality is shown in Figure 3.3. This process highlights the importance of stakeholder involvement for the achievement of viable results.

While there is a significant potential to develop bioenergy deployments that deliver more than bioenergy, several barriers exist that need to be addressed. Growing bioenergy crops in a landscape design has challenges of acceptance, policy, scientific and economic nature. It requires planning at a higher level than a farm unit, needs policy incentives to gain traction, and may have costs higher than growing crops in the business-as-usual format: the logistics costs of planting, growing, harvesting and collecting biomass from a fragmented land base made of buffers and subfield patches has implicit costs, in the process of being quantified, related to moving equipment and collecting stranded resources over a larger land surface. Importantly, end uses and omnivorous technologies need to be developed that can be crop-agnostic and handle the diverse products of a sustainable landscape. As these crops are grown in an engineered design to provide more than energy, another challenge that lies ahead is to quantify these additional services and value them to calculate the overall value of the goods provided, and the return on the investment, for a farmer and as a societal benefit, of producing these crops in this different than usual approach. Nevertheless, a growing thought movement towards the perennialization of landscapes, and the development of solid and certifiable systems for the accounting and valuation of these ecosystem services are witnesses to the opportunities that lie ahead for bioenergy crops to deliver far more than energy.

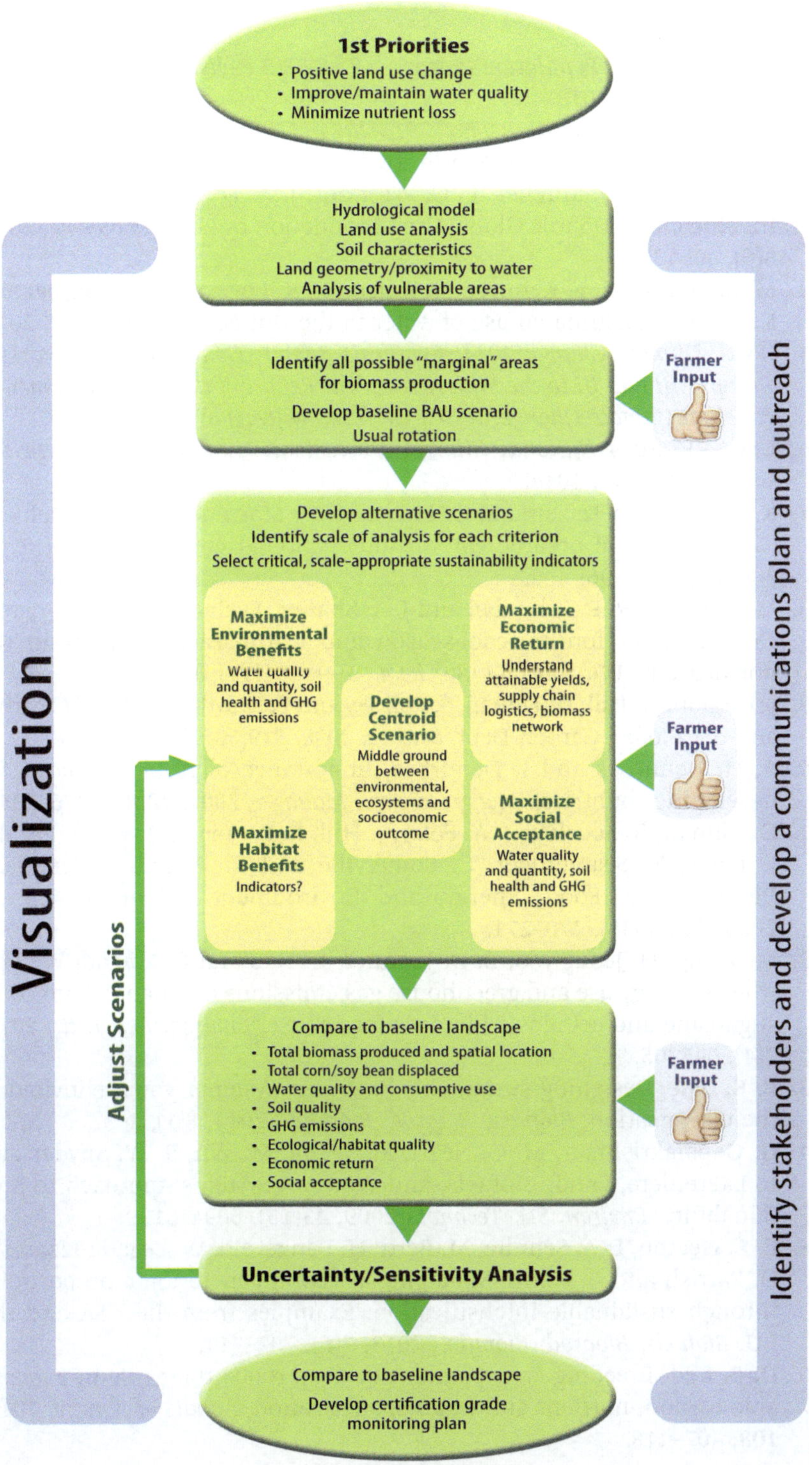

Figure 3.3 Example of a process to prioritize, design and monitor a sustainable bioenergy landscape to protect water quality.

References

1. UNDESA *World Population Prospects: The 2012 Revision*; United Nations: New York NY, 2013.
2. WRI *World Resources Report 2013-2014: Creating a Sustainable Food Future.: Interim Findings*; World Resources Insitute: 2013.
3. D. K. Ray, N. D. Mueller, P. C. West and J. A. Foley, Yield Trends Are Insufficient to Double Global Crop Production by 2050, *PLoS One*, 2013, **8**(6), e66428.
4. M. A. Maupin, J. F. Kenny, S. S. Hutson, J. K. Lovelace, N. L. Barber and K. S. Linsey, Estimated use of water in the United States in 2010, 2014.
5. IPCC, *Climate Change 2014: Synthesis Report. Contribution of Working Groups I, II and III to the Fifth Assessment Report of the Intergovernmental Panel on Climate Change*, IPCC: Geneva, Switzerlansd, 2014.
6. USGCRP, *2014 National Climate Assessment*; U.S. Global Change Research Program: 2014.
7. K. J. Anderson-Teixeira, S. C. Davis, M. D. Masters and E. H. Delucia, Changes in soil organic carbon under biofuel crops, *GCB Bioenergy*, 2009, **1**(1), 75–96.
8. M. D. Corre, R. R. Schnabel and J. A. Shaffer, Evaluation of soil organic carbon under forests, cool-season and warm-season grasses in the northeastern US, *Soil Biol. Biochem.*, 1999, **31**(11), 1531–1539.
9. J. Fargione, J. Hill, D. Tilman, S. Polasky and P. Hawthorne, Land Clearing and the Biofuel Carbon Debt, *Science*, 2008, **319**(5867), 1235–1238.
10. A. H. Demirbas and I. Demirbas, Importance of rural bioenergy for developing countries, *Energy Convers. Manage.*, 2007, **48**(8), 2386–2398.
11. D. Tilman, R. Socolow, J. A. Foley, J. Hill, E. Larson, L. Lynd, S. Pacala, J. Reilly, T. Searchinger, C. Somerville and R. Williams, Beneficial Biofuels—The Food, Energy, and Environment Trilemma, *Science*, 2009, **325**(5938), 270–271.
12. M. Wang, H. Jeongwoo, B. D. Jennifer, C. Hao and E. Amgad, Well-to-wheels energy use and greenhouse gas emissions of ethanol from corn, sugarcane and cellulosic biomass for US use, *Environ. Res. Lett.*, 2012, 7(4), 045905.
13. B. E. Dale, Designing sustainable bioenergy systems: you are invited to the conversation, *Biofuels, Bioprod. Biorefin.*, 2014, **8**(1), 1–3.
14. G. Gopalakrishnan, M. C. Negri, M. Wang, M. Wu, S. W. Snyder and L. LaFreniere, Land, Biofuels, and Water: A Systems Approach to Sustainability, *Environ. Sci. Technol.*, 2009, **43**(15), 6094–6100.
15. E. A. Heaton, L. A. Schulte, M. Berti, H. Langeveld, W. Zegada-Lizarazu, D. Parrish and A. Monti, Managing a second-generation crop portfolio through sustainable intensification: Examples from the USA and the EU, *Biofuels, Bioprod. Biorefin.*, 2013, 7(6), 702–714.
16. D. B. Kell, Breeding crop plants with deep roots: their role in sustainable carbon, nutrient and water sequestration, *Annals of Botany*, 2011, **108**, 407–418.

17. B. McGuire and S. Rupp, *Perennial Herbaceous biomass production and harvest in the Prairie pothole Region of the Northren Great Plains. Best Managemwent Guidelines to achieve sustainability of wildlife resources*, National Wildlife Federation, 2013.
18. (a) D. J. Cooper, D. R. D'amico and M. L. Scott, Physiological and morphological response patterns of Populus deltoides to alluvial groundwater pumping, *Environ. Manage.*, 2003, **31**(2), 0215–0226; (b) R. L. Scott, W. James Shuttleworth, D. C. Goodrich and T. Maddock III, The water use of two dominant vegetation communities in a semiarid riparian ecosystem, *Agric. For. Meteorol.*, 2000, **105**(1), 241–256.
19. E. A. Heaton, F. G. Dohleman and S. P. Long, Seasonal nitrogen dynamics of Miscanthus×giganteus and Panicum virgatum, *GCB Bioenergy*, 2009, **1**(4), 297–307.
20. Z. Ma, C. W. Wood and D. I. Bransby, Impacts of soil management on root characteristics of switchgrass, *Biomass Bioenergy*, 2000, **18**(2), 105–112.
21. M. A. Liebig, H. A. Johnson, J. D. Hanson and A. B. Frank, Soil carbon under switchgrass stands and cultivated cropland, *Biomass Bioenergy*, 2005, **28**(4), 347–354.
22. (a) D. E. Busch, N. L. Ingraham and S. D. Smith, Water uptake in woody riparian phreatophytes of the southwestern United States: a stable isotope study, *Ecol. Appl.*, 1992, 450–459; (b) Y. A. Kuzovkina and M. F. Quigley, Willows beyond wetlands: uses of Salix L. species for environmental projects, *Water, Air, Soil Pollut.*, 2005, **162**(1–4), 183–204; (c) M. Lubczynski, The hydrogeological role of trees in water-limited environments, *Hydrogeol. J.*, 2009, **17**(1), 247–259; (d) C. Negri, R. R. Hinchman and J. B. Wozniak, *Capturing a'mixed'contaminant plume: tritium phytoevaporation at Argonne National Laboratory's area 319*, Argonne National Laboratory, IL, US, 2000.
23. R. A. Holland, F. Eigenbrod, A. Muggeridge, G. Brown, D. Clarke and G. Taylor, A synthesis of the ecosystem services impact of second generation bioenergy crop production, *Renewable Sustainable Energy Rev.*, 2015, **46**, 30–40.
24. G. Gopalakrishnan, M. C. Negri and W. Salas, Modeling biogeochemical impacts of bioenergy buffers with perennial grasses for a row-crop field in Illinois, *GCB Bioenergy*, 2012, **4**(6), 739–750.
25. H. Ssegane, M. C. Negri, J. Quinn and M. Urgun-Demirtas, Multifunctional landscapes: Site characterization and field-scale design to incorporate biomass production into an agricultural system, *Biomass Bioenergy*, 2015, **80**(0), 179–190.
26. A. C. McBride, V. H. Dale, L. M. Baskaran, M. E. Downing, L. M. Eaton, R. A. Efroymson, C. T. Garten Jr, K. L. Kline, H. I. Jager, P. J. Mulholland, E. S. Parish, P. E. Schweizer and J. M. Storey, Indicators to support environmental sustainability of bioenergy systems, *Ecol. Indic.*, 2011, **11**(5), 1277–1289.

27. D. K. Lee, V. N. Owens and J. J. Doolittle, Switchgrass and Soil Carbon Sequestration Response to Ammonium Nitrate, Manure, and Harvest Frequency on Conservation Reserve Program Land, *Agron. J.*, 2007, **99**(2), 462–468.
28. D. K. Lee, J. J. Doolittle and V. N. Owens, Soil carbon dioxide fluxes in established switchgrass land managed for biomass production, *Soil Biol. Biochem.*, 2007, **39**(1), 178–186.
29. H. Blanco-Canqui, R. Lal and R. Lemus, Soil Aggregate Properties and Organic Carbon for Switchgrass and Traditional Agricultural Systems in the Southeastern United States, *Soil Sci.*, 2005, **170**(12), 998–1012.
30. P. Kahle, E. Hildebrand, C. Baum and B. Boelcke, Long-term effects of short rotation forestry with willows and poplar on soil properties, *Arch. Agron. Soil Sci.*, 2007, **53**(6), 673–682.
31. C. Bonin, R. Lal, M. Schmitz and S. Wullschleger, Soil physical and hydrological properties under three biofuel crops in Ohio, *Acta Agric. Scand., Sect. B*, 2012, **62**(7), 595–603.
32. M. J. Colloff, K. R. Pullen and S. A. Cunningham, Restoration of an Ecosystem Function to Revegetation Communities: The Role of Invertebrate Macropores in Enhancing Soil Water Infiltration, *Restor. Ecol.*, 2010, **18**, 65–72.
33. R. C. Schwartz, S. R. Evett and P. W. Unger, Soil hydraulic properties of cropland compared with reestablished and native grassland, *Geoderma*, 2003, **116**(1–2), 47–60.
34. M. J. Shipitalo, V. Nuutinen and K. R. Butt, Interaction of earthworm burrows and cracks in a clayey, subsurface-drained, soil, *Appl. Soil Ecol.*, 2004, **26**(3), 209–217.
35. M. R. Schmer, M. A. Liebig, K. P. Vogel and R. B. Mitchell, Field-scale soil property changes under switchgrass managed for bioenergy, *GCB Bioenergy*, 2011, **3**(6), 439–448.
36. Z. Qin, J. B. Dunn, H. Kwon, S. Mueller and M. M. Wander, Soil carbon sequestration and land use change associated with biofuel production: empirical evidence, *GCB Bioenergy*, 2015, DOI: 10.1111/gcbb.12237.
37. M. A. Sanderson, R. L. Reed, S. B. McLaughlin, S. D. Wullschleger, B. V. Conger, D. J. Parrish, D. D. Wolf, C. Taliaferro, A. A. Hopkins, W. R. Ocumpaugh, M. A. Hussey, J. C. Read and C. R. Tischler, Switchgrass as a sustainable bioenergy crop, *Bioresour. Technol.*, 1996, **56**(1), 83–93.
38. D. L. Gebhart, H. B. Johnson, H. S. Mayeux and H. W. Polley, The CRP increases soil organic carbon, *J. Soil Water Conserv.*, 1994, **49**(5), 488–492.
39. IPCC *Good Pracrice Guidance for land use, land use change and forestry. INtergovernmental Panel on Climate Change (IPCC), National Greenhouse Gas Inventories Programme*, Institute for GLobal Environmental Strategies, 2003.
40. C. T. Garten and S. D. Wullschleger, Soil Carbon Dynamics beneath Switchgrass as Indicated by Stable Isotope Analysis, *J. Environ. Qual.*, 2000, **29**(2), 645–653.

41. A. Monti, *Switchgrass, A valuable biomass Crop for Bioenergy*, Springer Verlag, London, 2012.
42. F. G. Dou, F. M. Hons, W. R. Ocumpaugh, J. C. Read, M. A. Hussey and J. P. Muir, Soil Organic Carbon Pools Under Switchgrass Grown as a Bioenergy Crop Compared to Other Conventional Crops, *Pedosphere*, 2013, **23**, 409–416.
43. C. S. Zan, J. W. Fyles, P. Girouard and R. A. Samson, Carbon sequestration in perennial bioenergy, annual corn and uncultivated systems in southern Quebec, *Agric., Ecosyst. Environ.*, 2001, **86**(2), 135–144.
44. E. M. Hansen, B. T. Christensen, L. S. Jensen and K. Kristensen, Carbon sequestration in soil beneath long-term Miscanthus plantations as determined by 13C abundance, *Biomass Bioenergy*, 2004, **26**(2), 97–105.
45. G. M. Richter, F. Agostini, M. Redmile-Gordon, R. White and K. W. T. Goulding, Sequestration of C in soils under Miscanthus can be marginal and is affected by genotype-specific root distribution, *Agric., Ecosyst. Environ.*, 2015, **200**(0), 169–177.
46. NRCS, U. *Planting and Managing Giant Miscanthus as a Biomass Energy crop*, Washington DC, 2011.
47. A. S. Gregory, C. P. Webster, C. W. Watts, W. R. Whalley, C. J. A. Macleod, A. Joynes, A. Papadopoulos, P. M. Haygarth, A. Binley, M. W. Humphreys, L. B. Turner, L. Skot and G. P. Matthews, Soil Management and Grass Species Effects on the Hydraulic Properties of Shrinking Soils, *Soil Sci. Soc. Am. J.*, 2010, **74**(3), 753–761.
48. C. J. A. Macleod, A. Binley, S. L. Hawkins, M. W. Humphreys, L. B. Turner, W. R. Whalley and P. M. Haygarth, Genetically modified hydrographs: what can grass genetics do for temperate catchment hydrology?, *Hydrol. Processes*, 2007, **21**(16), 2217–2221.
49. C. T. Garten Jr, J. L. Smith, D. D. Tyler, J. E. Amonette, V. L. Bailey, D. J. Brice, H. F. Castro, R. L. Graham, C. A. Gunderson, R. C. Izaurralde, P. M. Jardine, J. D. Jastrow, M. K. Kerley, R. Matamala, M. A. Mayes, F. B. Metting, R. M. Miller, K. K. Moran, W. M. Post III, R. D. Sands, C. W. Schadt, J. R. Phillips, A. M. Thomson, T. Vugteveen, T. O. West and S. D. Wullschleger, Intra-annual changes in biomass, carbon, and nitrogen dynamics at 4-year old switchgrass field trials in west Tennessee, USA, *Agric., Ecosyst. Environ.*, 2010, **136**(1–2), 177–184.
50. C. T. Garten and S. D. Wullschleger, Soil Carbon Inventories under a Bioenergy Crop (Switchgrass): Measurement Limitations, *J. Environ. Qual.*, 1999, **28**(4), 1359–1365.
51. K. D. Eggemeyer, T. Awada, F. E. Harvey, D. A. Wedin, X. Zhou and C. W. Zanner, Seasonal changes in depth of water uptake for encroaching trees Juniperus virginiana and Pinus ponderosa and two dominant C4 grasses in a semiarid grassland, *Tree Physiol.*, 2009, **29**(2), 157–169.
52. P. Crow and T. J. Houston, The influence of soil and coppice cycle on the rooting habit of short rotation poplar and willow coppice, *Biomass Bioenergy*, 2004, **26**(6), 497–505.

53. M. C. Negri, E. G. Gatliff, J. J. Quinn and R. R. Hinchman, Root Development and Rooting at Depths, in *Phytoremediation*, John Wiley & Sons, Inc., 2004, pp. 233–262.
54. A. Ferro, M. Gefell, R. Kjelgren, D. S. Lipson, N. Zollinger and S. Jackson, Maintaining Hydraulic Control Using Deep Rooted Tree Systems, in *Phytoremediation*, ed. D. Tsao, Springer, Berlin Heidelberg, 2003, vol. 78, pp. 125–156.
55. F. Ferchaud, G. Vitte, F. Bornet, L. Strullu and B. Mary, Soil water uptake and root distribution of different perennial and annual bioenergy crops, *Plant Soil*, 2015, **388**(1/2), 307–322.
56. D. Neukirchen, M. Himken, J. Lammel, U. Czypionka-Krause and H. W. Olfs, Spatial and temporal distribution of the root system and root nutrient content of an established Miscanthus crop, *Eur. J. Agron.*, 1999, **11**(3–4), 301–309.
57. A. Monti and A. Zatta, Root distribution and soil moisture retrieval in perennial and annual energy crops in Northern Italy, *Agric., Ecosyst. Environ.*, 2009, **132**(3–4), 252–259.
58. R. H. Skinner, W. Zegada-Lizarazu and J. P. Schmidt, Environmental impacts of switchgrass management for bioenergy production, in *Switchgrass, Green Energy and Technology*, ed. A. Monti, Springer-Verlag, London, U.K., 2012.
59. J. E. Gilley, B. Eghball, L. A. Kramer and T. B. Moorman, Narrow grass hedge effects on runoff and soil loss, *J. Soil Water Conserv.*, 2000, **55**(2), 190–196.
60. H. Blanco-Canqui, C. J. Gantzer, S. H. Anderson and E. E. Alberts, Grass Barriers for Reduced Concentrated Flow Induced Soil and Nutrient Loss, *Soil Sci. Soc. Am. J.*, 2004, **68**(6), 1963–1972.
61. B. Eghball, J. E. Gilley, L. A. Kramer and T. B. Moorman, Narrow grass hedge effects on phosphorus and nitrogen in runoff following manure and fertilizer application, *J. Soil Water Conserv.*, 2000, **55**(2), 172–176.
62. A. Rachman, S. H. Anderson, C. J. Gantzer and A. L. Thompson, Influence of Stiff-Stemmed Grass Hedge Systems on Infiltration, *Soil Sci. Soc. Am. J.*, 2004, **68**(6), 2000–2006.
63. Y. Kuzovkina and M. Quigley, Willows Beyond Wetlands: Uses of Salix L. Species for Environmental Projects, *Water, Air, Soil Pollut.*, 2005, **162**(1–4), 183–204.
64. R. Lemus, D. Parrish and O. Abaye, Nitrogen-Use Dynamics in Switchgrass Grown for Biomass, *BioEnergy Res.*, 2008, **1**(2), 153–162.
65. J. Toillon, B. Rollin, E. Dallé, M. Feinard-Duranceau, J.-C. Bastien, F. Brignolas and N. Marron, Variability and plasticity of productivity, water-use efficiency, and nitrogen exportation rate in Salix short rotation coppice, *Biomass Bioenergy*, 2013, **56**, 392–404.
66. H. G. Adegbidi, T. A. Volk, E. H. White, L. P. Abrahamson, R. D. Briggs and D. H. Bickelhaupt, Biomass and nutrient removal by willow clones in experimental bioenergy plantations in New York State, *Biomass Bioenergy*, 2001, **20**(6), 399–411.

67. L. P. Abrahamson, T. A. Volk, L. B. Smart and K. D. Cameron, *Shrub Willow Biomass Producers Handbook*, State University of New York, College of Environmental Science and Forestry, Syracuse, NY, 2010.
68. H. W. Zub and M. Brancourt-Hulmel, Agronomic and physiological performances of different species of Miscanthus, a major energy crop. A review, *Agron. Sustainable Dev.*, 2010, **30**(2), 201–214.
69. D. I. Bransby, S. B. McLaughlin and D. J. Parrish, A review of carbon and nitrogen balances in switchgrass grown for energy, *Biomass Bioenergy*, 1998, **14**(4), 379–384.
70. W. L. Stout and G. A. Jung, Biomass and Nitrogen Accumulation in Switchgrass: Effects of Soil and Environment, *Agron. J.*, 1995, **87**(4), 663–669.
71. Y. Huang, D. H. Rickerl and K. D. Kephart, Recovery of Deep-Point Injected Soil Nitrogen-15 by Switchgrass, Alfalfa, Ineffective Alfalfa, and Corn, *J. Environ. Qual.*, 1996, **25**(6), 1394–1400.
72. G. F. McIsaac, M. B. David and C. A. Mitchell, Miscanthus and Switchgrass Production in Central Illinois: Impacts on Hydrology and Inorganic Nitrogen Leaching, *J. Environ. Qual.*, 2010, **39**(5), 1790–1799.
73. P. V. Le, P. Kumar and D. T. Drewry, Implications for the hydrologic cycle under climate change due to the expansion of bioenergy crops in the Midwestern United States, *Proc. Natl. Acad. Sci. U. S. A.*, 2011, **108**(37), 15085–15090.
74. S. D. Wullschleger, E. B. Davis, M. E. Borsuk, C. A. Gunderson and L. R. Lynd, Biomass Production in Switchgrass across the United States: Database Description and Determinants of Yield, *Agron. J.*, 2010, **102**(4), 1158–1168.
75. W. Zegada-Lizarazu, S. D. Wullschleger, S. S. Nair and A. Monti, Crop Physiology, in *Switchgrass, Green Energy and Technology*, ed. A. Monti, Springer-Verlag, London U. K., 2012.
76. A. Lindroth, T. Verwijst and S. Halldin, Water-use efficiency of willow: Variation with season, humidity and biomass allocation, *J. Hydrol.*, 1994, **156**(1–4), 1–19.
77. R. L. Hall, S. J. Allen, P. T. W. Rosier and R. Hopkins, Transpiration from coppiced poplar and willow measured using sap-flow methods, *Agric. For. Meteorol.*, 1998, **90**(4), 275–290.
78. J. J. Opperman, G. E. Galloway, J. Fargione, J. F. Mount, B. D. Richter and S. Secchi, Sustainable Floodplains Through Large-Scale Reconnection to Rivers, *Science*, 2009, **326**(5959), 1487–1488.
79. G. M. Richter, A. B. Riche, A. G. Dailey, S. A. Gezan and D. S. Powlson, Is UK biofuel supply from Miscanthus water-limited? *Soil Use Manage.*, 2008, **24**(3), 235–245.
80. J. J. Mann, J. N. Barney, G. B. Kyser and J. M. Di Tomaso, Miscanthus × giganteus and Arundo donax shoot and rhizome tolerance of extreme moisture stress, *GCB Bioenergy*, 2013, **5**(6), 693–700.
81. J. H. Fike, D. J. Parrish, D. D. Wolf, J. A. Balasko, J. T. Green Jr, M. Rasnake and J. H. Reynolds, Long-term yield potential of

switchgrass-for-biofuel systems, *Biomass Bioenergy*, 2006, **30**(3), 198–206.
82. L. Rinehart, Switchgrass as a bioenergy crop, http://attra.ncat.org/attra-pub/PDF/switchgrass.pdf.
83. J. N. Barney, J. J. Mann, G. B. Kyser, E. Blumwald, A. Van Deynze and J. M. DiTomaso, Tolerance of switchgrass to extreme soil moisture stress: Ecological implications, *Plant Sci.*, 2009, **177**(6), 724–732.
84. (a) S. B. McLaughlin and L. Adams Kszos, Development of switchgrass (Panicum virgatum) as a bioenergy feedstock in the United States, *Biomass Bioenergy*, 2005, **28**(6), 515–535; (b) M. A. Sanderson, J. C. Read and R. L. Reed, Harvest Management of Switchgrass for Biomass Feedstock and Forage Production, *Agronomy J.*, 1999, **91**(1), 5–10.
85. S. R. Pezeshki, P. Anderson and F. D. Shields, Effects of soil moisture regimes on growth and survival of black willow (Salix nigra) posts (cuttings), *Wetlands*, 1998, **18**(3), 460–470.
86. N. Amlin and S. Rood, Comparative tolerances of riparian willows and cottonwoods to water-table decline, *Wetlands*, 2002, **22**(2), 338–346.
87. J. Yang, H. Zhao and T. Zhang, Heat and drought tolerance of two willow species, Salix gordejevii and Salix babylonica: A comparative study, *Isr. J. Plant Sci.*, 2004, **52**(4), 301–306.
88. M. Stauffer, C. Leyval, J. J. Brun, P. Leportier and J. Berthelin, Effect of willow short rotation coppice on soil properties after three years of growth as compared to forest, grassland and arable land uses, *Plant Soil*, 2014, **377**(1–2), 423–438.
89. K. M. Robinson, A. Karp and G. Taylor, Defining leaf traits linked to yield in short-rotation coppice Salix, *Biomass Bioenergy*, 2004, **26**(5), 417–431.
90. F. T. de Vries, E. Hoffland, N. van Eekeren, L. Brussaard and J. Bloem, Fungal/bacterial ratios in grasslands with contrasting nitrogen management, *Soil Biol. Biochem.*, 2006, **38**(8), 2092–2103.
91. A. W. Ratcliff, M. D. Busse and C. J. Shestak, Changes in microbial community structure following herbicide (glyphosate) additions to forest soils, *Appl. Soil Ecol.*, 2006, **34**(2–3), 114–124.
92. B. A. Robertson, P. J. Doran, E. R. Loomis, J. R. Robertson and D. W. Schemske, Avian Use of Perennial Biomass Feedstocks as Post-Breeding and Migratory Stopover Habitat, *PLoS One*, 2011, **6**(3), e16941.
93. B. Robertson, D. Landis, T. S. Sillett, E. Loomis and R. Rice, Perennial Agroenergy Feedstocks as En Route Habitat for Spring Migratory Birds, *BioEnergy Res.*, 2013, **6**(1), 311–320.
94. R. Isaacs, J. Tuell, A. Fiedler, M. Gardiner and D. Landis, Maximizing arthropod-mediated ecosystem services in agricultural landscapes: the role of native plants, *Front. Ecol. Environ.*, 2008, **7**(4), 196–203.
95. J. E. Losey and M. Vaughan, The Economic Value of Ecological Services Provided by Insects, *BioScience*, 2006, **56**(4), 311–323.

96. M. Gardiner, J. Tuell, R. Isaacs, J. Gibbs, J. Ascher and D. Landis, Implications of Three Biofuel Crops for Beneficial Arthropods in Agricultural Landscapes, *BioEnergy Res.*, 2010, **3**(1), 6–19.
97. B. Robertson, C. Porter, D. Landis and D. Schemske, Agroenergy Crops Influence the Diversity, Biomass, and Guild Structure of Terrestrial Arthropod Communities, *BioEnergy Res.*, 2012, **5**(1), 179–188.
98. T. D. Meehan, B. P. Werling, D. A. Landis and C. Gratton, Pest-Suppression Potential of Midwestern Landscapes under Contrasting Bioenergy Scenarios, *PLoS One*, 2012, **7**(7), e41728.
99. B. P. Werling, T. D. Meehan, B. A. Robertson, C. Gratton and D. A. Landis, Biocontrol potential varies with changes in biofuel–crop plant communities and landscape perenniality, *GCB Bioenergy*, 2011, **3**(5), 347–359.
100. B. P. Werling, T. L. Dickson, R. Isaacs, H. Gaines, C. Gratton, K. L. Gross, H. Liere, C. M. Malmstrom, T. D. Meehan, L. Ruan, B. A. Robertson, G. P. Robertson, T. M. Schmidt, A. C. Schrotenboer, T. K. Teal, J. K. Wilson and D. A. Landis, Perennial grasslands enhance biodiversity and multiple ecosystem services in bioenergy landscapes, *Proc. Natl. Acad. Sci.*, 2014, **111**(4), 1652–1657.
101. R. L. Rowe, D. Goulson, C. P. Doncaster, D. J. Clarke, G. Taylor and M. E. Hanley, Evaluating ecosystem processes in willow short rotation coppice bioenergy plantations, *GCB Bioenergy*, 2013, **5**(3), 257–266.
102. L. J. Thomson and A. A. Hoffmann, Pest management challenges for biofuel crop production, *Curr. Opin. Environ.*, 2011, **3**(1–2), 95–99.
103. J. Gutierrez-Lopez, H. Asbjornsen, M. Helmers and T. Isenhart, Regulation of soil moisture dynamics in agricultural fields using strips of native prairie vegetation, *Geoderma*, 2014, **226–227**, 238–249.
104. M. Liebman, M. J. Helmers and L. A. Schulte, Integrating conservation with biofueld feedstock production, in *Managing Agricultural Landscapes for Environmental Quality II: Achieving More Effective Conservation*, ed. P. Nowak and M. Schnepf, Soil and Water Conservation Society, Ankeny IA, 2010.
105. T. D. Meehan, C. Gratton, E. Diehl, N. D. Hunt, D. F. Mooney, S. J. Ventura, B. L. Barham and R. D. Jackson, Ecosystem-service tradeoffs associated with switching from annual to perennial energy crops in riparian zones of the US Midwest, *PloS one*, 2013, **8**(11), e80093.
106. N. Jordan, L. Schulte, C. Williams, D. Mulla, D. Pitt, C. S. Slotterback, R. Jackson, D. Landis, B. Dale, D. Becker, M. Rickenbach, M. Bobby and H. Bringi, Landlabs: An Integrated approach to creating agricultural enterprises that meet the triple bottom line, *J. High. Educ. Outreach Engage.*, 2013, **17**(4).

CHAPTER 4

Butanol Production by Fermentation: Efficient Bioreactors

ADRIANO P. MARIANO,[a] THADDEUS C. EZEJI*[b] AND NASIB QURESHI*[c]

[a] Laboratory of Optimization, Design and Advanced Control – LOPCA, School of Chemical Engineering, University of Campinas – UNICAMP, Campinas, SP, Brazil; [b] The Ohio State University, Department of Animal Sciences, Ohio State Agricultural Research and Development Center, 305 Gerlaugh Hall, 1680 Madison Avenue, Wooster, OH 44691, USA; [c] United States Department of Agriculture,[†] National Center for Agricultural Utilization Research, Bioenergy Research Unit, Peoria, IL 61604, USA

*Email: ezeji.1@osu.edu; Nasib.Qureshi@ars.usda.gov

4.1 Introduction

Bio-based *n*-butanol (hereafter referred to as butanol) is traditionally produced *via* the acetone–ethanol–butanol (ABE or solvent) process in which ABE are produced during fermentation of sugars by solventogenic *Clostridium* species.[1] As one of the oldest fermentation processes employed for the commercial production of chemicals, the ABE process has a long

[†]Mention of trade names or commercial products in this article is solely for the purpose of providing scientific information and does not imply recommendation or endorsement by the United States Department of Agriculture. USDA is an equal opportunity provider and employer.

RSC Green Chemistry No. 43
Commercializing Biobased Products: Opportunities, Challenges, Benefits, and Risks
Edited by Seth W. Snyder

Published by the Royal Society of Chemistry, www.rsc.org

history that dates back to 19th century and its industrial significance was recognized in the early 20th century. Production of butanol was discovered by Pasteur in 1861 and in 1912, Chaim Weizmann (recognized as the father of ABE fermentation) discovered a microorganism called *Clostridium acetobutylicum*, which is able to ferment sugars or starch to ABE.[2] The first commercial ABE fermentation plant was built in Terra Haute, Indiana (USA), in 1918 by the Commercial Solvents Corporation. This venture supplied butanol for the production of butyl acetate, the primary component of paint lacquers. With the expiration of the Weizmann patent for the ABE process in 1936, ABE plants were built throughout the United States, and in other countries including Great Britain, Canada, France, Japan, India, China, Australia, South Africa, Taiwan, Egypt, Brazil, and the former Soviet Union, thus making the ABE process a major industrial fermentation process and recognized second in importance and history only to yeast-based ethanol fermentation.

Due to increasing substrate costs and the availability of much cheaper petroleum-derived butanol, most of the industrial fermentation plants liquidated between the 1950s and 1960s.[3] In the 1980s, butanol plants in South Africa, Egypt, and the Soviet Union also ceased operations. Interestingly, due to political and economic constraints, the Chinese butanol industry expanded rapidly after the 1950s, reaching a peak of about 30 plants with an annual solvent production capacity of 170 000 tons mainly from fermentation of corn starch. The majority of these plants liquidated in the 2000s due to strong competition from petroleum-derived butanol, whose demand was boosted by Chinese economic development and market reforms.[4] The majority of the plants ceased operation in the aftermath of the 2008 financial crisis during which the price of butanol declined from US$2400/ton to US$1000/ton. The feedstock for these plants was corn starch and/or molasses and by 2010 the prices of butanol increased again. During recent years, it has been shown that replacement of corn starch substrate with lignocellulosic biomass substrate would cut down the production cost of butanol from approximately US$1900 per ton to US$1300 per ton[5–7] thus increasing profit margins.

While commercial production of bio-based butanol in China has declined significantly, global research and development efforts toward developing superior butanol-producing microbes together with advanced process technologies continued mainly due to rising oil prices and environmental concerns such as greenhouse gas emissions. It is noteworthy that butanol is recognized by its "drop-in" characteristics and is a superior blend stock that can be added to gasoline, diesel, biodiesel, and ethanol. Additionally, butanol can also be catalytically upgraded to jet fuel. The current chemical market for butanol is approximately 3 million ton per year, which is used for the production of a wide range of polymers, plastics, solvent for paints, and chemical stabilizers.

The majority of the bio-based butanol plants in China employ batch bioreactors for fermentation followed by product recovery by distillation.

These production and recovery methods lack energy efficiency and high productivity. To address these issues new inventions have been developed around the world, including generation of superior microbial strains with improved butanol yield and product tolerance, and novel processes for simultaneous butanol production and recovery. It is believed that these novel processes will play a major role in reviving the butanol fermentation industry.[6] The process development that the authors of this chapter refer to include continuous fermentation in high productivity reactors along with simultaneous product recovery using energy efficient alternate product recovery technologies. It is estimated that a two-fold increase in productivity could reduce capital investment and operating costs by approximately 20%.[5] With this view, the major focus of this chapter is to highlight the development of novel continuous reactor systems around the globe. These reactor systems include free cell continuous, immobilized cell continuous, and cell recycle continuous reactors. Additionally, some technologies on simultaneous product recovery will also be presented.

4.2 Traditional Technology: Batch Fermentation

Industrially, butanol was traditionally produced in batch reactors (150 000 to 750 000 liter tanks) as a mixture of ABE after a fermentation time of 36 to 72 h, which was recovered by distillation. Corn starch in the United States and sugarcane molasses in South Africa were the typical feedstocks. In some operations, a set of 20 or more series bioreactors were connected. The first reactor was fed with 60 g L^{-1} sterile sugar/liquefied corn starch solution and inoculated with actively growing *Clostridium* cells. The product of the first reactor containing microbial cells was fed to the second reactor and so on. The temperature of the reactors was maintained at 35–37 °C. Details of inoculum preparations and operation of industrial size reactors can be found elsewhere.[2]

In the ABE process, fermentation goes through two distinct metabolic phases. The first phase, called acidogenic, involves production of acetic and butyric acids. These acids are secreted by *Clostridium* species such as *C. beijerinckii* and *C. acetobutylicum* into the medium causing the pH of the medium to drop to acidic range. The second phase, called solventogenic, starts after the "pH-breakpoint" and results in the re-assimilation of acetic and butyric acids for ABE production. Depending on the microorganism, ABE mass yield ranges from 0.30 to 0.40 g g^{-1} and total ABE accumulates to a concentration of 20 g L^{-1} or less (usual ratio A–B–E 3 : 6 : 1). Butanol is toxic to the culture and inhibition can occur at concentrations as low as 5 g L^{-1} and a microbial cell concentration higher than 4 g L^{-1} is rarely achieved. Besides butanol, high sugar concentration (exceeding 100 g L^{-1}) can also be inhibitory to the microorganism. For example, maximum cell concentration of *C. beijerinckii* BA101 is attained at glucose concentrations between 60 and 100 g L^{-1}, and cell growth is severely inhibited above 160 g L^{-1} glucose. For this culture, a long lag phase for cell growth was observed which ranged from a few hours to 20 hours when glucose concentration ranged from 60 to 160 g L^{-1}, respectively.[8]

Although the operation of batch reactors is simple with a low risk of contamination, the operation mode results in low ABE or butanol productivity due to butanol toxicity and downtime such as charging, sterilization, lag phase, emptying the reactor, and cleaning. As a result, reactor productivity in the batch operation rarely exceeds 0.5–0.6 g L h^{-1} (slow fermentation). For illustration, Table 4.1 shows productivity and other important performance parameters such as product concentration and yield for batch fermentations conducted on a laboratory scale using different feedstocks and microbial strains.

Batch ABE fermentation technology was the process of choice in the last century due to ease of operation. However, as indicated, it resulted in low ABE productivities, which results in the requirement of large fermentors and hence is an uneconomic process. Therefore, efforts were made to develop novel fermentation technologies with improved productivity for this fermentation. One of the techniques to improve productivity is the use of continuous fermentation systems that result in elimination of lag time, and time to empty reactors, cleaning and refilling them for the next fermentation thus reducing process time considerably and hence improving productivity. These continuous bioreactors can be categorized into three reactor types namely suspended cell, immobilized cell, and cell recycle continuous bioreactors. Furthermore, a combination of continuous bioreactors with simultaneous product recovery systems is expected to improve the economics of butanol production. A detailed description of these technologies is presented in the following paragraphs, tables and figures.

4.3 Continuous Systems

4.3.1 Free Cell Continuous Bioreactors

In continuous systems the reactor is initiated in a batch mode and microbial cell growth is allowed until exponential growth phase is attained. While microbial cells are in exponential growth phase, the reactor is fed continuously with fermentation medium and product stream is withdrawn at the same flow rate, thus maintaining a constant volume in the reactor.

Although trials with single-stage continuous bioreactors have been partially successful, scale-up of the process for industrial butanol production is challenging due to the complex physiology of *Clostridium* cultures such as oscillating production levels of ABE and culture instability. It is interesting to note that while high ABE productivities can be achieved at various time periods, ABE production is rarely steady. Another common issue found in long-time continuous operations is the degeneration of the culture over time during which genetic modification of the *Clostridium* culture may occur.[2,19]

Furthermore, ABE concentration in single-stage continuous reactors is generally lower than values achieved in batch operations. The design of an

Table 4.1 Performance parameters of representative batch fermentations obtained in laboratory scale experiments using different feedstocks and *Clostridium* strains.

Feedstock	Culture	Initial sugar concentration (g L^{-1})	Sugar utilization (g L^{-1})	ABE (g L^{-1})	ABE yield (g g^{-1})	ABE prod. (g L h^{-1})	Ref.
Glucose	*C. beijerinckii* P260	62	48.9	20.1	0.41	0.28	9
Glucose	*C. acetobutylicum*	66	60.0	18.0	0.30	0.30	10
Glucose	*C. saccharo-perbutylacetonicum* N1-4 ATCC 13 564	90	55.8	21.2	0.38	0.26	11
Xylose	*C. acetobutylicum* ATCC 824	30	27.2	7.52	0.28	0.25	12
Soy molasses	*C. beijerinckii* BA101	34.7	34.5	10.7	0.31	0.11	13
Wheat straw	*C. beijerinckii* P260	41.9	41.4	13.4	0.32	0.14	14
Sago starch	*C. saccharobutylicum* DSM 13 864	50	50	16.4	0.35	0.59	15
Corn fiber	*C. beijerinckii* IB4	30.3	26.4	9.5	0.31	0.13	16
Corn stover	*C. acetobutylicum* ATCC 824	50[a]	41	12.4	0.30	0.17	17
Barley straw	*C. beijerinckii* P260	58.8[a]	21.8	7.1	0.33	0.10	18

[a]Mixed with laboratory-grade sugar.

industrial continuous process should target at least the final concentration obtained in batch reactors (~20 g L^{-1}) in order to improve economics. However, at such ABE concentrations, cultures are very unstable leading to an increase of acid formation.

To overcome the problems of the single-stage continuous bioreactors, two- or multi-stage continuous fermentation systems have been developed. In two-stage bioreactors, the inhibitory effect of butanol can be decreased by allowing the successive acidogenic and solventogenic fermentations in the separate reactors. Consequently, cells are exposed to ABE concentrations comparable to values attained in batch reactors only in the second stage bioreactors, resulting in more stable solvent production.[1,20,21] The benefits of this strategy are clearly illustrated in Table 4.2, which presents the performance of representative single, double, and multi-stage continuous fermentations conducted on the laboratory scale.

During middle of the last century, multi-stage free-cell continuous systems transitioned from laboratory benches to pilot and commercial plants. A good example is the former Dokshukino plant in USSR (Russia now) in which a multi-stage free-cell continuous fermentation system demonstrated efficiency and cost savings during a four-week continuous operation on pilot plant and large scales.[30] Use of multi-stage continuous fermentations was also exercised in China. In fact, the continuous operation mode has been used in most of their butanol plants and productivity in the continuous multi-stage bioreactors (generally six to eight tanks with 200–500 m^3 volume capacity per tank running continuously for 7 to 21 days) was usually 20% to 50% higher than batch operations.[4,7] In the US, the biotech startup company Green Biologics has been developing an alternative approach in which one bacterial strain is used to produce butyric acid in the first stage and a second bacterial strain converts the acid into butanol in the second stage.[5]

Similar to the Russian experience, Chinese plants also encountered contamination problems. It is difficult to avoid contamination problems in continuous operations. In an effort to fully take advantage of continuous operation in particular for productivity and energy efficiency, coupling of immobilized cell reactors, cell recycle units, and in-line product recovery have been evaluated. Important technical aspects about these emerging technologies are presented in the following sections.

4.3.2 Immobilized Cell Continuous Reactors

Although continuous free cell systems offer higher productivity than batch systems, high cell concentration cannot be retained as there is no means to retain the cells in the reactors and cell washout occurs at high dilution rates. For this reason, methods to retain high cell concentration and operate the reactor at high flow rates have been developed. Specifically, the two major systems employ "cell immobilization" and "cell recycle" techniques.[31] While this subsection will focus on the cell immobilization, the cell recycle system will be discussed in Subsection 4.3.3. By using cell immobilization methods, cell concentrations in excess of 50–70 g L^{-1} can

Table 4.2 Performance parameters of representative free cell continuous fermentations obtained in laboratory scale experiments using different feedstocks and *Clostridium* strains.

Feedstock	Culture	Initial sugar concentration (g L^{-1})	Number of stages	Dilution rate (h^{-1})	ABE (g L^{-1})	ABE yield (g g^{-1})	ABE prod. (g L h^{-1})	Ref.
Lab-grade glucose	*C. saccharo-perbutylacetonicum* N1-4 (ATCC 13 564)	50	1	0.20	9.27	—	1.85	22
Saccharified degermed corn (with P2 nutrients)	*C. beijerinckii* BA 101	60	1	0.03	9.7	—	0.29	23
Soluble corn starch	*C. beijerinckii* BA 101	30	1	0.02	9.8	—	0.20	24
Lab-grade glucose (with addition of butyrate)	*C. beijerinckii* NCIMB 8052	60	1	0.04	9.4	0.24[a]	0.22[a]	25
Gelatinized sago starch	*C. saccharobutylicum* DSM 13 864	30	1	0.05	9.10	0.30	0.46	15
Lab-grade glucose	*C. acetobutylicum* DSM 1731	60	2	0.125/0.03	18.2	—	—	1
Lab-grade glucose	*C. acetobutylicum* DSM 792	60	2	0.13/0.13	15.4	0.25	1.93	16
Lab-grade glucose	*C. acetobutylicum* ATCC 824	60	2	0.08/0.04	21	0.36	0.56	21
Lab-grade glucose	Packed-bed reactor in series with a free cell reactor	60	2	0.57/0.18	9.27	0.23	1.24	27
Lab-grade glucose	*C. beijerinckii* NRRL B592	50	2	0.12/0.022	15	0.25	0.27	28
Cane molasses	*C. saccharo-butylicum* DSM 13 864	58	4	0.15/0.15/ 0.125/0.1	13.7	0.32	0.44	29
Corn stover hydrolysate	*C. saccharo-butylicum* DSM 13 864	52	4	0.15/0.15/ 0.15/0.15	11.4	0.41	0.43	29

[a]Value refers to butanol and acetone. Fresh medium contained 2 g L^{-1} sodium butyrate.

be achieved. Such reactors can be operated at high flow rates, resulting in a 20 to 40 fold increase in productivity as compared to batch reactors.[32] Additionally, immobilized cell continuous reactors are more stable with longer life spans than free cell systems.[26,33,34]

There are three categories of cell immobilization techniques, namely adsorption, entrapment, and covalent bond formation. Their details and advantages and disadvantages are as follows:

(i) Adsorption is a simple technique in which microbial cells adsorb onto surfaces of adsorption materials such as wood shavings, bonechar, glass wool, clay brick, wood pulp, and sugarcane bagasse.[35–38] The adsorption of microbial cells onto the aforementioned materials occurs through interactions involving van der Waals forces and ionic and hydrogen bonds. Notably, this technique does not require any chemicals for attaching microbial cells onto the support materials. The adsorption process can either be conducted *in situ* in a single stage in which the reactor is packed with adsorbent materials followed by inoculation with the culture and feeding with nutrient solutions or two to more stages in which column reactors packed with adsorbent materials are connected in series and operated.
(ii) Entrapment is a process in which microbial cells are trapped inside gel layers such as calcium-alginate, polyacrylamide, and carrageenan. In this technique, cells grown in batch reactors are harvested by centrifugation. The microbial cell pellets are transferred to a gel suspension/solution and mixed. The mixture is passed through a needle to form gel beads. The gel beads containing microbial cells may require further treatment such as hardening or polymerization. The gel beads are either packed in a column reactor or suspended in a continuous stirred tank reactor (CSTR).
(iii) Covalent bonding immobilization is facilitated by covalent bond formation between an activated support and microbial cells in the presence of a binding agent. Compared to physical adsorption, stronger microbial cell-support binding is formed, which reduces cell loss due to detachment from the support. Depending on the microorganism, some of the binding agents (chemicals such as γ-amino propyl triethoxy silane, isocyanate, hydroxyethyl acrylate, sodium periodate, and cyanuric chloride) may be inhibitory to the culture.[39]

Development of new technologies for large scale ABE fermentation has largely focused on adsorption and entrapment techniques. Although the majority of the academic initiatives are still at the laboratory scale levels, the technology commercializing company (Cobalt Technologies Inc.) is developing an immobilized cell continuous reactor, currently at pilot scale, for the industrial scale production of butanol.

Whereas adsorption and entrapment techniques offer significant improvement in butanol productivity, laboratory experiments have demonstrated that the former is superior to the latter (Table 4.3). In addition, the

Table 4.3 Performance parameters of representative immobilized cell continuous fermentations obtained in laboratory scale experiments using different feedstocks and reactor design.

Feedstock	Cell immobilization technique, support, and reactor type	Initial sugar concentration ($g L^{-1}$)	Sugar utilization ($g L^{-1}$)	Dilution rate (h^{-1})	ABE ($g L^{-1}$)	ABE yield ($g g^{-1}$)	ABE prod. ($g L h^{-1}$)	Ref.
Defibered-sweet-potato-slurry	Adsorption, ceramic D-21 beads, four packed-bed columns	39.7	39.7	0.129	7.73	0.19	1.0	40
Lab-grade glucose	Adsorption, polyester sponge strips, trickle-bed reactor with two serial columns[a]	60	45	0.272	15.5	0.34	4.2	41
Lab-grade glucose (with addition of butyrate)	Adsorption, spiral wound fibrous matrix, packed-bed column	50	42	0.9	8.9	0.42	4.6[b]	42
Lab-grade glucose	Adsorption, clay brick, packed-bed column	60	20.4	2.0	7.9	0.38	15.8	34
Lab-grade glucose	Adsorption, clay brick, packed-bed column	67.5	30.5	0.32	6.29	0.30	2.01	43
Lab-grade glucose	Adsorption, porous polyvinyl alcohol particles, stirred tank	60	50.2	0.04	22.1	0.44	0.40[b]	25
Lactose from cheesy whey permeate	Adsorption, bonechar, packed-bed column	50	15	1.0	—	0.32	5.8	44
Lab-grade glucose	Adsorption, bricks, packed-bed column	60	54.5	0.054	14.3	0.24	0.48[b]	45
Wood chips (spruce) with addition of glucose	Adsorption, wood pulp, packed-bed column	53	29	0.64	7.5	0.27	4.9	36
Lab-grade glucose	Adsorption, sugarcane bagasse, packed-bed column	60	14.5	0.60	3.5	0.25	2.12	38
Lab-grade glucose	Entrapment, calcium alginate gel, stirred tank	10	8	0.4	1.75	0.2[b]	0.7[b]	46
Lab-grade glucose	Entrapment, calcium alginate gel, stirred tank	60	—	1.0	3.94	0.21	4.0	26
Lab-grade glucose	Entrapment, k-carrageenan gel, stirred tank	60	—	0.69	4.0	0.18	2.8	26
Lab-grade glucose	Entrapment, chitosan gel, two stirred tanks in series	60	—	0.53	2.7	0.17	1.4	26

[a]Trickle-bed reactor is a packed-bed reactor with a low feed flow rate from the reactor top.
[b]Accounting for butanol only.

adsorption process is a simpler technique and has economic advantages over the entrapment technique as it can employ low-cost residues for microbial cell support and no chemical addition is necessary to achieve cell immobilization. In addition, mass transfer resistance for the transportation of substrate and products between fermentation broth and microbial cells is reduced (in the adsorption technique) because the biofilm generated after the attachment of the microbial cells to the support allows direct contact between broth and microbial cells. Mass transfer in the entrapment technique causes inhibition of encapsulated cells by slowing down the rate at which butanol leaves the entrapped cells, thus hampering the biocatalytic efficiency of the microorganism.[26] For these reasons, adsorption can be considered the most suitable cell immobilization technique for the ABE fermentation.

Different configurations of immobilized cell reactors, such as stirred tank, packed-bed, and fluidized-bed reactors, have been evaluated for ABE fermentation. An illustrative representation of each reactor type is shown in Figure 4.1 and additional details are given as follows:

(i) In the stirred tank reactor, while mixing provides a homogeneous fermentation broth, the mixing process may result in high shear forces which can cause damage to both support and microbial cells.
(ii) The packed-bed reactor consists of a column packed with immobilized cells through which substrate solution flows. It is operated in a plug-flow mode with minimum back mixing. This reactor has some unique advantages due to its simplicity and low cost of operation. The impact of shear forces on microbial cells is minimized because stress caused by the fluid phase velocity is minimal. However, substrate utilization by the fermenting microorganisms and productivity can be affected by mass transfer limitations such as blockage due to excess microbial cell production and accumulation of gases (gas hold-up)

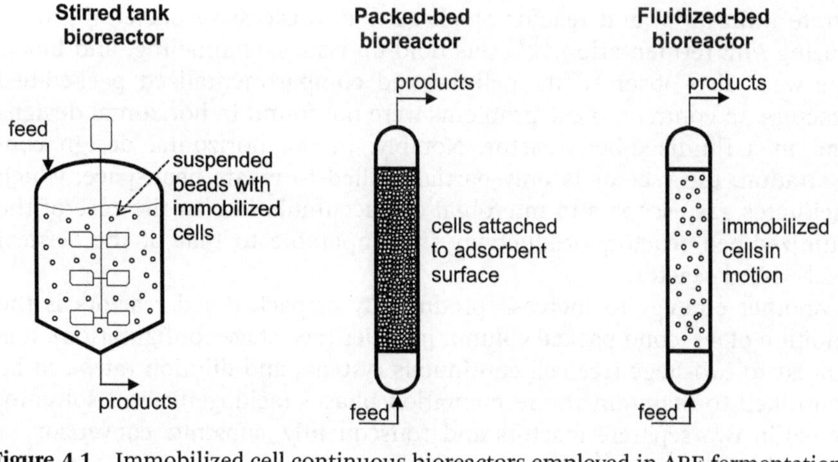

Figure 4.1 Immobilized cell continuous bioreactors employed in ABE fermentation.

which reduces the working volume of the reactor. In addition to these drawbacks, the presence of different physiological/morphological types of microbial cells in the packed-bed reactor may hinder its performance. Laboratory scale studies suggest that the majority of the microbial cells in the packed-bed reactors may be inert, *i.e.* they neither grow nor produce ABE, and in this manner, only a relatively small fraction of microbial cells may be effectively producing ABE. Common fluctuations in "steady state" conditions found in these reactors could be a result of variations in the relative proportions of different types of microbial cells.[44]

(iii) In the fluidized-bed reactors, immobilized cells are maintained in motion by a continuous flow of the feed solution and stream recirculation from the top of the reactor to the bottom. This reactor provides conditions that are intermediate to those of the stirred tank and packed-bed reactor. Mixing in the fluidized bed is better than in packed-bed reactors and the shear rate is low compared to the stirred tank reactor.[39]

Reactor design plays an important role in the production of fuels and chemicals by fermentation. In fermentation systems characterized by gas production and product inhibition, especially in ethanol and butanol fermentation, reactors that offer low gas hold-up and exhibit no backmixing (plug-flow behavior) are expected to have better performance. Notably, horizontal partially-packed-bed reactors have higher productivity than vertical packed-bed reactors for the production of ethanol by immobilized yeast cells. The horizontal design minimized gas hold-up and product inhibition.[47–50] Since ABE and ethanol fermentation are similar in terms of gas hold-up and product inhibition, it is expected that reactor design may also influence productivity in butanol fermentation.

Indeed, a vertical packed-bed reactor with cells adsorbed onto bonechar suffered from problems of high gas hold-up, product inhibition, low substrate utilization, and reactor blockage due to excessive bacterial growth during ABE fermentation.[33,44] Gas hold-up issues, channeling, and blockage were also observed in inclined and compartmentalized packed-bed reactors. In contrast, these problems were not found in horizontal designs and in a fluidized-bed reactor. Notably, in the horizontal design configuration, the reactor is only partially filled to create headspace, which facilitates gas escape and microbial cell accumulation. In the case of the fluidized-bed reactor, productivity is comparable to that of the vertical packed-bed reactor.

Another strategy to increase productivity of packed-bed reactors is the addition of a second packed column in series (two-stage configuration). It is similar to two-stage free cell continuous systems and dilution rates can be controlled to maintain the fermentation phases (acidogenic and solventogenic) in two separate reactors and consequently, substrate conversion to butanol is enhanced.[33,34,51]

4.3.3 Cell Recycle Continuous Reactors

In an immobilized state, microbial cells may experience substrate, nutrient, and product diffusion limitation, resulting in cell inactivation or death, especially in the innermost cell layers, due to starvation and accumulation of toxic butanol. With smaller numbers of cells that may effectively partake in the substrate conversion, reactor productivity may decline as a result. Another common problem with packed-bed reactors is blockage due to excessive cell growth. Although immobilized cell reactors may offer lower investment and operating costs, several studies have focused on the development of cell recycle systems using synthetic membrane (hereafter referred to as membrane) technologies as a means of preventing cells from being carried out with the outflow stream.

In cell recycle systems, the reactor is initiated in batch mode and cell growth is allowed till the end of the exponential phase. Prior to attaining the stationary phase, the fermentation broth is circulated through a filtration membrane system. The membrane allows the aqueous product solution to pass through while retaining the cells, accumulating to levels around 100 g L^{-1}. As both reactor inflow and product (permeate) removal flow are equal, a constant volume is maintained in the reactor. A small bleed stream (<10% of dilution rate) is necessary to keep cells productive, and to avoid accumulation of unknown inhibitory components.[22,52] A schematic diagram of the cell recycle system is shown in Figure 4.2.

In a free cell continuous reactor, the specific cell growth rate is equal to the dilution rate as can be demonstrated by a steady-state mass balance. By introducing cell recycling, it is the bleed dilution rate that balances the specific growth rate in the steady-state regime. Thus, while the concentration of cells in the fermentor is simply controlled by the bleed flow rate, the total dilution rate mainly controls substrate conversion.[53] Productivity generally increases many-fold relative to free cell systems, achieving values as high as approximately 10 g L h^{-1}.[22,54] The performance of representative cell recycle continuous fermentations conducted on the laboratory scale is presented in Table 4.4.

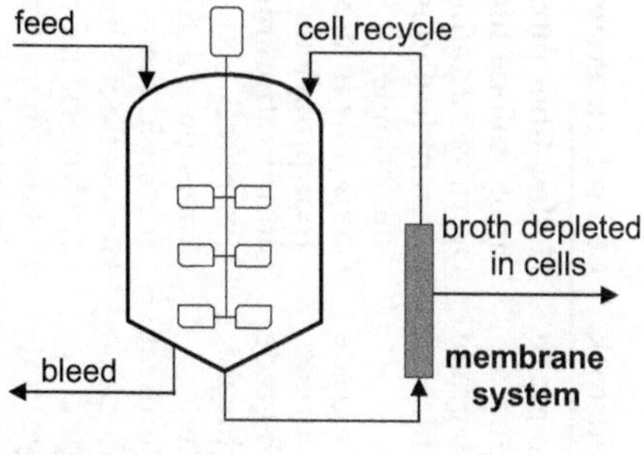

Figure 4.2 Cell recycle continuous reactor.

Table 4.4 Performance parameters of representative cell recycle continuous fermentations obtained in laboratory scale experiments using different membranes and feedstocks.

Feedstock	Cell recycle system	Initial sugar conc. (g L^{-1})	Sugar utilization (g L^{-1})	Dilution rate (h^{-1})[a]	Cell conc. (g L^{-1})	ABE (g L^{-1})	ABE yield (g g^{-1})	ABE prod. (g L h^{-1})	Ref.
Lab-grade glucose	Hollow-fiber ultrafilter (polysulfone membrane)	45	45	0.5 (0.435–0.065)	20	13	0.29	6.5	55
Lab-grade glucose	Overflow ultrafiltration system (cellulose-triacetate membrane)	60–90	—	0.10 (—)	13.1	22	0.32	1.4	54
Lab-grade glucose	Cross-flow microfiltration (polypropylene membrane)	60	—	0.64 (—)	8	8.4	0.20	5.4	56
Lab-grade glucose	Ultrafiltration module (mineral)	70	10–65	0.33 (—)	14	9.3–20.5	—	1.3	52
Whey permeate	Cross-flow microfiltration (ceramic membrane)	60	—	0.41 (0.39–0.02)	11.8–12.8	7.1	0.31	2.9	57
Lab-grade glucose	Hollow-fiber ultrafilter (polymeric membrane)	50	38	0.85 (0.74–0.11)	33	8.6	—	7.5	22
Glycerol	Hollow-fiber ultrafilter (polymeric membrane)	60	35	0.9 (0.9–0)	408[c]	9.2	—	8.3	58
Lab-grade xylose	Hollow-fiber ultrafilter (polymeric membrane)	50	22	0.78 (—)	17.4	5.9	—	3.3[b]	59
Lab-grade glucose	Cross-flow ultrafilter (polymeric membrane)	60	15	0.29 (—)	22	3.4	0.23	1.0	60

[a]Values between parentheses are the permeate and bleeding dilution rates, respectively.
[b]Accounting for butanol only.
[c]OD_{600}.

The filtration process in the cell recycle continuous fermentation employs the pressure gradient between internal and external surfaces of the filter to facilitate mass transfer, retains microbial cells and allows cell free fermentation broth to pass through the filter pores. The retention characteristics of the membrane are greatly influenced by the pore diameter. The membranes that can be used for cell separation include microfiltration and ultrafiltration.

An important aspect of membranes is their chemical and physical structure. The majority of commercially available separation membranes are made of polymeric materials such as polypropylene, polytetrafluoroethylene (PTFE), polysulfone, and cellulose-acetate. A newer generation of separation membranes are made up of ceramics. Ceramic membranes have high thermal stability, fine chemical stability and biocompatibility, and good erosion resistance. These properties permit the use of strong cleaning agents such as steam sterilization, backflushing, and ultrasonication. Moreover, ceramic membranes are less susceptible to biological degradation and aging. However, they are relatively expensive and intrinsically brittle. More specific details on membrane separation technologies can be found elsewhere[61] as well as in Chapters 6 and 7. Although different types of filtration membranes have been developed and several of them have been investigated for ABE fermentation, high membrane cost and fouling still remain the major obstacles for the use of this technology in commercial fermentations including ABE.

4.3.4 Continuous Bioreactors and Simultaneous Product Recovery

As seen in previous sections, marked improvements in reactor productivity have been achieved with novel technologies operated in continuous mode. However, these high productivity values are often associated with low product concentration in the fermentation broth and low sugar conversion. Such dilute product concentration in the fermentation broth results in excessive energy consumption to recover butanol from the product stream, and disposal of large volumes of stillage is cost and environmentally prohibitive. For illustration, an increase in the concentration of butanol from 13 to 36 g L^{-1} could cut energy consumption in a downstream distillation unit by ~50% (from 25 to 12 MJ kg^{-1} Bu*t*OH) and wastewater footprint by 60% (from 72 to 29 L stillage/L butanol).[62] Notably, sugar loss in the effluent should be avoided since substrate cost is a key economic element for biofuel production, accounting for 50–75% of the production cost.

A possible solution to the problem of incomplete utilization of substrate is to recycle the reactor effluent to achieve complete substrate utilization. Towards alleviating problems associated with product inhibition, coupling

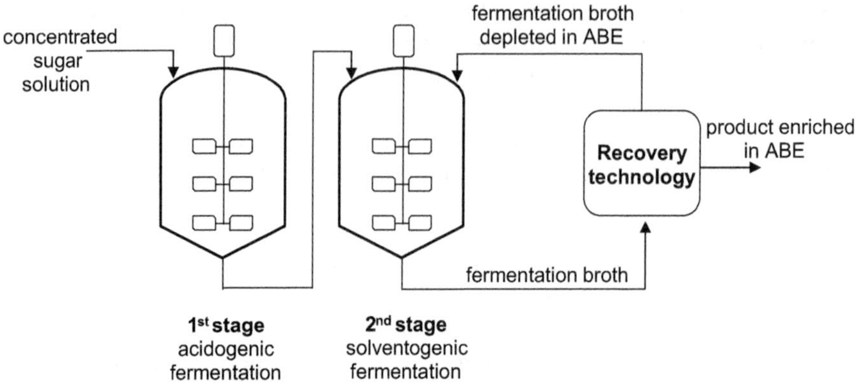

Figure 4.3 A two-stage free cell continuous bioreactor with product recovery integrated in the second fermentation.

an effective product recovery unit to the reactor to continuously remove formed products will relieve product inhibition and consequently, enable complete utilization of a concentrated feed media (Figure 4.3). Indeed, significant research efforts on the development of simultaneous fermentation and *in situ* product recovery technologies for ABE fermentation, including adsorption, gas stripping, pervaporation, perstraction, reverse osmosis, liquid–liquid extraction, and vacuum (flash) fermentation, have been made during the last two decades. Some of these technologies are also gaining attention for scale-up, especially liquid–liquid extraction and vacuum (flash) fermentation. Details on principles, advantages and drawbacks of the aforementioned *in situ* butanol recovery technologies can be found elsewhere.[31,63,64]

Representative studies on continuous bioreactors integrated with product recovery are shown in Table 4.5. For most of these technologies, higher sugar utilization and improved ABE concentration have been obtained, thus making the butanol fermentation process more energy- and cost-efficient. In such systems, two or more consecutive fermentors with or without cell immobilization or cell recycle may be used to separate the acidogenic and solventogenic phases. The solventogenic reactor can be kept at a lower dilution rate since the integrated product recovery will prevent the cells from being exposed to high levels of butanol (Figure 4.3).

Figure 4.4 presents the major features of each bioreactor type presented in this chapter, highlighting the evolution from batch technology to cutting-edge continuous bioreactors. The exciting and ongoing history of the development of efficient bioreactor systems for fermentative butanol production can be summarized even as we aspire to develop efficient bioreactor systems with high productivity, utilization and conversion of high sugar feed concentrations, and generation of concentrated product streams.

Table 4.5 Performance parameters of representative continuous bioreactors with product recovery obtained in laboratory scale experiments using different recovery technologies and feedstocks.

Feedstock	Reactor type, product recovery technology	Initial sugar concentration ($g L^{-1}$)	Sugar utilization ($g L^{-1}$)	Overall dilution rate (h^{-1})	ABE ($g L^{-1}$)	ABE yield ($g g^{-1}$)	ABE prod. ($g L h^{-1}$)	Ref.
Whey permeate	Two-stage immobilized cell reactor, Gas-stripping	40	40.9	0.43	12.3	0.35	2.6	57
Whey permeate	Single-stage fluidized-bed reactor, Gas-stripping	49.6	15.3	1.37	54	0.4	5.1	65
Whey permeate	Single-stage packed-bed reactor, Liquid–liquid extraction	60	60	0.35	5.5	0.36	1.9	66
Lab-grade glucose	Packed-bed reactor, in series with a free cell reactor Pervaporation (integrated to 2nd stage)	60	52	0.12	13.9	0.28	1.72	27
Lab-grade glucose	Single-stage packed-bed reactor, Pervaporation	60	60	2.0	11.9–49.2	0.44	16.2	67
Lab-grade glucose	Single-stage free cell reactor, Gas stripping	100	112.5[d]	0.03	36.5–102	0.41	0.92	68
Lab-grade glucose	Single-stage free cell reactor, Pervaporation	40	—	0.075	—	—	2.34	69
Glucose[a]	Single-stage cell recycle reactor, Flash fermentation	140	134	0.25	47	0.33	11.7	70
Lab-grade *n*-butyric acid/glucose	Two-stage free cell reactor, Gas-stripping (integrated to 2nd stage)	10.5–60[b]	9.8–56[b]	0.025	15.7[c]	0.16[c]	0.61	71
Lab-grade glucose	Two-stage immobilized cell reactor, Liquid–liquid extraction (integrated to first stage)	60	42	0.20	14	0.35	2.5	38
Lab-grade glucose	Two-stage free cell reactor, Pervaporation (integrated to 2nd stage)	126	112	0.009	115	0.27	0.30	72
Lab-grade xylose/glucose	Two-stage cell recycle reactor, Pervaporation (integrated to 2nd stage)	150	130	0.017	38.4	0.29	0.65	60

[a]Results obtained from mathematical modeling and simulation.
[b]*n*-butyric acid – glucose ($g L^{-1}$).
[c]In relation to butanol.
[d]Total sugars consumed.

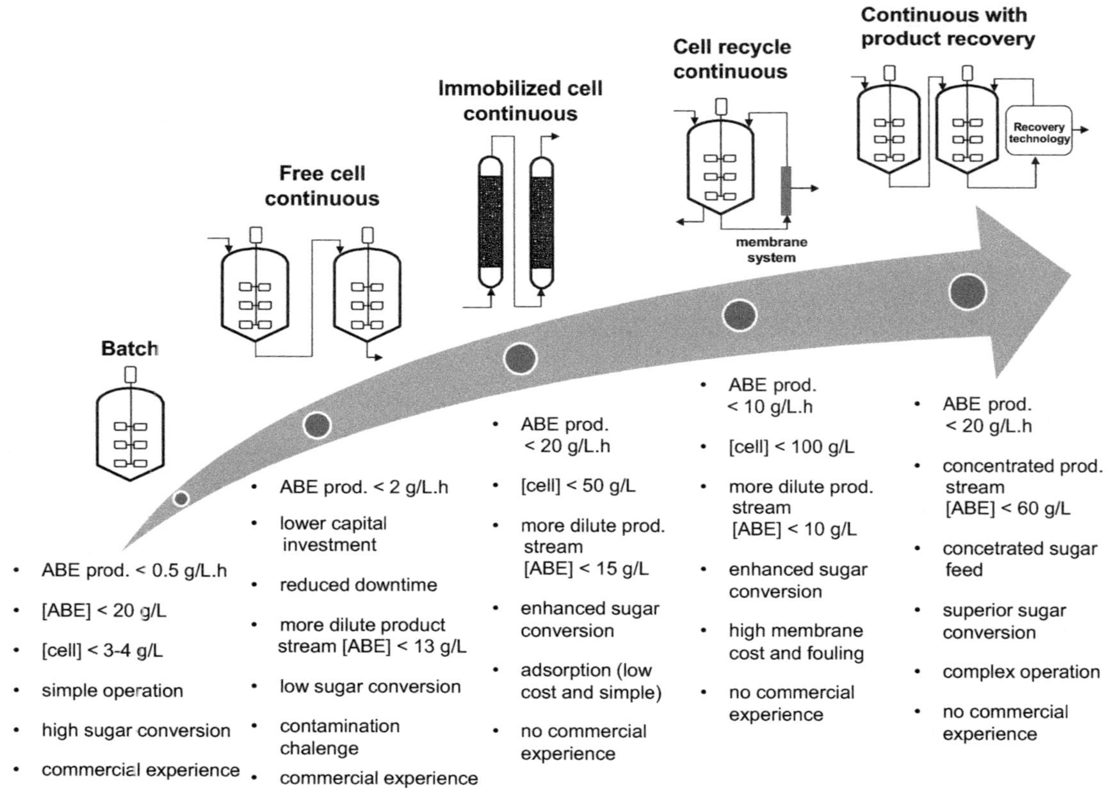

Figure 4.4 Comparative analysis of bioreactors for butanol production by fermentation.

4.4 Perspectives and Concluding Remarks

Whereas the economic and socio-environmental need for the utilization of low-cost lignocellulosic biomass is common to the biofuels industry, butanol production by fermentation is particularly plagued by low process efficiency. These are mainly corroborated by low butanol yield, which impacts productivity and operational cost. For the sake of comparison, yeast-based ethanol fermentation generates a product concentration nearly ten times greater (120–150 g L^{-1}) than that of butanol. The transition towards utilization of lignocellulosic feedstock will further exacerbate the problem due to the toxic effects of microbial inhibitors generated during the pretreatment and hydrolysis of lignocellulosic biomass to monomeric sugars (lignocellulosic biomass hydrolysates). Owing to economic constraints, purification of lignocellulosic biomass hydrolysates to remove microbial inhibitors prior to fermentation may not be feasible.

This scenario clearly suggests that while the novel fermentation technologies presented in this chapter are gaining attention in the re-emerging butanol industry, more work is needed to fine-tune these technologies or develop new ones that will adequately address the limitations of existing technologies. Indeed, these technological advances in bioreactor design and operations will pose challenges to engineers during industrial scale-up as reported in recent commercialization endeavors. The development of new advanced bioreactors or fine-tuning existing ones to address the limitations of existing technologies discussed in this chapter will indeed be a necessary piece in the intricate puzzle for making commercial production of butanol economically feasible.

References

1. H. Bahl, W. Andersch and G. Gottschalk, Continuous production of acetone and butanol by *Clostridium acetobutylicum* in a two-stage phosphate limited chemostat, *Eur. J. Appl. Microbiol. Biotechnol.*, 1982, **15**, 201–205.
2. D. T. Jones and D. R. Woods, Acetone-Butanol Fermentation Revisited, *Microbiol. Rev.*, 1986, **50**, 484–524.
3. P. Dürre, New insights and novel developments in clostridial acetone/butanol/isopropanol fermentation, *Appl. Microbiol. Biotechnol.*, 1998, **49**, 639–648.
4. J.-S. Chiao and Z-h Sun, History of the Acetone-Butanol-Ethanol fermentation industry in China: development of continuous production technology, *J. Mol. Microbiol. Biotechnol.*, 2007, **13**, 12–14.
5. E. M. Green, Fermentative production of butanol—the industrial perspective, *Curr. Opin. Biotechnol.*, 2011, **22**, 1–7.
6. Y. Jiang, J. Liu, W. Jiang, Y. Yang and S. Yang, Current status and prospects of industrial bio-production of n-butanol in China, *Biotechnol. Adv.*, 2014, Oct 27. pii: S0734-9750(14)00156-6.

7. Y. Ni and Z. Sun, Recent progress on industrial fermentative production of acetone–butanol–ethanol by *Clostridium acetobutylicum* in China, *Appl. Microbiol. Biotechnol.*, 2009, **83**, 415–423.
8. N. Qureshi and H. P. Blaschek, Recent advances in ABE fermentation: hyper-butanol producing *Clostridium beijerinckii* BA101, *J. Ind. Microbiol. Biotechnol.*, 2001, **27**, 287–291.
9. N. Qureshi, C. S. Badal and M. A. Cotta, Butanol production from wheat straw hydrolysate using *Clostridium beijerinckii*, *Bioprocess Biosyst. Eng.*, 2007, **30**, 419–427.
10. A. M. Junelles, R. Janati-Idrissi, H. Petitdemange and R. Gay, Iron effect on acetone-butanol fermentation, *Curr. Microbiol.*, 1988, **17**, 299–303.
11. A. Ishizaki, S. Michiwaki, E. Crabbe, G. Kobayashi, K. Sonomoto and S. Yoshino, Extractive acetone-butanol-ethanol fermentation using methylated crude palm oil as extractant in batch culture of *Clostridium saccharoperbutylacetonicum* Nl-4 (ATCC 13564), *J. Biosci. Bioeng.*, 1999, **87**, 352–356.
12. M. Wayman and S. Yu, Acetone-butanol fermentation of xylose and sugar mixtures, *Biotechnol. Lett.*, 1985, 7, 255–260.
13. N. Qureshi, A. Lolas and H. P. Blaschek, Soy molasses as fermentation substrate for production of butanol using *Clostridium beijerinckii* BA101, *J. Ind. Microbiol. Biotechnol.*, 2001, **26**, 290–295.
14. N. Qureshi, B. C. Saha, R. E. Hector, S. R. Hughes and M. A. Cotta, Butanol production from wheat straw by simultaneous saccharification and fermentation using *Clostridium beijerinckii*: Part I—Batch fermentation, *Biomass Bioenergy*, 2008, **32**, 168–175.
15. S. T. Liew, A. Arbakariya, M. Rosfarizan and A. R. Raha, Production of solvent (acetone-butanol-ethanol) in continuous fermentation by *Clostridium saccharobutylicum* DSM 13864 using gelatinised sago starch as a carbon source, *Malays. J. Microbiol.*, 2006, **2**, 42–50.
16. T. Guo, Y. Tang, Q. Zhang, T. Du, D. Liang, M. Jiang *et al.*, *Clostridium beijerinckii* mutant with high inhibitor tolerance obtained by low-energy ion implantation, *J. Ind. Microbiol. Biotechnol.*, 2012, **39**, 401–407.
17. L. Wang and H. Chen, Increased fermentability of enzymatically hydrolyzed steam-exploded corn stover for butanol production by removal of fermentation inhibitors, *Process Biochem.*, 2011, **46**, 604–607.
18. N. Qureshi, B. C. Saha, B. Dien, R. E. Hector and M. A. Cotta, Production of butanol (a biofuel) from agricultural residues: part I—use of barley straw hydrolysate, *Biomass Bioenergy*, 2010, **34**, 559–565.
19. T. C. Ezeji, N. Qureshi and H. P. Blaschek, Industrially relevant fermentations, in *Handbook on Clostridia*, ed. P. Dürre, Taylor and Francis, New York, NY, 2005, ch. 36, pp. 797–812.
20. C. Godin and J. M. Engasser, Improved stability of the continuous production of acetone-butanol by *Clostridium acetobutylicum*, *Biotechnol. Lett.*, 1988, **6**, 389–392.

21. C. Godin and J. M. Engasser, Two-stage continuous fermentation of *Clostridium acetobutylicum*: effects of pH and dilution rate, *Appl. Microbiol. Biotechnol.*, 1990, **33**, 269–273.
22. Y. Tashiro, K. Takeda, G. Kobayashi and K. Sonomoto, High production of acetone–butanol–ethanol with high cell density culture by cell-recycling and bleeding, *J. Biotechnol.*, 2005, **120**, 197–206.
23. T. Ezeji, N. Qureshi and H. P. Blaschek, Production of acetone–butanol–ethanol (ABE) in a continuous flow bioreactor using degermed corn and *Clostridium beijerinckii*, *Process Biochem.*, 2007, **42**, 34–39.
24. T. C. Ezeji, N. Qureshi and H. P. Blaschek, Continuous butanol fermentation and feed starch retrogradation: butanol fermentation sustainability using *Clostridium beijerinckii* BA101, *J. Biotechnol.*, 2005, **115**, 179–187.
25. S.-M. Lee, M. O. Cho, C. H. Park, Y.-C. Chung, J. H. Kim, B.-I. Sang and Y. Um, Continuous butanol production using suspended and immobilized *Clostridium beijerinckii* NCIMB 8052 with supplementary butyrate, *Energy Fuels*, 2008, **22**, 3459–3464.
26. Ch. Frick and K. Schügerl, Continuous acetone-butanol production with free and immobilized *Clostridium acetobutylicum*, *Appl. Microbiol. Biotechnol.*, 1986, **25**, 186–193.
27. J. R. Gapes, D. Nimcevic and A. Friedl, Long-term continuous cultivation of *Clostridium beijerinckii* in a two-stage chemostat with on-line solvent removal, *Appl. Environ. Microbiol.*, 1996, **62**, 3210–3219.
28. O. Mutschlechner, H. Swoboda and J. R. Gapes, Continuous two-stage ABE-fermentation using *Clostridium beijerinckii* NRRL B592 operating with a growth rate in the first stage vessel close to its maximal value, *J. Mol. Microbiol. Biotechnol.*, 2000, **2**, 101–105.
29. Y. Ni, Z. Xi, Y. Wang and Z. Sun, Continuous butanol fermentation from inexpensive sugar-based feedstocks by *Clostridium saccharobutylicum* DSM 13864, *Bioresour. Technol.*, 2013, **129**, 680–685.
30. D. Nimcevic and J. R. Gapes, The Acetone-Butanol Fermentation in Pilot Plant and Pre-Industrial Scale, *J. Mol. Microbiol. Biotechnol.*, 2000, **2**, 15–20.
31. T. C. Ezeji, N. Qureshi and H. P. Blaschek, Bioproduction of butanol from biomass: from genes to bioreactors, *Curr. Opin. Biotechnol.*, 2007a, **18**, 220–227.
32. N. Qureshi and T. C. Ezeji, Butanol, 'a superior biofuel' production from agricultural residues (renewable biomass): recent progress in technology, *Biofuels, Bioprod. Biorefin.*, 2008, **2**, 319–330.
33. N. Qureshi and I. S. Maddox, Continuous solvent production from whey permeate using cells of *Clostridium acetobutylicum* immobilized by adsorption onto bonechar, *Enzyme Microb. Technol.*, 1987, **9**, 668–671.
34. N. Qureshi, J. Schripsema, J. Lienhardt and H. P. Blaschek, Continuous solvent production by *Clostridium beijerinckii* BA101 immobilized by adsorption onto brick, *World J. Microbiol. Biotechnol.*, 2000, **16**, 377–382.

35. N. Qureshi, Solvent Production, in *Encyclopedia of Microbiology*, ed. M. Schaechter, Elsevier, Oxford, 2009, pp. 512–528.
36. S. A. Survase, E. Sklavounos, G. Jurgens, A. Van Heiningen and T. Granström, Continuous acetone–butanol–ethanol fermentation using SO_2–ethanol–water spent liquor from spruce, *Bioresour. Technol.*, 2011, **102**, 10996–11002.
37. S. A. Survase, A. Van Heiningen and T. Granström, Continuous biocatalytic conversion of sugar mixture to acetone–butanol–ethanol by immobilized *Clostridium acetobutylicum* DSM 792, *Appl. Microbiol. Biotechnol.*, 2012, **93**, 2309–2316.
38. S. B. Bankar, S. A. Survase, R. S. Singhal and T. Granström, Continuous two stage acetone–butanol–ethanol fermentation with integrated solvent removal using *Clostridium acetobutylicum* B 5313, *Bioresour. Technol.*, 2012, **106**, 110–116.
39. S.-T. Yang, *Bioprocessing for Value-added Products from Renewable Resources*, Elsevier Academic Press, Amsterdam, 2007, pp. 376–385.
40. H. R. Badr, R. Toledo and M. K. Hamdy, Continuous acetone–ethanol–butanol fermentation by immobilized cells of *Clostridium acetobutylicum*, *Biomass Bioenergy*, 2001, **20**, 119–132.
41. C.-H. Park, M. R. Okos and P. C. Wankat, Acetone-butanol-ethanol (ABE) fermentation in an immobilized cell trickle bed reactor, *Biotechnol. Bioeng.*, 1988, **34**, 18–29.
42. W.-C. Huang, D. E. Ramey and S.-T. Yang, Continuous production of butanol by *Clostridium acetobutylicum* immobilized in a fibrous bed reactor, *Appl. Biochem. Biotechnol.*, 2004, **113**, 887–898.
43. N. Qureshi, P. Karcher, M. Cotta and H. P. Blaschek, High-productivity continuous biofilm reactor for butanol production (effect of acetate, butyrate, and corn steep liquor on bioreactor performance), *Appl. Biochem. Biotechnol.*, 2004, **113**, 713–721.
44. N. Qureshi and I. S. Maddox, Reactor design for the ABE fermentation using cells of *Clostridium acetobutylicum* immobilized by adsorption onto bonechar, *Bioprocess Eng.*, 1988, **3**, 69–72.
45. H.-W. Yen, R.-J. Li and T.-W. Ma, The development process for a continuous acetone–butanol–ethanol (ABE) fermentation by immobilized *Clostridium acetobutylicum*, *J. Taiwan Inst. Chem. Eng.*, 2011, **42**, 902–907.
46. C. Förberg, S.-O. Enfors and L. Haggström, Control of immobilized, non-growing cells for continuous production of metabolites, *Eur. J. Appl. Microbiol. Biotechnol.*, 1983, **17**, 143–147.
47. T. Shiotani and T. Yamane, A horizontal packed bed bioreactor to reduce CO_2 holdup in the continuous production of ethanol by immobilized yeast cells, *Eur. J. Appl. Microbiol. Biotechnol.*, 1981, **13**, 96–101.
48. G. K. Chotani and A. Constantinides, Immobilized cell cross flow reactor, *Biotechnol. Bioeng.*, 1984, **26**, 217–220.
49. K. Toda, H. Ohtake and T. Asakura, Ethanol production in horizontal bioreactor, *Appl. Microbiol. Biotechnol.*, 1986, **24**, 97–101.

50. L. R. Junena, M. Terasawa, T. Yamane and S. Shimizu, Continuous ethanol fermentation using flocculent yeast entrapped in horizontal parallel flow bioreactor system, *Biotechnol. Lett.*, 1986, **8**, 431–436.
51. S. A. Survase, E. Sklavounos, G. Jurgens, A. Van Heiningen and T. Granström, Continuous production of isopropanol and butanol using *Clostridium beijerinckii* DSM 6423, *Appl. Microbiol. Biotechnol.*, 2011, **91**, 305–1313.
52. E. Ferras, M. Minier and G. Goma, Acetonobutylic fermentation: Improvement of performance by coupling continuous fermentation and ultrafiltration, *Biotechnol. Bioeng.*, 1986, **28**, 523–533.
53. A. P. Mariano, C. B. B. Costa, M. R. Wolf Maciel, F. Maugeri Filho, D. I. P. Atala, D. F. Angelis and R. Maciel Filho, Dynamics and control strategies for a butanol fermentation process, *Appl. Biochem. Biotechnol.*, 2010, **160**, 2424–2448.
54. D. Schlote and G. Gottschalk, Effect of cell recycle on continuous butanol-acetone fermentation with *Clostridium acetobutylicum* under phosphate limitation, *Appl. Microbiol. Biotechnol.*, 1986, **24**, 1–5.
55. P. Pierrot, M. Fick and J. M. Engasser, Continuous acetone-butanol fermentation with high productivity by cell ultrafiltration and recycling, *Biotechnol. Lett.*, 1986, **8**, 253–256.
56. A. S. Afschar, H. Biebl, K. Schaller and K. Schugerl, Production of acetone and butanol by *Clostridium acetobutylicum* in continuous culture with cell recycle, *Appl. Microbiol. Biotechnol.*, 1985, **22**, 394–398.
57. B. M. Ennis and I. S. Maddox, Production of solvents (ABE fermentation) from whey permeate by continuous fermentation in a membrane bioreactor, *Bioprocess Eng.*, 1989, **4**, 27–34.
58. A. Malaviya, Y. S. Jang and S. Y. Lee, Continuous butanol production with reduced byproducts formation from glycerol by a hyper producing mutant of *Clostridium pasteurianum*, *Appl. Microbiol. Biotechnol.*, 2012, **93**, 1485–1494.
59. J. Zheng, Y. Tashiro, T. Yoshida, M. Gao, Q. Wang and K. Sonomoto, Continuous butanol fermentation from xylose with high cell density by cell recycling system, *Bioresour. Technol.*, 2013, **129**, 360–365.
60. U. Kayaalp, Two stage ABE fermentation with *in situ* pervaporation and high cell density, Master's thesis, Faculdade de Ciências e Tecnologia, Universidade Nova de Lisboa, 2013.
61. R. D. Noble and S. A. Stern, *Membrane Separations Technology: Principles and Applications*, Elsevier, 3rd edn, 1995, pp. 16–19.
62. A. P. Mariano and R. Maciel Filho, Improvements in biobutanol fermentation and their impacts on distillation energy consumption and wastewater generation, *BioEnergy Res.*, 2012, **5**, 504–514.
63. B. M. Ennis, N. Qureshi and I. S. Maddox, In-line toxic product removal during solvent production by continuous fermentation using immobilized *Clostridium acetobutylicum*, *Enzyme Microb. Technol.*, 1987, **9**, 672–675.

64. L. M. Vane, Separation technologies for the recovery and dehydration of alcohols from fermentation broths, *Biofuels, Bioprod. Biorefin.*, 2008, **2**, 553–588.
65. N. Qureshi and I. S. Maddox, Integration of continuous production and recovery of solvents from whey permeate: use of immobilized cells of *Clostridium acetobutylicum* in a fluidized bed reactor coupled with gas stripping, *Bioprocess Eng.*, 1991, **6**, 63–69.
66. N. Qureshi and I. S. Maddox, Continuous production of acetone-butanol-ethanol using immobilized cells of *Clostridium acetobutylicum* and integration with product removal by liquid-liquid extraction, *J. Ferment. Bioeng.*, 1995, **80**, 185–189.
67. J. Lienhardt, J. Schripsema, N. Qureshi and H. P. Blaschek, Butanol production by *Clostridium beijerinckii* BA101 in an immobilized cell biofilm reactor, *Appl. Biochem. Biotechnol.*, 2002, **98–100**, 591–598.
68. T. C. Ezeji, N. Qureshi and H. P. Blaschek, Microbial Production of a Biofuel (Acetone-Butanol-Ethanol) in a Continuous Bioreactor: Impact of Bleed and Simultaneous Product Removal, *Bioprocess Biosyst. Eng.*, 2013, **36**, 109–116.
69. P. Izák, K. Schwarz, W. Ruth, H. Bahl and U. Kragl, Increased productivity of *Clostridium acetobutylicum* fermentation of acetone, butanol, and ethanol by pervaporation through supported ionic liquid membrane, *Appl. Microbiol. Biotechnol.*, 2008, **78**, 597–602.
70. A. P. Mariano, D. F. Angelis, D. I. P. Atala, F. Maugeri Filho, M. R. Wolf Maciel and R. Maciel Filho, An alternative process for butanol production: continuous flash fermentation, *Chem. Prod. Process Model.*, 2008, **3**, A.34.
71. H. Richter, N. Qureshi, S. Heger, B. Dien, M. A. Cotta and L. T. Angenent, Prolonged conversion of n-butyrate to n-butanol with *Clostridium saccharoperbutylacetonicum* in a two-Stage continuous culture with in-situ product removal, *Biotechnol. Bioeng.*, 2012, **109**, 913–921.
72. W. Van Hecke, P. Vandezande, S. Claes, S. Vangeel, H. Beckers, L. Diels and H. De Wever, Integrated bioprocess for long-term continuous cultivation of *Clostridium acetobutylicum* coupled to pervaporation with PDMS composite membranes, *Bioresour. Technol.*, 2012, **111**, 368–377.

CHAPTER 5

Catalysis's Role in Bioproducts Update

KIM MAGRINI-BAIR,* DEREK R. VARDON AND GREGG T. BECKHAM

National Renewable Energy Laboratory, 15223 Denver West Parkway, Golden, CO 80401, USA
*Email: Kim.Magrini@nrel.gov

5.1 Introduction

North America is leading the global bio-based and sustainable chemicals industry. Consumer demands for greener products are driving retailers, developers, and government policy makers to replace petrochemicals and chemicals of concern with safer bio-alternatives. Bio-based chemicals are currently receiving an unexpected boost from the shale gas boom for drop-in bio-based C_3–C_5 chemicals. Additionally, bio-based chemicals are not impacted by volatile natural gas and oil prices, thus offering the potential for a more stable or even lower price than their petrochemical analogs.[1] Consumers will soon notice that the origin of everyday products is changing with items such as clothing, shoes, water and soda bottles, and even automobile tires being manufactured from plant-based rather than petroleum-based materials. This quiet revolution has been steadily moving to the market place. Several bio-based intermediates approaching commercialization use plant-based raw materials to produce polyester, spandex (Lycra), synthetic rubber, and nylon.[2] Critical criteria for adopting these biomaterials and precursors are that they must be indistinguishable from their petroleum-based counterparts, cost the same or less, and be available in

RSC Green Chemistry No. 43
Commercializing Biobased Products: Opportunities, Challenges, Benefits, and Risks
Edited by Seth W. Snyder

Published by the Royal Society of Chemistry, www.rsc.org

sufficient and sustained quantities. The rationale for moving to bio-based feedstocks is their potential lower cost especially if derived from waste biomass where the cost is essentially free, smaller carbon footprint, and the absence of toxic trace elements that can occur with petroleum-based feedstocks. Catalysts, key components of industrial processing of these feedstocks, are required to form both intermediates and finished products. This chapter reviews and updates recent developments of the role catalysts play in bioproducts development, deployment, and commercialization using process options[3] and the US Department of Energy's (DOE) report on the top value added chemicals from biomass[4] as the baseline to measure biomolecule conversions and catalyst improvements.

5.2 Catalyst Considerations

Designing heterogeneous catalysts for converting bio-based feedstocks, carbohydrates, lipids, and lignins, creates new challenges and opportunities for catalyst technology as until now most development has focused on producing fuels and chemicals from crude oil and natural gas. Because crude oil and natural gas-based feedstocks are not predominantly reactive, they require catalysts that activate them for further reaction *via* high-temperature cracking, which forms shorter chain molecules that contain unsaturated C–C bonds, which can be used for facile catalytic transformations. In contrast, bio-based renewable feedstocks naturally contain numerous oxygen-containing functional groups that can be used for chemical transformations, which in turn, require new catalysts that can perform selective chemistry in the presence of multiple reactive functional groups (hydroxyl, carbonyl, carboxylic acid, ester, and/or ether) within a reactant molecule. An associated challenge is to develop chemistry and catalysts that work with fermentation-derived, dilute, aqueous mixtures of oxygenated chemicals (platform molecules) rather than the current petroleum-derived non-aqueous, non-oxygenated feedstocks and to avoid energy intensive, expensive and wasteful concentration and purification steps.

> *One outcome from this significant difference in feedstock chemistry is that well established heterogeneous catalyst technology developed for hydrocarbon feedstocks, which are efficient for selective hydrogenation, selective oxidation, isomerization, and reforming, has not been designed for the dehydration, decarboxylation, decarbonylation, hydrogenolysis, esterification, and ketonization reactions that are needed for converting bio-based feedstocks to desirable products.*[5]

Another critical challenge created by these multiple feedstock chemical functionalities is the reduced volatility of the compounds at reaction temperatures. This property requires condensed phase reaction systems, which is in marked contrast to the primarily gas-phase reaction systems used for hydrocarbon processing. The resulting liquid–solid interfaces place new demands on catalytic materials relative to catalytic properties, catalyst stability and transport properties, which are further intensified when the condensed phase solvent is water. As well, due to their high oxygen content, bio-based feedstocks are hydrogen deficient relative to hydrocarbons, so three-phase reaction systems comprising a solid catalyst, liquid solvent–reactant mixture and hydrogen gas will generally be required to generate products that replace those from hydrocarbons.[5]

Lignocellulosic biomass, however, is the most abundant and bio-renewable resource with significant potential for the sustainable production of chemicals and fuels. Catalytic hydrolysis, solvolysis, liquefaction, pyrolysis, gasification, hydrogenolysis and hydrogenation are the major processes currently used to achieve these biomass conversions. Catalytic hydrolysis is conducted with acid catalysts comprising inorganic and organic acids and various solid acids such as sulfonated carbon, zeolites, heteropoly acids and oxides. Liquefaction and fast pyrolysis of biomass are primarily conducted over catalysts of tailored acidity/basicity. Gasification is conducted over supported transition and noble metal catalysts. Reaction conditions, solvents and catalysts are the primary factors that affect the yield and composition of target products. Most processes, however, produce complex mixtures, which both complicates and adds cost to upgrading and separation. An emerging technique is to integrate hydrolysis, liquefaction or pyrolysis with hydrogenation over multifunctional solid catalysts to convert lignocellulosic biomass to value-added fine chemicals and bio-hydrocarbon fuels with emerging catalysts based on supported transition metal catalysts and zeolite-related materials. Even so, significant technological barriers exist to commercializing bio-based products.[6]

5.3 DOE's Top Value Added Chemicals from Biomass Revisited

In 2004, the US DOE identified a set of biomass-derived compounds most suited to replace petroleum-derived chemicals.[4] These twelve building block chemicals (containing multiple functional groups for conversion) could be produced from sugars by biological or chemical processes and subsequently converted to many high-value bio-based chemicals or materials. Reviewing the available literature showed that while biological processes accounted for most of the routes from plant feedstocks to these building blocks, chemical processes predominated in the conversion of building blocks to molecular derivatives and intermediates. These candidate chemicals are shown in Table 5.1 (unshaded entries). In 2010, Bozell and Petersen[7] reviewed the status of these chemicals using criteria comprising the number of literature references, applicability to multiple products, direct substitution for a petrochemical, use for high volume products, potential as a platform chemical, scale up or commercialization extent, and applicability as a biorefinery building block. These molecules are shown in shaded entries in

Table 5.1 Updated evaluation of the DOE's top 15 value added chemicals from biomass.[4] Evaluation criteria included commercialization extent and degree of activity from the research sector. Bold entries added by the 2010 Bozell review of the DOE's top ten chemicals.[7]

DOE's top 15 biomass-derived compounds	Source	Downstream chemicals and materials	Key uses and products	Commercial now	In 10 years
Succinic acid (fumaric, malic)	**Bacterial glucose fermentation, chemical oxidation of 1,4-butanediol**	**1,4-Butanediol, tetrahydrofuran, γ-butyrolactone, maleic anhydride, pyrrolidones**	**Solvents, polyesters, polyurethanes, nylon, food and beverage acidity control, fabrics, inks, paints**	**Now**	**Yes**
Methanol, ethanol	**Gasification then catalytic upgrading of unrecyclable MSW**	**Ethylene, dimethyl acetate, dimethyl ether, acetic acid**		**Soon**	**Yes**
2,5-Furandicarboxylic acid	Chemical glucose dehydration, 5-hydroxymethylfurfural oxidation	2,5-Dihydroxymethyl-furan2,5-bis(amino methyl)tetrahydrofuran	Polyethylene terephthalate analogs, polyamides (nylon, plastic bottles, fabrics, carpet fiber)	Soon	Yes
3-Hydroxypropanoic acid	Glycerol or glucose bacterial fermentation	1,3-Propanediol, acrylic acid, methyl acrylate, acrylamide	Polytrimethyhlene terephthalate, acrylate polymers, carpet fibers, paints and adhesives, superabsorbent polymers, contact lenses	No	Yes
Glycerol	**Chemical or enzymatic transformation of vegetable oils**	**Propylene and ethylene glycol, 1,3-propanediol, glyceric and lactic acid, acetol, acrolein, epichlorhydrin**	**Polyesters, butanol, soaps, cosmetics, food, deicing fluids, pharmaceuticals, coatings, carpet fiber**	**Yes**	**Yes**
Sorbitol	**Corn syrup derived glucose hydrogenation, bacterial fermentation (in development)**	**Isosorbide, propylene and ethylene glycol, glycerol, lactic acid, alkanes**	**Sweeteners, toothpaste, chewing gum, polyethylene terephthalate analogs, fuel enhancers, deicing agents, water treatment**	**Yes**	**Yes**

Xylitol (arabinitol)	**Xylose hydrogenation, extraction from lignocellulose, bacterial fermentation (in development)**	**Propylene and ethylene glycol, glycerol, xylaric acid, furfural**	**Sweeteners, gum, cough drops and medicine, deicing fluids, new polyesters**	**Yes**	**Yes**
Levulinic acid	**Acid catalyzed dehydration of sugars**	**2-Methyltetrahydrofuran, γ-valerolactone, 1,4-pentanediol, acetylacrylic and diphenolic acids, caprolactam, adiponitrile, pyrrolidones**	**Fuel additives, solvents, acrylate polymers, BPA-free polycarbonates, polyesters, polyamides, pharmaceuticals, herbicides, plastic bottles/ containers**	**No**	**Maybe**
Itaconic acid	Fungal glucose fermentation	4-Methyl-γ-butyrolactone, 3-methyltetrahydrofuran, pyrrolidones	Styrene-butadiene copolymers, polyitaconic acid, rubber, plastics, paper and architectural coatings	No	Maybe
3-Hydroxybutyrolactone	Multistep synthesis from starch	3-Hydroxytetrahydrofuran, acrylate-lactone, 3-aminotetrahydrofuran	Solvents, synthetic intermediates for pharmaceuticals, polyurethane fiber analogs, new polymers	No	Maybe
Glutamic acid	Bacterial glucose fermentation	1,5-Pentanediol, glutaric acid, 5-amino-1-butanol	Polyesters, nylon analogs, glutamate flavor enhancers, fabrics, plastics	No	Maybe
Glucaric acid	Starch or glucose oxidation by nitric acid or bleach	Lactones, polyhydroxypolyamides, adipic acid	Solvents, nylon analogs, branched polyesters, fabrics, plastics, detergents	No	Maybe
Aspartic acid	Enzymatic amination of fumaric acid, fermentation route	2-Amino-1,4-butanediol, 3-aminotetrahydrofuran, aspartic anhydride, amino-γ-valerolactone	Aspartame, polyaspartate, sweeteners, chelating agents for water treatment, superabsorbent polymers	No	No
Furans (furfural)	**Selective dehydration of carbohydrates; selective oxidation of alcohols**				
Biohydrocarbons (isoprene)	**Thermochemical, biochemical processing**				

Table 5.1. In 2014, C&EN[2] reviewed once again these top twelve chemicals for commercialization potential and they now provide the basis for reviewing progress in catalytic transformations to bio-based chemicals and materials.

5.4 Process Option for Biomass Conversion to Bioproducts

Within the framework of a biorefinery, three process options for converting renewable feedstocks to bioproducts are proposed by Gallezot[3] and shown in Figure 5.1. Waste lignin and biomass can form chemicals and fuels *via* thermal deconstruction and catalytic upgrading. Examples are (1) gasifying biomass to syngas followed by catalytic upgrading to chemicals *via* gas phase or aqueous phase catalysis and (2) pyrolyzing these molecules to smaller species that can then be upgraded usually by extraction then catalytic upgrading. This approach generally requires multi-step syntheses. The platform molecule approach uses fermentation and hydrolysis to produce chemical products and fuels and again is generally a multi-step synthesis route. One pot conversions are the most desirable route to bio-products if catalyst selectivity and activity can be maintained in dilute aqueous environments characteristic of this approach.

The rest of this chapter assesses new routes using these processing platforms through the lens of emerging catalysts being developed for efficient bio-molecule conversion.

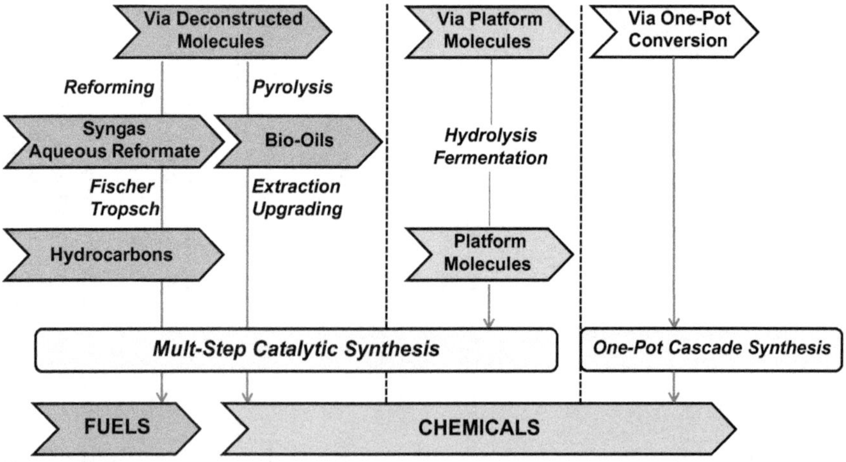

Figure 5.1 Process options for converting carbohydrates and lignocellulosic feedstocks to bioproducts in a biorefinery framework.
Reproduced from ref. 3 with permission from the Royal Society of Chemistry.

5.5 Biomass to Products through Deconstructed Molecules

5.5.1 Biomass-derived Syngas Upgrading

Biomass and its intermediates and derivatives can be thermally converted to syngas (CO and H_2) by using existing industrial processes of steam or autothermal reforming. Produced syngas is then converted to hydrocarbons and fuels *via* Fischer–Tropsch synthesis or other emerging catalytic processes currently under development. This approach is relatively insensitive to feedstock type. Recent work conducted at the National Renewable Energy Laboratory (NREL) demonstrated mixed alcohol synthesis from oak-derived syngas generated from indirect steam gasification. Two 150 hour continuous runs demonstrated the technology at the pilot scale converting up to 0.5 ton of oak feedstock per day.[8] Reforming and mixed alcohol synthesis catalysts were developed for this work with a nickel-alumina fluidizable catalyst used for tar reforming (US Patent 7 915,196 B2, 20 11[9]), a Johnson Matthey fixed-bed Rh-based catalyst for residual methane reforming,[10] and a K–Co–Mo-sulfide mixed alcohol synthesis catalyst jointly developed with Dow.[11] Economic assessment of ethanol production from these pilot tests is documented in an Aspen® process model.[12] Biomass-derived methanol as a platform chemical is currently being commercialized in Canada by Enerkem using non-recyclable municipal solid waste as a feedstock for gasification to syngas followed by catalytic upgrading to methanol. Longer term goals are converting methanol to ethanol as a platform chemical for launching varied conversion schemes to fuels, chemicals and materials.[13]

5.5.2 Biomass Pyrolysis Products Upgrading

Although lignin is the second-most abundant biopolymer in plant cell walls, its use in biochemical conversion of lignocellulosic biomass is typically relegated to combustion for heat and energy due to its inherent heterogeneity and recalcitrance.[14] As cellulosic ethanol biorefineries come online globally, the amount of waste lignin will dramatically increase, warranting new strategies for lignin valorization.[15] Recent work at NREL demonstrated an integrated approach that employs biological "funneling" of depolymerized lignin aromatics with chemical catalysis to produce renewable fuels and direct petrochemical replacements from lignin. In nature, metabolic pathways have evolved in certain microorganisms to convert a wide array of lignin-derived aromatic compounds to intracellular polymers,[16] or with genetic modification, extracellular dicarboxylic acids. By first applying this biological funneling approach to address the heterogeneity of lignin, subsequent separation and catalytic processes can be greatly simplified, providing a new path towards lignin valorization.

In a recent demonstration of this approach, *Pseudomonas putida* KT2440 was utilized to biologically funnel lignin-derived aromatics from alkaline pretreated corn stover[17] and produce intracellular medium chain length polyhydroxyalkanoic acids (*mcl*-PHAs).[18] *mcl*-PHAs have direct application as renewable polymers or as a feedstock for hydrocarbon fuels following thermal depolymerization and catalytic upgrading. After producing and recovering *mcl*-PHAs from a lignin-derived liquor stream, heat (250 °C) was applied under an inert atmosphere to depolymerize and dehydrate *mcl*-PHAs, resulting in C_6–C_{14} alkenoic acids. A Pt–Re bimetallic catalyst, previously used for unsaturated fatty acid deoxygenation,[19] was then utilized to hydrogenate and deoxygenate these alkenoic acids in hydrothermal media, producing linear and cyclic hydrocarbons with one carbon less than the parent acid, as shown in Figure 5.2. As in conventional petrochemical processing, the application of bimetallic catalysts can enhance reactivity, improve selectively, and extend catalyst stability under the harsh reaction conditions typically employed for biomass upgrading. Ongoing efforts in bimetallic catalyst material development include alloying noble metals with low-cost secondary metals,[20] tuning structural properties to increase the density and geometry of active crystallite sites,[21] as well as exploring earth abundant metal oxides to reduce material costs[22,23] given the potential for deactivation when processing biomass-derived substrates.

Moving forward, efforts will be required to investigate the sustained performance of commercially available catalytic materials, determine the influence of trace impurities from upstream biological and separation processes on catalyst stability, and apply next-generation catalytic materials to

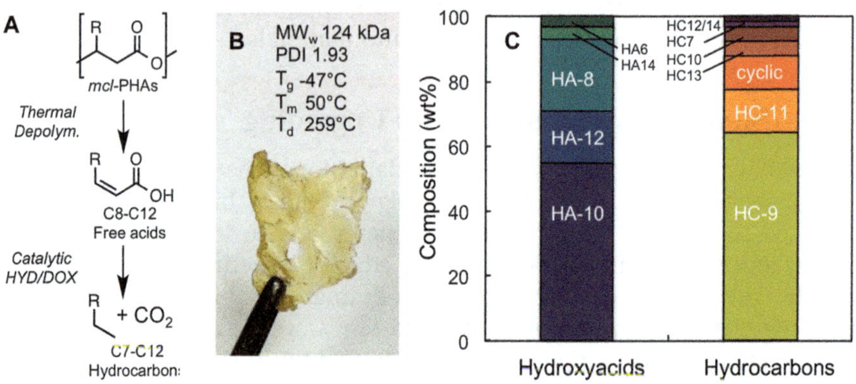

Figure 5.2 Thermal depolymerization and catalytic deoxygenation of *mcl*-PHAs derived from lignin. (A) Example of a thermal-catalytic upgrading pathway to alkenoic acids and hydrocarbon fuels. (B) *mcl*-PHA physicochemical properties derived from the biological conversion of corn stover alkaline pretreated liquor. Properties include weight average molecular weight (MW_w), polydispersity index (PDI), glass transition temperature (T_g), melting point (T_m), and decomposition temperature (T_d). (C) Initial *mcl*-PHA hydroxy acid (HA) composition and hydrocarbon (HC) distribution after thermal depolymerization and catalytic deoxygenation. Adapted from Linger *et al.* with permission from ref. 18. Copyright 2014 National Academy of Sciences.

facilitate chemistries under mild conditions with high activity, selectivity, yield, and stability. Furthermore, feedback from integrated techno-economic analysis and environmental impact studies will be key to identifying and overcoming technological barriers for integrated biological funneling and chemical catalysis for lignin valorization. [24]

5.6 Biomass to Products through Platform Chemicals

The following section assesses progress of the top twelve biomass-derived molecules towards demonstration and commercialization.

5.6.1 Succinic Acid

Succinic acid, identified in numerous publications as the top (bio) platform molecule, is available from the bioconversion of glucose at concentrations as high as approximately 6 wt%.[25] Figure 5.3 shows the conversion routes of succinic acid as a platform chemical. Esterification reactions are one of the most useful transformations for organic acids,[26] especially for a dicarboxylic acid as the diester can be used as an intermediate in the manufacture of polymers, fine chemicals, perfumes, plasticizers and solvents.[26,27,28,29]

Many acid catalysts have been reported in these reactions,[30,31] although only a few authors have dealt with the esterification of either succinic acid[32] or its cyclic anhydride.[31,32] Budarin[33] developed a new family of tunable mesoporous carbonaceous catalysts derived from starch that can be used for efficient aqueous phase conversions. Tuning the hydrophobicity-hydrophilicity balance of mesoporous starbons by carbonization at different

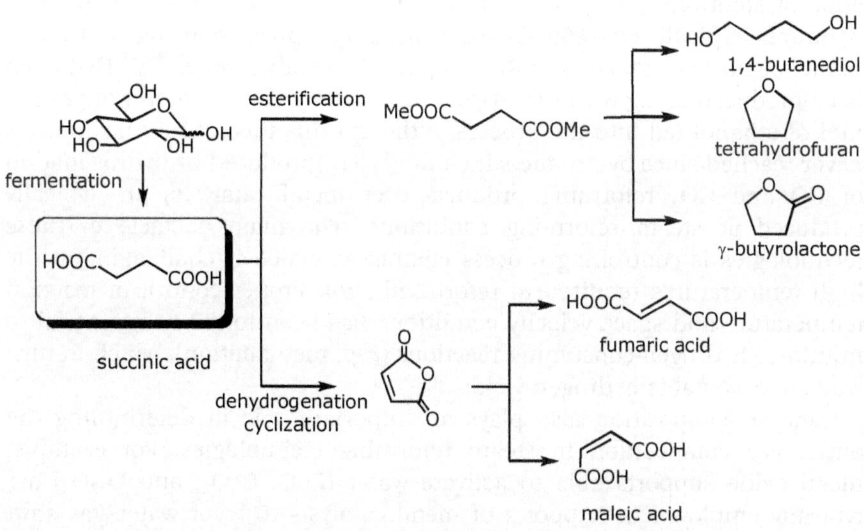

Figure 5.3 Succinic acid as a platform chemical from Bozell and Petersen.[7] Reproduced from ref. 7 with permission from the Royal Society of Chemistry.

temperatures (250–750 °C) followed by simple sulfuric acid treatment yielded a series of porous solid acids that functioned in dilute aqueous solutions. The reaction of succinic acid in aqueous alcohol to the diester significantly exceeded that of other solid acid catalysts comprising zeolites, sulfated zirconias and clays, though zeolite-beta was the most promising catalyst for diester formation, whereas sulfated zirconia and montmorillonite KSF gave the highest observed selectivities to the monoester. Designing catalysts for use in these dilute aqueous solutions is a critical need for efficient bio-based "building block" conversions.

Another promising heterogeneous conversion route first dehydrates succinic acid to succinic anhydride, then hydrogenates the anhydride to g-butyrolactone with a ruthenium silica catalyst, followed by butyrolactone hydrogenation to 1,4-butanediol with a copper-zinc catalyst and gives yields greater than 70%.[34] Other recent work showed high activity for levulinic and succinic acid hydrogenation using supported noble-metal catalysts (Ru, Pd or Pt) and the corresponding Re-promoted catalysts. Both acids were converted into lactones or diols depending on catalyst and reaction conditions with the highest selectivity to 1,4-pentanediol of 82% achieved at 140 °C with a 1.9% Ru–3.6% Re/C catalyst.[35]

5.6.2 Ethanol

Apart from its use as a biofuel, ethanol possesses enormous potential as a chemical feedstock. A number of valuable chemicals can be derived from ethanol by using continuous-flow catalytic technologies. Steam reforming is the most studied continuous catalytic process of ethanol in which aqueous ethanol solutions are gasified into hydrogen-rich mixtures at high temperatures (typically 600–800 °C) and atmospheric pressure using Ni, Co and noble metals (*e.g.*, Pt, Pd, and Rh) supported on stable oxides.[36,37] Hydrogen is co-produced along with CO_2 with a maximum yield of 6 moles of gas per mol of ethanol fed into the process, although this theoretical maximum is never reached since by-products including CH_4 (produced by hydrogenation of CO and CO_2 reforming products over metal catalysts) are typically produced at steam reforming conditions. The main obstacle of these technologies is controlling process chemistry, which is challenging at the high temperatures required to reform ethanol. Proper control of reaction temperature and space velocity conditions has been found to be crucial to minimize hydrogen-consuming reactions (*e.g.*, methanation), which in turn leads to acceptable hydrogen yields.[38]

Catalyst composition also plays an important role in determining the outlet gas composition in steam reforming technologies. For example, metal–oxide supports able to activate water (ZnO, CeO_2, and La_2O_3) are typically employed as supports of metal catalysts to favor water–gas shift processes (WGS, $CO + H_2O \rightarrow CO_2 + H_2$) which lead to purified H_2 streams with very low levels of CO that are potentially useful in fuel cell applications. These materials also possess basic characteristics that avoid coke-forming polymerization reactions (typically catalyzed on acidic sites) and improve catalyst stability.

5.6.3 2,5-Furandicarboxylic Acid

2,5-Furandicarboxylic acid (FDCA) is a potentially significant renewable building block because it can substitute terephthalic acid (TPA) in the production of polyesters and other current polymers containing an aromatic component. Avantium has announced plans to produce 40 ton/year of FDCA monomer at a pilot plant in the second half of 2011 (Avantium starts biopolyester pilot plant ICIS Green Chemicals). The Avantium process for preparing polyester comprises transesterifying FDCA with ethylene glycol in the presence of a tin-based esterification catalyst. Polycondensation and solid state polymerization result in films, fibers, and shaped materials.[39]

The primary technical barrier in the production and use of FDCA is the development of an effective and selective dehydration process from sugars. The control of sugar dehydration could be a very powerful technology, leading to a wide range of additional, inexpensive building blocks, but it is not yet well understood. Currently, dehydration processes using hydroxymethylfurfural (HMF) as an intermediate are generally non-selective, unless, immediately upon their formation, the unstable intermediate products can be transformed into more stable materials such as methoxymethylfurfural (MMF). Required R&D will include development of selective dehydration systems and catalysts. FDCA formation will require development of cost effective and industrially viable oxidation technology that can operate in tandem with required dehydration processes.[4] Despite its chemical stability (342 °C melting point), FDCA undergoes reactions typical for carboxylic acids, such as halogen substitution to give carboxylic dihalides, diester formation and the formation of amides.[40] The versatility of FDCA is also seen in the number of derivatives available *via* relatively simple chemical transformations. Selective reduction can lead to partially hydrogenated products, such as 2,5-dihydroxymethylfuran, and fully hydrogenated materials, such as 2,5-bis(hydroxymethyl)tetrahydrofuran. The potential applications of furan-based building blocks for polymer applications have been extensively reviewed by Gandini and Belgacem.[41] Examples of polyesters, polyamides and polyurethanes have been described in extensive literature reports.

5.6.4 3-Hydroxypropanoic Acid

3-Hydroxypropionic acid (HPA) is a beta hydroxy acid that is water soluble and dehydrates to acrylic acid when distilled. Figure 5.4 shows the production routes for HPA and its derivatives. It is used in the industrial production of various chemicals such as acrylates and can be produced by engineered microbes. University of Minnesota researchers recently produced poly(3-hydroxypropionic acid) from 3-hydroxypropanoic acid in a single pot reaction using ring-opening polymerization. This method produces a higher

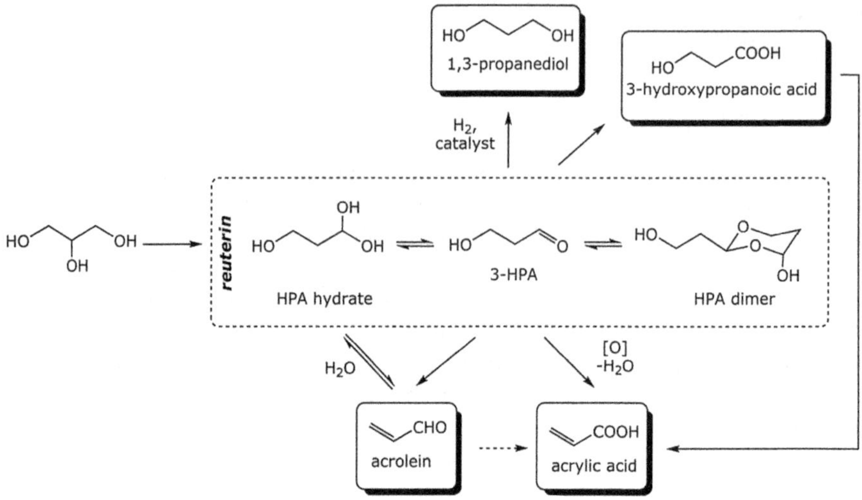

Figure 5.4 Production of 3-HPA and related derivatives from Bozell and Petersen.[7] Reproduced from ref. 7 with permission from the Royal Society of Chemistry.

molecular weight which makes the polymer more structurally sound using a process with lower toxicity than competing technologies.[42] A novel approach to HPA conversion vaporized an aqueous solution of HPA followed by fixed-bed catalytic gas phase conversion to acrylic acid thus avoiding the issue of recovering product from the precursor solution.[43]

5.6.5 Glycerol

Glycerol is not a carbohydrate, but structurally it can be considered as a "mini-sugar", in that transformations characteristic of glycerol may be applied to carbohydrates. More importantly, glycerol is a significant material because of its ready availability and strong potential to become a primary building block for the biorefinery. Technology for its manufacture is established, and processes for its conversion into higher value materials have received significant research attention recently. In addition to biodiesel, higher value products from glycerol could provide an important revenue stream and a reduction of dependence on subsidies for an industry that faces overcapacity issues on the fuel side. Biodiesel-derived glycerol could be burned for process fuel value if the market becomes saturated, but higher value uses provide expanded production of bio-based chemicals from an inexpensive feedstock. Some projections indicate costs as low as \$0.11 per kg for crude glycerine solutions. Several technologies have emerged as candidates for conversion of glycerol into chemicals particularly in glycerol reduction, dehydration and fermentation.[7]

Catalytic hydrogenolysis converts glycerol into a family of derivatives, including ethylene glycol, propylene glycol, acetol and lactic acid. Suppes has reported a selective hydrogenolysis of glycerol to propylene glycol, a commercial material with an annual production of over 450 million kg.[44] Reduction of glycerol at 200 psi H_2 and 200 °C over a copper chromite catalyst gives propylene glycol selectivities of nearly 90% at 65% conversion. The mechanism of the reaction is suggested to be an initial dehydration of glycerol to acetol, followed by selective reduction of the carbonyl group. Suppes' route also offers product control. By altering the process conditions, acetol can be made as the primary product in greater than 90% selectivity by eliminating the hydrogen and using reactive distillation to convert glycerol to acetol.[45] Industrial interest in bio-based propylene glycol has been significant. Senergy has licensed Suppes' technology and is reportedly piloting the process. Archer Daniels Midland (ADM) announced that a new propylene glycol plant will be starting in 2010. The plant will have a capacity of 100 000 tonnes per year and is closely linked to ADM's existing biodiesel production.

Alternative reductions of glycerol give different product profiles. Treatment of glycerol with hydrogen in the presence of Ru/C or Pt/C leads to mixtures of ethylene and propylene glycol. Carrying out the reduction in the presence of hydroxide bases induces formation of lactic acid as the major product.[46]

Catalytic and thermal dehydration of glycerol provides several derivatives. Depending on the conditions used, dehydration occurs *via* loss of a primary hydroxyl group, leading to hydroxypropionaldehyde and acrolein, or at the secondary hydroxyl group, leading to hydroxyacetone. Acrolein has received recent attention as a precursor to acrylic acid, a high volume chemical with an annual production of 1.2×10^9 kg. Acrolein is produced in 86% selectivity at 70% conversion by treating glycerol in hot compressed water for 8 s under supercritical conditions in the presence of H_2SO_4 or $Zn(SO_4)_2$ promoters.[47] Glycerol dehydration has also been carried out in the gas phase over supported acid catalysts. Acrolein was formed with 65% selectivity at 100% conversion over 15 wt%WO_3/ZrO_2.[48]

5.6.6 Sorbitol

Sugar alcohols are promising intermediates for the production of hydrocarbons as drop-in products for the petrochemical refinery. Huber and Dumesic[49] have reported the chemical conversion of sorbitol to light alkanes *via* aqueous phase reforming as shown in Figure 5.5. Pt/Al_2O_3 is the preferred catalyst for the transformation, and promotes conversion of sorbitol into hexane at 50% selectivity. The remainder of the sugar is converted to lighter materials. The bifunctional catalyst induces several reactions. Sorbitol is dehydrated on the catalyst's acidic sites, and the resulting intermediates are hydrogenated on the metal sites. Through several dehydration and reduction cycles, sorbitol is converted to hexane. Reforming of the sorbitol on metal sites leads to the formation of CO_2 and H_2, which are converted to methane. Light hydrocarbons result from hydrogenolysis of the sorbitol.

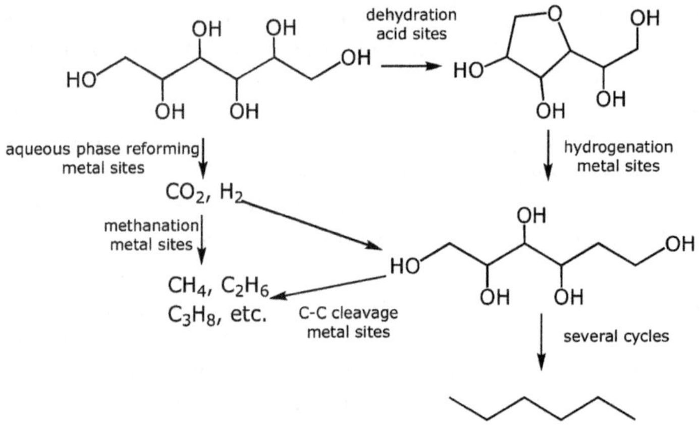

Figure 5.5 Proposed intermediates in the aqueous phase reforming of carbohydrates to light alkanes.[3]
Reproduced from ref. 3 with permission from the Royal Society of Chemistry.

5.6.7 Levulinic Acid

Levulinic acid is of interest as a primary biorefinery building block and platform chemical because of its simple and relatively high yield production from acid treatment of C_6 sugars. However, its isolation and purification can be complicated by the presence of intractable materials.

High yield industrial production of levulinic acid from C_6 polysaccharides has been achieved by Biofine Renewables using a proprietary two reactor system to minimize conditions suitable for side product formation.[50] Reactive extraction of levulinic acid from aqueous media with a family of different solvents containing Amberlite™ LA-2 has been investigated as a way to improve isolation and purification.[51]

The use of levulinic acid as a platform chemical continues to be studied. Manzer[52] has reported elegant, high yield catalytic transformations of levulinic acid into substituted pyrrolidones, lactones and levulinate esters. Supported heteropoly acids (HPAs) have been investigated as catalysts for the conversion of levulinic acid into diphenolic acid, a potential green replacement for bisphenol A in the production of polycarbonates.[53,54]

5.6.8 Itaconic Acid

Because itaconic acid has a structure similar to maleic anhydride, it can also be used for the production of unsaturated polyester resins. While not a drop-in substitute, it is a good replacement chemical.[55] DSM is developing a route for 100% bio-based polyester composites and in May 2014 published a patent for producing bio-based polyester composites from itaconic acid. The invention discloses a recombinant yeast cell which is capable of producing (overexpressing) one or more of 4-methyl itaconate or 1-methyl itaconate.[6]

Recent work demonstrated a 50% methacrylic acid yield with greater than 80% selectivity by decarboxylating itaconic acid using heterogeneous catalysts based on Pd, Pt and Ru. The reaction occurred in water at 200–250 °C without any external added pressure, conditions significantly milder than those described previously for the same conversion with better yield and selectivity. The decarboxylation process is also applicable to citric acid, a more widely available bio-based feedstock, and leads to the production of methacrylic acid in one pot with 41% selectivity. Aconitic acid, the intermediate compound in the pathway from citric acid to itaconic acid, was also used successfully as a substrate.[56]

5.6.9 3-Hydroxybutyrolactone

3-Hydroxybutyrolactone can be made from starch oxidation though catalysts and processes need further improvement to become economical. Little work outside of academia has been done with this bio-molecule.

5.6.10 Glutamic Acid

PNNL has done some work on glutamic acid.[57] Their technology hydrogenates glutamic acid to pyroglutaminol and prolinol.[58] Laboratory scale amounts of both products are available for license. PNNL says that "biomass feedstocks are increasingly in higher demand across the petrochemical industry as manufacturers research ways to make chemical products from something more environmentally friendly than petroleum". Glutamic acid is a ready platform for conversion to value added products. Among those are pyroglutaminol and prolinol, which may benefit a number of applications.

5.6.11 Glucaric Acid

Montana-based renewable chemicals company Rivertop Renewables is scaling up its oxidation process for glucaric acid-based products from the lab to pilot manufacturing. So far, it has produced "numerous" batches of glucaric acid-based products at approximately 385 kg per run with manufacturing partner DTI, the company said. Product made in this initial phase of contract manufacturing will be used to fulfil Rivertop's commercial contracts for bio-based corrosion inhibitors.[59]

5.6.12 Aspartic Acid

Not much work with aspartic acid has occurred since 2011, when Flexible Solutions announced that their aspartic acid[60] from sugar had reached commercial operating status at their Taber Alberta facility. Flexible stated that the next step is to increase production at the new plant. Production ramp up is expected to take several months given that this process is a biochemical process as well as the first of its kind in the world. The company also stated that for competitive reasons, they will not disclose details concerning production or production volumes though their initial target is 5000 metric tons per year.

Aspartic acid production at FSI's Taber plant in Alberta will be delayed by 3 to 5 months, slowing down the planned introduction of bio-sourced polyaspartic acid. However, the delay will not affect normal business and organic growth in the company's NanoChem division, which uses aspartic acid to produce thermal polyaspartate.[59]

5.7 Biomass to Products through One-pot Reactions

Bio-molecule conversion to products in a one-pot conversion is an attractive and economic process option if catalysts and products can be readily separated from the reaction mixture. Two elements that can enable one-pot reactions are catalyst site isolation and substrate-selective catalysis.[61] While substrate-selective catalysis is generally conducted with enzymes, catalyst site isolation can be achieved with sequentially linked columns containing anchored catalysts through which reagents flow with subsequent product formation and elution. Alternatively, a one-pot, one-catalyst system is the goal for this type of synthesis. BioAmber reportedly transformed its bio-based succinic acid into 1,4-butanediol (BDO) and tetrahydrofuran (THF) in a single catalytic step. BDO and THF are used to make engineering plastics, polyurethanes, biodegradable polyesters, spandex and other specialty chemicals.[62]

In principle, the ample hydroxyl groups in the structure of cellulose make it an ideal feedstock for the production of industrially important polyols such as ethylene glycol (EG), according to the atom economy rule. However, effectively depolymerizing cellulose under mild conditions is challenging due to the intra- and intermolecular hydrogen bonding network. As well, controlling product selectivity is complicated by the thermal instabilities of cellulose-derived sugars. Wang and Zhang[63] developed a one-pot catalytic conversion of cellulose to EG using a cascade reaction in which (1) acid catalyzes the hydrolysis of cellulose to water-soluble oligosaccharides and glucose; (2) produced oligosaccharides and glucose undergo C–C bond cleavage to form glycolaldehyde with a tungsten catalyst; then (3) hydrogenation of glycolaldehyde by a transition metal catalyst produces the end product ethylene glycol. Due to the instabilities of glycolaldehyde and cellulose-derived sugars, the reaction rates should be $r_1 \ll r_2 \ll r_3$ in order to achieve a high yield of EG. Tuning the molar ratio of tungsten to transition metal and changing the reaction temperature successfully optimizes this reaction.

5.8 Conclusions

The goal for catalyst development using bio-based feedstocks is to design new, efficient, selective and robust materials for the dehydration, decarboxylation, decarbonylation, hydrogenolysis, esterification, and ketonization reactions that are required for converting these oxygenated bio-based feedstocks to desirable renewable products. Summarizing the work done on developing bio-based products in the last decade is straightforward as incremental improvements in catalysis have moved bio-based polymers and their associated products to commercialization: ADM, PepsiCo, and Dow are deploying bio-based bottles, films and packaging to the marketplace. Aqueous phase bio-molecule processing is also developing rapidly as are the catalysts that can function in aqueous solutions of dilute feedstocks. One-pot processes are desirable if catalysts and products can be recovered economically. Finally, developments in catalyst technology that efficiently convert biomass-derived syngas and pyrolysis liquids and gases to both fuels and chemicals continue to improve in the harsh conditions required for these processes. The next decade should show market penetration of bio-based products as associated catalysts and processes move from development to demonstration and potential commercialization.

References

1. www.infocvastinc.com/bio-chem.
2. M. M. Bomgardner. C&EN, October 27, 2014, pp. 10–14.
3. P. Gallezot, Process options for converting renewable feedstocks to bioproducts, *Green Chem.*, 2014, **9**, 295–302.
4. T. Werpy and G. Petersen: *Top Value Added Chemicals from Biomass. Volume I – Results of Screening for Potential Candidates from Sugars and Synthesis Gas*, Produced by the Staff at Pacific Northwest National Laboratory (PNNL); National Renewable Energy Laboratory (NREL), Office of Biomass Program (EERE), 2004.
5. B. S. Shanks, *Design of Heterogeneous Catalysts for the Conversion of Biorenewable Feedstocks, in Catalysis for the Conversion of Biomass and its Derivatives*, ed. Malte Behrens, Abhaya K. Datye, Max Planck Research Library for the History and Development of Knowledge, 2013.
6. C. H. Zhou, X. Xia, C. X. Lin, D. S. Tongs and J. Beltramini, Catalytic conversion of lignocellulosic biomass to fine chemicals and fuels, *Chem. Soc. Rev.*, 2011, **40**(11), 5588–5617.
7. J. J. Bozell and G. Petersen, Technology development for the production of bio-based products from biorefinery carbohydrates—the US Department of Energy's Top 10 revisited, *Green Chem.*, 2010, **12**, 539–554.
8. R. L. Bain, K. A. Magrini-Bair, J. E. Hensley, W. S. Jablonski, K. M. Smith, K. R. Gaston and M. M. Yung, Pilot Scale Production of Mixed Alcohols from Wood, *Ind. Eng. Chem. Res.*, 2014, **53**(6), 2204–2218.
9. Y. O. Parent, K. A. Magrini, S. M. Landin and M. A. Ritland, Attrition Resistant Fluidizable Reforming Catalyst, *PCT Int. Appl.*, 2007, p. 43, Issued March 10, 2011, *US Pat.* 7,915,196 B2, Issued March 29, 2011.
10. A. M. Steele, S. Poulston, K. A. Magrini-Bair and W. S. Jablonski, Catalytic syngas purification from model biomass gasification streams, *Catal. Today*, 2013, **214**, 74–81.

11. J. E. Hensley, S. Pylypenko and D. R. Ruddy, Deactivation and stability of K-CoMoSx mixed alcohol synthesis catalysts, *J. Catal.*, 2014, **309**, 199–208.
12. A. Dutta, J. Hensley, R. Bain, K. Magrini, E. Tan, G. Apanel, D. Barton, P. Groenendijk, D. Ferrari, W. Jablonski and C. Carpenter, Techno-economic analysis for the production of mixed alcohols via indirect gasification of biomass based on demonstration experiments, *Ind. Eng. Chem. Res.*, 2014, **53**(30), 12149–12159.
13. http://edmonton.ca/for_residents/garbage_recycling/biofuels-facility.aspx.
14. S. P. S. Chundawat, G. T. Beckham, M. E. Himmel and B. E. Dale, Deconstruction of Lignocellulosic Biomass to Fuels and Chemicals, *Annu. Rev. Chem. Biomol. Eng.*, 2011, **2**, 121–145.
15. A. J. Ragauskas *et al.*, Lignin Valorization: improving lignin processing in the biorefinery, *Science*, 2014, **344**, 1246843.
16. G. Fuchs, M. Boll and J. Heider, Microbial degradation of aromatic compounds — from one strategy to four, *Nat. Rev. Microbiol.*, 2011, **9**, 803–816.
17. E. M. Karp *et al.*, Alkaline pretreatment of corn stover: bench-scale fractionation and stream characterization, *ACS Sustainable Chem. Eng.*, 2014, **2**, 1481–1491.
18. J. G. Linger *et al.*, Lignin valorization through integrated biological funneling and chemical catalysis, *Proc. Natl. Acad. Sci.*, 2014, **111**, 12013–12018.
19. D. R. Vardon *et al.*, Hydrothermal catalytic processing of saturated and unsaturated fatty acids to hydrocarbons with glycerol for in situ hydrogen production, *Green Chem.*, 2014, 1507–1520.
20. D. M. Alonso, S. G. Wettstein and J. A. Dumesic, Bimetallic catalysts for upgrading of biomass to fuels and chemicals, *Chem. Soc. Rev.*, 2012, **41**, 8075–8098.
21. G. A. Somorjai, H. Frei and J. Y. Park, Advancing the Frontiers in Nanocatalysis, Biointerfaces, and Renewable Energy Conversion by Innovations of Surface Techniques, *J. Am. Chem. Soc.*, 2009, **131**, 16589–16605.
22. T. Prasomsri, M. Shetty, K. Murugappan and Y. Román-Leshkov, Insights into the catalytic activity and surface modification of MoO_3 during the hydrodeoxygenation of lignin-derived model compounds into aromatic hydrocarbons under low hydrogen pressures, *Energy Environ. Sci.*, 2014, 7, 2660–2669.
23. T. Prasomsri, T. Nimmanwudipong and Y. Román-Leshkov, Effective hydrodeoxygenation of biomass-derived oxygenates into unsaturated hydrocarbons by MoO_3 using low H_2 pressures, *Energy Environ. Sci.*, 2013, **6**, 1732–1738.
24. R. Davis *et al.*, Process design and economics for the conversion of lignocellulosic biomass to hydrocarbons: Dilute-acid and enzymatic deconstruction of biomass to sugars and biological conversion of sugars to hydrocarbons, *NREL. Tech. Rep.*, 2013.

25. A. Koutinas, R. Wang, C. Webb, Top Value chemicals from biomass, ed. T. Werpy and G. Petersen, http://www.eere.energy.gov/biomass/pdfs/35523.pdf, unpublished results.
26. (a) J. Otera, *Esterification*, Wiley-VCH, Weinheim, Germany, 2003; (b) J. Mulzer, *Comprehensive Organic Synthesis*, ed. B. M. Trost and L. Fleming, Pergamon Press, Oxford, 1991.
27. Y. Liu, E. Ranucci, M. S. Lindblad and A. C. Albertsson, New Biodegradable Polymers from Renewable Sources: Polyester-carbonates Based on 1,3-Propylene-co-1,4-Cyclohexanedimethylene Succinate, *J. Polym. Sci., Polym. Chem.*, 2001, **39**, 2508–2519.
28. D. Shekhawat, K. Nagarajan, J. E. Jackson and D. J. Miller, Formation of citraconic anhydride via condensation of dialkyl succinates and formaldehyde, *Appl. Catal., A*, 2002, **223**, 261–273.
29. Diversified Natural Products, Inc. http://www.dnpworld.com.
30. (a) B. R. Jermy and A. Pandurangan, *Appl. Catal., A*, 2005, **288**, 25–33; (b) B. Rabindram Jermy and A. Pandurangan, *J. Mol. Catal. A*, 2005, **237**, 146–154; (c) F. Tataroglu Seyidov, Y. Mansoori and N. Goodarzi, *J. Mol. Catal. A*, 2005, **240**, 186–190; (d) Y. Mansoori, F. Seyidov Tataroglu and M. Sadaghian, *Green Chem.*, 2005, 7, 870–873; (e) T. A. Peters, N. E. Benes, A. Holmen and J. T. F. Keurentjes, *Appl. Catal., A*, 2006, **297**, 182–188; (f) J. Lilja, J. WärnS, T. Salmi, L. J. Petterson, J. Ahlkvist, H. Grénman, M. Rçnnholm and D. Y. Murzin, *Chem. Eng. J.*, 2005, **115**, 1–12; (g) J. Lilja, J. Aumo, T. Salmi, D. Y. Murzin, P. Mäki-Arvela, M. Sundell, K. Ekman, R. Peltonen and H. Vainio, *Appl. Catal., A*, 2002, **228**, 253–267; (h) T. Okuhara, *Chem. Rev.*, 2002, **102**, 3641–3665.
31. (a) H. J. Bart, J. Reidetschl, K. Schatka and A. Lehmann, *J. Int. Chem. Kinet.*, 1994, **26**, 1013–1021; (b) C. R. Reddy, P. Iyengar, G. Nagendrappa and B. S. Jai Prakash, *J. Mol. Catal. A*, 2005, **229**, 31–37.
32. C. R. Reddy, P. Iyengar, G. Nagendrappa and B. S. Jai Prakash, Esterification of dicarboxylic acids to diesters over M^{n+}-montmorillonite clay catalysts, *Catal. Lett.*, 2005, **101**, 87–91.
33. V. Budarin, R. Luque, D. J. Macquarrie and J. H. Clark, Towards a bio-based industry: benign catalytic esterifications of succinic acid in the presence of water, *Chem. - Eur. J.*, 2007, **13**, 6914–6919.
34. M.-S. Kim, S.-H. Chung, H.-J. Eom and K.-Y. Lee, *Catalytic hydrogenation of bio-based succinic acid for the production of 1, 4-butanediol through the indirect pathway*, Abstract 2103 AIChE Annual Meeting, November 3-8, 2013, San Francisco, CA.
35. Louis Corbel-Demailly, Bao-Khanh Ly, D.-P. Minh, B. Tapin, C. Especel, F. Epron, A. Cabiac, E. Guillon, M. le Besson and C. Pinel, Heterogeneous catalytic hydrogenation of bio-based levulinic and succinic acids in aqueous solutions, *ChemSusChem*, 2013, **6**, 2388–2395.
36. A. Haryanto, S. Fernando, N. Murali and S. Adhikari, Current status of hydrogen production techniques by steam reforming of ethanol: A review, *Energy Fuels*, 2005, **19**, 2098–2106.

37. M. Ni, D. Y. C. Leung and M. K. H. Leung, A review on reforming bio-ethanol for hydrogen production, *Int. J. Hydrogen Energy*, 2007, **32**, 3238–3247.
38. J. C. Serrano-Ruiz, R. Luque, J. M. Campelo and A. A. Romero, Continuous-flow processes in heterogeneously catalyzed transformations of biomass derivatives into fuels and chemicals, *Challenges*, 2012, **3**, 114–132.
39. E. DeJong, M. A. Dam, L. Sipos, G.-J. M. Gruter, Furandicarboxylic acid (FDCA), A versatile building block for a very interesting class of polyesters, in *Bio-based Monomers, Polymers, and Materials*, ed. P. Smith *et al.*, ACS Symposium Series, American Chemical Society, Washington, DC, 2012.
40. J. Lewkowski, Synthesis, chemistry and applications of 5-hydroxymethylfurfural and its derivatives, *ARKIVOC*, 2001, 17–54.
41. A. Gandini and N. M. Belgacem, *Prog. Polym. Sci.*, 1997, **22**, 1203–1379; E. DeJong, M. A. Dam, L. Sipos and G.-J. M. Gruter, Furandicarboxylic acid (FDCA), a versatile building block for a very interesting class of polyesters, in *Bio-based Monomers, Polymers, and Materials*, ed. P. Smith *et al.*, ACS Symposium Series, American Chemical Society, Washington, DC, 2012.
42. http://license.umn.edu/technologies/z05135_biodegradable-polyester-produced-from-non-toxic-renewable-monomer).
43. D. Decoster, S. Hoyt and S. Roach, Dehydration of 3-hydroxypropanoic acid to acrylic acid, WO 2013192451 A1, December 27, 2013.
44. M. A. Dasari, P. P. Kiatsimkul, R. Sutterlin and G. J. Suppes, Low-pressure hydrogenolysis of glycerol to propylene glycol, *Appl. Catal., A*, 2005, **281**, 225–231.
45. C. W. Chiu, M. A. Dasari, G. J. Suppes and W. R. Sutterlin, Dehydration of glycerol to acetol via catalytic reactive distillation, *AIChE J.*, 2006, **52**, 3543–3548.
46. E. P. Maris, W. C. Ketchie, M. Murayama and R. J. Davis, Glycerol hydrogenolysis on carbon-supported PtRu, and AuRu bimetallic catalysts, *J. Catal.*, 2007, **251**, 281–294.
47. M. Watanabe, T. Lida, Y. Aizawa, T. M. Aida and H. Inomata, Acrolein synthesis from glycerol in hot-compressed water, *Bioresour. Technol.*, 2007, **98**, 1285–1290.
48. S. H. Chai, H. P. Wang, Y. Liang and B. Q. Xu, Sustainable production of acrolein: investigation of solid acid–base catalysts for gas-phase dehydration of glycerol, *Green Chem.*, 2007, **9**, 1130–1136.
49. G. W. Huber and J. A. Dumesic, An overview of aqueous-phase catalytic processes for production of hydrogen and alkanes in a biorefinery, *Catal. Today*, 2006, **111**, 119–132.
50. S. W. Fitzpatrick, Feedstocks of the future: Renewables for the production of chemicals and materials, *ACS Symp. Ser.*, 2006, **921**, 271–287.
51. H. Uslu, S. I. Kirbaslar and K. L. Wasewar, Reactive extraction of levulinic acid by Amberlite LA-2 extractant, *J. Chem. Eng. Data*, 2009, **54**, 712–718.

52. L. E. Manzer, Biomass derivatives: A sustainable source of chemicals, *ACS Symp. Ser.*, 2006, **921**, 40–51.
53. J. J. Bozell, L. Moens, D. C. Elliott, Y. Wang, G. G. Neuenscwander, S. W. Fitzpatrick, R. J. Bilski and J. L. Jarnefeld, Production of levulinic acid and use as a platform chemical for derived products, *Resour. Conserv. Recycl.*, 2000, **28**, 227–239.
54. http://www.agranet.com/portal2/renchem/home.jsp?template=newsarticle&artid=20017947399&pubid=ag134.
55. R. Singh, B. Erickson and P. Winters, Bio-Based Strategy: Food and Non-Food Market, in *Bio-Based Composites for High-Performance Materials: From Strategy to Industrial Application*, ed. W. Smitthipong, R. Chollakup and M. Nardin, CRC Press, Taylor and Francis Group, Boca Raton, FL, 2015.
56. J. Le Notre, S. C. M. Witte-van Dijk, J. M. van Haveren, E. L. Scott and J. P. M. Sanders, Synthesis of bio-based methacrylic acid by decarboxylation of Itaconic acid and citric acid catalyzed by solid transition-metal catalysts, *ChemSusChem*, 2014, 7, 2712–2720.
57. http://techportal.eere.energy.gov/technology.do/techID=257.
58. http://www.biofuelsdigest.com/bio-based/2013/01/08/top-molecules-the-does-12-top-bio-based-list-whats-worked-out.
59. http://www.agranet.com.portal12/renchem/home.jsp?template=newsarticle&artid=20017947399&pubid=ag134.
60. http://www.biofuelsdigest.com/bdigest/2011/11/14/flexible-solutions-to-make-aspartic-acid-from-sugar/.
61. S. J. Broadwater, S. L. Roth, K. E. Price, M. Kobašlija and D. T. McQuade, One-pot multi-step synthesis: a challenge spawning innovation, *Org. Biomol. Chem.*, 2005, 2899–2906.
62. http://www.bio-amber.com/bioamber/en/technology#conversion_technology.
63. A. Wang and T. Zhang, One-pot conversion of cellulose to ethylene glycol with multifunctional tungsten-based catalysts, *Acc. Chem. Res.*, 2013, **46**(7), 1377–1386.

CHAPTER 6

Separations Technologies for Biobased Product Formation—Opportunities and Challenges

BHANENDRA SINGH,[†] ANJU KUMARI[†] AND SAURAV DATTA*

Department of Biotechnology, Indian Institute of Technology Roorkee, Roorkee, Uttarakhand, India
*Email: sdattfbt@iitr.ac.in

6.1 Introduction

The limited availability of fossil fuels and considerable success in obtaining valuable fuels/chemicals from renewable feedstocks has enabled a paradigm shift in the science and technological world. As a result, recent years have witnessed a major thrust moving towards a sustainable, biobased economy using a biorefinery concept. The biorefinery concept is based on obtaining a broad spectrum of biofuels and value-added biobased products from renewable resources, analogous to the petroleum refinery concept.[1] Several policies, such as the Energy Independence and Security Act, 2007,[2] and the National Bioeconomy Blueprint,[3] were announced to provide enough impetus for adopting this concept. It has two major components—environmentally benign conversion technologies and efficient separations

[†]These authors have equal contribution.

RSC Green Chemistry No. 43
Commercializing Biobased Products: Opportunities, Challenges, Benefits, and Risks
Edited by Seth W. Snyder

Published by the Royal Society of Chemistry, www.rsc.org

technologies. Biological conversion of renewable resources by fermentation is an important process to produce biofuels and a wide range of value-added products (Chapter 4). In this process, microorganisms utilize various feedstocks, such as sugars, crops, animal residues, food waste and sewage sludge to produce fuels/chemicals through their metabolic activities. Conversion is followed by separation (recovery) of the target product(s). Separations technologies are important as they contribute to 30–50% of the overall production cost. Due to the complexities of the biological solution, a comprehensive separations strategy rather than a single separation technique is implemented for the recovery of the target product.

Although biofuel is the major product of a biorefinery in terms of bulk volume, at present it suffers from limited economic viability (Chapter 15). In contrast, value-added biobased products obtained from the biorefinery are able to provide higher financial incentives to the commercial investors for pursuing the biorefinery concept. Hence, there is a surging interest for commercializing value-added biobased products, which function as precursors or building blocks for industrially relevant downstream chemicals.[4] Particularly, oxygenated biobased species, such as alcohols, organic acids and furans, due to their reactive nature, are of great interest. For example, furan-2,5-dicarboxylic acid (FDCA) is designated as an alternative to terephthalate acid for downstream production of polyethylene terephthalate (PET). A distinct advantage of biomass-derived oxygenated species could be attributed to the fact that oxygen is already present in the original structure of the biomass, whereas, for petroleum-derived oxygenated species, oxygen needs to be introduced by chemical reaction.[5] Biomass is more atom-efficient for producing oxygenated products. This renders biomass-derived oxygenated species as front runners for commercialized biobased products.

It is evident from this discussion that in order to make the biorefinery concept a success, an economically viable separations/recovery of the biobased product is absolutely critical. Separations is a vast subject of engineering. With growing interest in biofuels and biobased products, the field has become interdisciplinary with participation of researchers from different scientific and technological backgrounds. There are dedicated books discussing theories, principles and engineering applications of separations.[6–9] The vast information available is often overwhelming for interdisciplinary researchers. Apart from that, the choice of separations technology for a particular process is not so trivial. The mechanism of separations varies from one technique to another based on the thermodynamic properties of the components involved as well as on their physical states. For example, density difference is the driving force of separation for sedimentation, which is conducted for solid–liquid separations. On the other hand, chemical potential and electrical potential are the driving forces for adsorption (liquid–liquid or gas–gas or gas–liquid separations) and electrodialysis (liquid–liquid separation), respectively. Few separations technologies use combinations of physical and chemical forces, *e.g.* a combination of chemical reaction followed by density difference enables separation by precipitation. Knowledge of all these factors is important as they dictate the design parameters of the process.[10]

We embrace this opportunity and have designed this chapter focusing on separations technologies that are relevant to biobased product formation. Both fundamental aspects of separations technologies and their state-of-the-art applications for biobased product formation are emphasized. Since this is an emerging area of research and commercialization, none of the technologies are devoid of limitations. Therefore, we discuss the technical challenges and potential solutions associated with the technologies. We believe that this chapter will function as a guideline for identifying suitable separations schemes for biobased product formation.

6.2 Fundamentals of Separations Technologies

We discuss the fundamental theories and principles of separations technologies that are relevant to biobased product formation from renewable resources, namely adsorption, crystallization, L–L extraction and membrane separations. Conventional separations technologies, which are widely used, but are associated with energy and environmental restrictions, are deliberately omitted. For example, distillation is one of the most popular techniques for dewatering and recovery of volatile compounds, however, due to its energy intensity, alternative technologies are emerging. We discuss other emerging separations technologies, such as membrane separations and extraction, as alternatives to distillation.

6.2.1 Adsorption

In biorefineries, adsorption is used for selective recovery of the target bioproduct (adsorbate) from a mixture containing unwanted components (or for the removal of impurities from the target bioproduct) using a solid surface (adsorbent). Activated carbon, molecular sieves (zeolites), activated alumina, silica gel and synthetic polymeric resin beads are some of the commercially available adsorbents. Polymeric resin beads, because of their versatility in surface functionality, are widely used for the recovery of bioproducts from processing solutions. Both uncharged and charged (cation exchange or anion exchange) resin beads are implemented for adsorption of target species. Adsorption takes place due to a selective interaction between the adsorbate molecules and the porous solid structure of the adsorbent. The interaction could be due to electrostatic attraction or Van der Waals force, which leads to physisorption. A stronger interaction, such as chemical bond formation, leads to chemisorption. For a particular separation, adsorbents are selected based on the following properties:

(i) *Selectivity*: ability of the adsorbent to selectively capture the target species from a mixture. This property depends on the pore size of the adsorbents, existence of affinity between the adsorbent and adsorbate, and rate of diffusion of the adsorbate molecules into the porous structure of the adsorbent.

(ii) *Capacity*: the higher the capacity of the adsorbent, the smaller the equipment size, leading to a cost efficient process. This property depends on the porous surface area of the adsorbent.

(iii) *Reversibility*: to reduce the cost of the process, reusability of the adsorbent is desirable. Reversibility depends on the type of interaction and is achieved by a regeneration step.

(iv) *Structural integrity*: the ability of the adsorbent to tolerate chemical, mechanical and thermal forces without compromising the structural integrity.

For industrial applications, a continuous adsorption process in a fixed-bed column is widely used (Figure 6.1). Here, adsorbent materials, such as resin beads, are packed in a column. Feed is fed from the bottom end of the column, while treated solution is removed from the top end of the column. Adsorption results in depletion of solute in the feed solution as it advances through the column. Overall, solute concentrations in the feed solution as well as in the adsorbent decrease along the height of the column, and also vary as a function of time for a particular location. Feed flow is stopped at the saturation of the bed (equilibrium). This is followed by a washing step to remove residual feed solution and non-specifically adsorbed solute from the

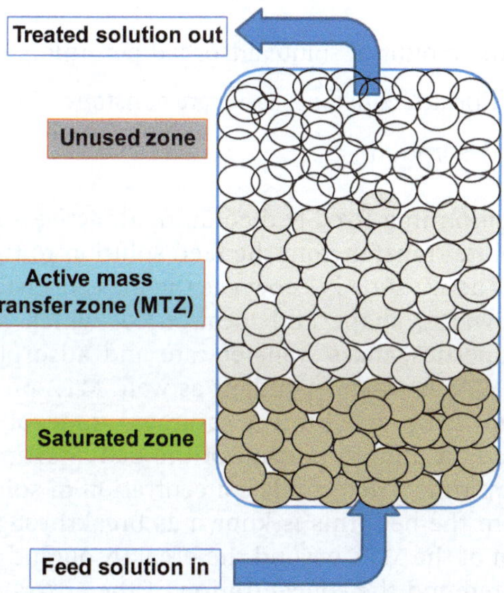

Figure 6.1 Schematic of a fixed-bed adsorption column showing three distinct zones of a typical adsorbent bed: saturated zone at equilibrium with the feed solution, active mass transfer zone (MTZ), where mass transfer takes place, and the unused portion of the bed that is yet to be in contact with the adsorbate (solute) molecules.

column. Then, the bound solute is recovered by desorbing from the adsorbent in a separate elution step.

The equilibrium relationship between adsorbate concentration in solution and in a solid adsorbent for a particular adsorbate–adsorbent pair is known as adsorption isotherm. The adsorption isotherm is important for design calculations in a fixed-bed column. It is determined from separate batch experiments with varying initial solute concentrations in feed solution. The Langmuir adsorption isotherm and Freundlich adsorption isotherm are two popular forms of expressing equilibrium relationships.[9] The Langmuir adsorption isotherm is mathematically represented as follows:

$$q = q_{\mathrm{m}} \frac{KC}{1 + KC} \tag{6.1}$$

It could also be written as:

$$\frac{1}{q} = \frac{1}{Kq_{\mathrm{m}}} \frac{1}{C} + \frac{1}{q_{\mathrm{m}}} \tag{6.2}$$

C = concentration of solute in solution at equilibrium

q = amount of solute adsorbed per unit amount of adsorbent at equilibrium

q_{m} = maximum amount of solute adsorbed per unit amount of adsorbent

$K = \frac{K_{\mathrm{a}}}{K_{\mathrm{d}}}$ = equilibrium constant; K_{a} = rate constant of adsorption; K_{d} = rate constant of desorption

During adsorption in a fixed-bed column, an active zone of the column exists, where solute transfer from the feed solution to the solid adsorbent takes place. This active zone is known as the mass transfer zone (MTZ), the characteristics (width, shape and location) of which are dependent on equilibrium properties, mass transfer rate and adsorption kinetics. The location of the MTZ is time-dependent as well. MTZ progresses along the bed as a "wave" leaving behind a saturated zone of the bed and encountering ahead an unused portion of the bed. Starting from time zero, when for the first time a detectable concentration of solute appears in the exit effluent from the bed, this is known as breakthrough time. With further progression of the MTZ beyond the breakthrough time, the whole bed becomes saturated and the concentration of the adsorbed solute remains in equilibrium with the concentration of the solute in feed solution. This is known as equilibrium time. Information obtained from MTZ analysis is used for column capacity determination and scale-up design as demonstrated next.[7]

Total capacity of the fixed - bed column, if the entire bed comes to equilibrium capacity (saturation) with the feed (expressed in time equivalent),

$$t_t = \int_0^{\infty} \left(1 - \frac{C}{C_0}\right) dt \tag{6.3}$$

Used capacity of the fixed - bed column up to the breakthrough time (t_b) (expressed in time equivalent),

$$t_u = \int_0^{t_b} \left(1 - \frac{C}{C_0}\right) dt \tag{6.4}$$

Total height of the bed $= H_t$; height of used bed up to breakthrough time $= H_b$

$$\frac{H_t}{H_u} = \frac{t_t}{t_u} \tag{6.5}$$

$$\text{Height of unused bed} = H_{unb} = \left(1 - \frac{t_u}{t_t}\right) H_t \tag{6.6}$$

C_0 = concentration of solute in feed solution
C = concentration of solute in solution at any axial position in the bed

By determining H_{unb} from a laboratory-scale experiment (under similar conditions to the larger scale experiment) and calculating the height of bed needed to achieve the required capacity at breakthrough point, H_b, the total height of the full-scale bed, H_t, could be then obtained from:[7]

$$H_t = H_u + H_{unb} \tag{6.7}$$

Diffusion governed slow operation, low capacity, high pressure drop, lack of selectivity, expensive adsorbents and cost of regeneration are the prominent challenges associated with adsorption. Significant advancement in materials science is able to overcome some of these challenges.

6.2.1.1 *Adsorption by Principles of Ion Exchange (IEX)*

Adsorption using ion exchange (IEX) adsorbents has worldwide industrial applications. IEX is a separations process, where ions in a solution are exchanged with similar kinds of mobile ions present in a solid phase. The solid phase consists of a charged base material (fixed ions) surrounded by the mobile counterions to maintain electroneutrality. IEX resins (solid phase) are categorized according to the type of mobile ions, such that the cation exchange (CEX) resin contains mobile cations and anion exchange (AEX) resin contains mobile anions. Typically, a fixed-bed column, containing IEX beads is used for contacting a process solution containing ionizable species, which are exchanged with the mobile ions of the solid IEX resin beads. For example, a cation exchange resin (modified polystyrene-derivative) with sulfonic acid as a fixed charge and Na^{+} as the mobile counterion is used to remove Ca^{+2} or Hg^{+2} present in the process solution using the IEX process. pH has a strong influence on the ionic properties of a solution, and therefore, it is an important parameter for the IEX-based separations process. For efficient recovery of anions, the operating pH should be greater than the pK_a of the target species, such that the species remains primarily in an ionized form. Similarly, efficient recovery of cations requires lower operating pH than the pK_a of the target species, such that it remains primarily in an ionized form. IEX resins are categorized as weak and strong based on their ionization capability. Weak resins operate over a narrow range of pH, whereas strong resins operate over a broader range of pH. So, weak and strong acid CEX resins and weak and strong base AEX resins are available. Besides, based on the morphology, microporous (gel-type) and macroporous (macro-reticular) resin beads are available.

The IEX capacity of the resins increases with the number of mobile ions present, which in turn depends on the available surface area of the solid matrix. Thus, higher surface area resin beads ensure higher ion uptake capacity. Higher ion uptake capacity leads to higher water sorption (resin swelling due to water sorption), which aids in enhancing ion transport through the solid matrix. In contrast, a higher degree of cross-linking leads to lower water sorption and reduced ion transport. However, a higher degree of cross-linking provides structural integrity to the resin. It is necessary, therefore, to trade off between ion uptake capacity and the structural integrity of the resin depending on the application.

6.2.2 Crystallization

A crystal is a solid form of a substance with orderly and repetitive arrangements of atoms and molecules. Crystallization is an equilibrium-driven solid–liquid separations process, where mass transfer of a solute from a homogeneous solution takes place to form the solid crystals. Crystallization is conducted by manipulating certain process variables such that the solute concentration becomes greater than the solubility at a particular temperature, *i.e.* by achieving supersaturation. Supersaturation is achieved either by cooling the solution (batch/tank type crystallizers) or by evaporating the solvent (evaporator crystallizers) or by a combination of both (adiabatic vacuum crystallizers). The choice depends on the nature of the solubility curve (solubility as a function of temperature) of the solute–solvent system. Under supersaturation conditions, a solute continues to form pure crystals until the process solution becomes saturated and terminates the crystallization process (equilibrium is attained). The size and shape of the crystals, along with the yield and purity, are important parameters for determining the performance efficiency of a crystallization process.

Crystallization involves the formation of a new solid phase from a homogeneous solution, therefore, understanding crystal formation and the growth mechanism is important for designing and operating crystallizers. The mechanism consists of two major steps—nucleation (nucleus or growth site formation) and crystal growth. Nucleation occurs either by a homogeneous mechanism due to molecular aggregation or by a contact mechanism due to interference of a contacting agent (container walls or agitator surface). Crystal growth has two steps—convective transport of the solute molecules from the bulk supersaturated solution to the crystallization site and incorporation of the solute molecules in the growing crystal embodiment. Once a nucleation site is available, crystal growth initiates and this sequence of nucleation and crystal growth continues until saturation is achieved. Formation of new crystals independent of the presence of existing crystals is known as primary nucleation, whereas formation of crystals in the presence of existing crystals is known as secondary nucleation. Theoretical analysis of the crystallization process, *i.e.* combination of nucleation and crystal growth, could be mathematically represented as follows:[6,9]

$$B_{\text{pn}} = A\left[\exp\frac{-16\pi\sigma^3 v_{\text{m}}^2}{3k^3T^3\{\ln(s+1)\}^2}\right] \tag{6.8}$$

$$B_{\text{sn}} = k_N w^b M_T^j s^p \tag{6.9}$$

$$\frac{\mathrm{d}m_{\text{c}}}{\mathrm{d}t} = K_{\text{l}}A_{\text{c}}(C - C_{\text{s}}) = \frac{A_{\text{c}}(C - C_{\text{s}})}{\frac{1}{k_{\text{l}}} + \frac{1}{k_{\text{r}}}} \tag{6.10}$$

B_{pn} = rate of primary nucleation; B_{sn} = rate of secondary nucleation
A = frequency factor; σ = interfacial tension; v_{m} = volume of a molecule; k = Boltzmann constant
T = absolute temperature; $s = \frac{(C - C_{\text{s}})}{C_{\text{s}}}$ = degree of supersaturation
C = bulk solute concentration in supersaturated solution
C_{s} = solute concentration at saturation; w = agitation rate
M_T = mass of crystal per unit volume of suspension
k_N, b, j, p are experimentally determined parameters
m_{c} = mass of a crystal; A_{c} = surface area of crystal; K_{l} = overall mass transfer coefficient
k_{l} = mass transfer coefficient for the convective transport from bulk to the crystallization site
k_{r} = mass transfer coefficient for incorporation into the crystal

Although crystallization is widely used in industry, major challenges associated with it, including higher processing time (slow, equilibrium-driven process) and energy intensiveness to achieve saturation, provide opportunities for further improvement.

6.2.3 Liquid–Liquid (L–L) Extraction

In L–L extraction, a solute is transferred from a solution phase in feed solvent to an extracting solvent phase by intimate contact between the two solvent phases. After extraction, the treated feed solvent phase, depleted of the solute, is known as the raffinate phase and the extracting solvent phase, enriched in the solute, is known as the extract phase. Typically, sufficient time is allowed so that equilibrium is attained between the raffinate and extract phases. Therefore, L–L extraction is an equilibrium-driven process and mass transfer plays an important role during the interphase transport of the solute molecules like the other separations technologies.

The selection criteria for the solvent used in L–L extraction processes are the following:

(i) *Selectivity*: the preference of the solute towards the extracting solvent compared to the feed solvent (career solvent). Solvents with higher selectivity are better candidates for extraction.

The selectivity of the solute for the extracting solvent over the feed solvent is represented by partition coefficient (distribution coefficient), K, which is expressed as:

$$K = \frac{C_{\mathrm{E}}}{C_{\mathrm{R}}} \tag{6.11}$$

C_{E} = equilibrium solute concentration in extract; C_{R} = equilibrium solute concentration in raffinate

(ii) *Miscibility*: for an ideal extraction, the feed and the extracting solvents should be completely immiscible. In certain cases, partially miscible solvents are also used for extraction.

(iii) *Density*: after extraction, in order to separate the raffinate and extract phases, a density difference between the feed and the extracting solvent is critical. The larger the density difference, the easier the separation.

The principles of different configurations of L–L extraction processes are described next.[9,11]

6.2.3.1 *Batch Extraction for Completely Immiscible Solvents*

Batch mode L–L extraction is typically conducted in a mixer–settler type arrangement. It is a two-step approach, where contact between the two phases takes place in a mixer followed by the separation of the phases in a settler based on the density difference. Sufficient time is allowed to attain equilibrium. Two important parameters for batch extraction are as follows:

$$\text{Extraction factor, } F = \frac{\text{amount of solute in extract}}{\text{amount of solute in raffinate}};$$

$$\text{Fraction extracted, } e = \frac{\text{amount of solute in extract}}{\text{amount of solute in initial feed}}$$

F and e can be mathematically expressed as follows:

$$F = \frac{EC_E}{RC_R} = \frac{KE}{R} \tag{6.12}$$

$$e = \frac{EC_E}{RC_{R_0}} = \frac{F}{1+F} \tag{6.13}$$

C_E = equilibrium solute concentration in extract; C_R = equilibrium solute concentration in raffinate
C_{R_0} = solute concentration in initial feed
R = volume of feed *i.e.* volume of raffinate; E = volume of extract

For a non-linear equilibrium relationship, the equation for the equilibrium line is obtained from the relationship $C_E = f(C_R)$, whereas the equation for the operating line is obtained from an overall material balance and is expressed as follows:

$$\text{Equation for operating line, } C_E = \left(\frac{R}{E}\right)(C_{R_0} - C_R) \tag{6.14}$$

C_E and C_R are then determined from the coordinates of the point of intersection between the equilibrium and operating lines on a C_E *vs.* C_R plot.

6.2.3.2 Continuous Single-stage Extraction for Completely Immiscible Solvents

In this case, the feed and the extracting solvent phases are continuously introduced into the mixer and are continuously removed from the settler. Sufficient time is allowed to attain equilibrium.

$$F = \frac{Q_E C_E}{Q_R C_R} = \frac{KQ_E}{Q_R} \tag{6.15}$$

$$e = \frac{Q_E C_E}{Q_R C_{R_0}} = \frac{F}{1+F} \tag{6.16}$$

Q_E = flow rate of the extracting solvent; Q_R = flow rate of the feed or raffinate

For a non-linear equilibrium relationship, the equation for the equilibrium line is obtained from the relationship $C_E = f(C_R)$. The equation for the operating line is obtained from an overall material balance and is expressed as follows:

$$\text{Equation for the operating line, } C_E = \left(\frac{Q_R}{Q_E}\right)(C_{R_0} - C_R) \tag{6.17}$$

C_E and C_R are then determined from the coordinates of the point of intersection between the equilibrium and operating lines on C_E *vs.* C_R plot.

6.2.3.3 Countercurrent Continuous Multi-stage Extraction for Completely Immiscible Solvents

This is a more industrially relevant mode of operation. It is assumed that the raffinate and the extract streams coming out from each stage are in equilibrium as demonstrated in Figure 6.2.

It is also assumed that the partition coefficient (K) is independent of the solute concentration. Various parameters are mathematically expressed as follows:

$$F = \frac{KQ_E}{Q_R} \tag{6.18}$$

For n number of stages, e can be expressed in terms of F as,

$$e = \frac{Q_E C_{E_n}}{Q_R C_{R_{n+1}}} = \frac{Q_R C_{R_{n+1}} - Q_R C_{R_1}}{Q_R C_{R_{n+1}}} = \frac{F(F^n - 1)}{F^{n+1} - 1} \tag{6.19}$$

Also, from the material balance for nth stage, one can write:

$$C_{R_{n+1}} = (1 + F + F^2 + \cdots\cdots + F^n)C_{R_1} = \left(\frac{F^{n+1} - 1}{F - 1}\right)C_{R_1} \tag{6.20}$$

For a non-linear equilibrium relationship, the equation for the equilibrium line is obtained from the relationship $C_{Ei} = f(C_{Ri})$. The equation for the operating line is obtained from an overall material balance and is expressed as follows:

$$\text{Equation for operating line, } C_{E_n} = \left(\frac{Q_R}{Q_E}\right)(C_{R_{n+1}} - C_{R_1}) \tag{6.21}$$

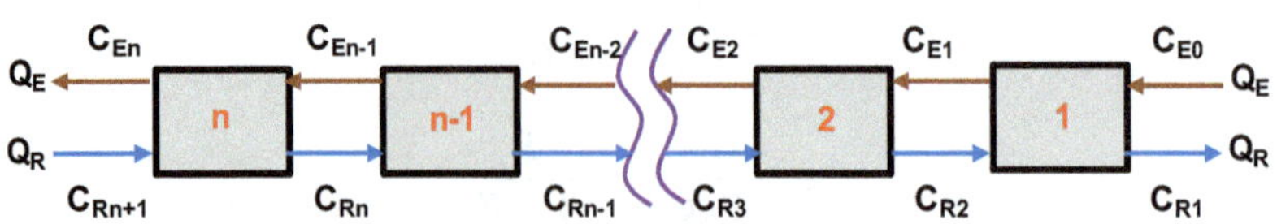

Figure 6.2 Schematic of a counter-current continuous multi-stage extraction process. The raffinate and extract streams exiting each stage are in equilibrium, *i.e.* $(C_E)_i$ is in equilibrium with $(C_R)_i$.

The required number of stages (n) is then determined graphically from a C_E *vs.* C_R plot.

6.2.3.4 *Supercritical Fluid Extraction (SFE)*

A substance above its critical temperature and pressure is transformed into a supercritical state. Supercritical fluid extraction (SFE) is an emerging technique of separation, where a supercritical fluid is used for extraction instead of an ordinary fluid. Water ($T_{cp} = 374\,°C$ and $P_{cp} = 217$ atm) and CO_2 ($T_{cp} = 31\,°C$ and $P_{cp} = 73$ atm) are two major solvents used for SFE. Supercritical fluids possess favorable properties of both gaseous substance and liquid substance that are extremely beneficial for extraction.[11] These are the ability to penetrate a substance easily (gas) and the ability to dissolve a substance easily (liquid), which leads to significantly higher diffusivity and solvent capacity, respectively. Besides, the viscosity of a substance in a supercritical state is substantially smaller than that in the liquid state. Another distinct advantage of SFE is the sharp increase in the solubility of the solutes in a supercritical fluid with marginal increase in pressure. Overall, favorable mass transfer and thermodynamic characteristics of SFE result in increased process efficiency. Besides, the increasing demand for SFE in pharmaceutical, food and beverage, flavor and fragrance industries, and biorefineries could also be attributed to the environmentally friendly nature of the supercritical fluid compare to toxic organic solvents.

Although L–L extraction is efficient in concentrating the target species, regeneration and recycling of the used solvent is an economically challenging process. Typically, energy-intensive processes, such as evaporation or distillation, are used to recover the extracted species and recycle the solvent, making it less attractive to industry. The requirement of a higher processing time due to a slow, equilibrium-governed operation is another major concern. Besides, higher chances of cross-contamination among the solvents and potential hazards from harmful solvents limit its use. Advanced extraction processes, such as membrane-assisted extraction or ionic liquid (IL) green-solvent based extraction, are emerging to overcome these challenges.

6.2.4 Membrane Separations Technologies

Membrane separation technologies are becoming increasingly popular due to several advantages including versatility (different choices of membrane materials and flow configurations), higher productivity (pressure-driven convective flow—lower mass transfer resistance compared to diffusion), ease of scaling up, lower chemical usage and lower energy consumption. In membrane separations technologies, solute molecules are either selectively transported through or selectively retained by a semi-permeable membrane based on size or solubility or electrical properties of the solute molecules. Most commonly used membrane separations technologies for bioproducts formation are microfiltration (MF), reverse osmosis (RO) and electrodialysis (ED).[12–14] Wide applications of membrane separations technologies are hindered by some major challenges notably flux decline due to fouling of the membrane and compatibility issues with the membrane materials. However, the advent of improved membrane materials and advancement in process engineering are able to solve these challenges to a large extent.

6.2.4.1 *Microfiltration (MF)*

Conventional membrane separations technologies are classified based on the pore size of the membranes. MF membranes have a pore diameter range of 0.05–2 μm and are extensively used for bacterial cell removal, enzyme recovery, clarification of beverages/juices and affinity purification of biomolecules. Typically, a *trans*-membrane pressure drop range of 1–10 psi is used for MF. The driving force for mass transfer is pressure gradient and the selective transport of molecules through MF membranes occurs based on a sieving mechanism. Cellulose acetate (CA), polyethersulfone (PES), polyvinylidenefluoride (PVDF) and nylon are a few of the common MF membrane materials.

For conventional membrane operations, such as MF, UF and RO, the following parameters are profoundly important and are used for performance evaluation:

$$\text{Solvent flux, } J_\text{w} = \frac{\text{solvent flow rate}}{\text{membrane external area}}$$

$$\text{Rejection, } R = \frac{\text{conc. of solute in feed} - \text{conc. of solute in permeate}}{\text{conc. of solute in feed}}$$

$$\text{Selectivity of solute A over B, } S_{\text{A/B}} = \frac{\text{conc. of solute A in permeate/conc. of solute B in permeate}}{\text{conc. of solute A in feed/conc. of solute B in feed}}$$

Transport through MF membranes is mathematically expressed as:

$$J_\text{w} = A \times \Delta P \tag{6.22}$$

J_w = solvent flux; ΔP = transmembrane pressure drop

Permeability, A, of a particular solute through a membrane is determined by equating the above flux equation with flux equation derived from Hagen–Poiseuille's equation. For this, membrane pores are assumed to be straight, cylindrical conduits through which laminar flow takes place during separation.

$$A = \frac{J_\text{w}}{\Delta P} = \frac{\varepsilon r^2}{8\mu L} \tag{6.23}$$

ε = membrane porosity = $\dfrac{\text{porous surface area}}{\text{total membrane area}} = \dfrac{\pi r^2 N_P}{A_m}$; r = pore radius; A_m = total membrane area; N_P = number of pores; L = length of the pores; μ = solution viscosity

Again, from the resistance model, solvent flux is expressed as follows:

$$J_w = \frac{\Delta P}{\mu R_m}; \quad R_m = \text{membrane resistance} \tag{6.24}$$

6.2.4.2 *Reverse Osmosis (RO)*

RO membranes have pore diameter range of 1–5 Å and *trans*-membrane pressure drop range of 400–1000 psi. RO is the other extreme of the membrane spectrum in comparison to MF. Cellulose acetate (CA) and polyamides (PA) are commonly used as RO membrane materials. Because of the virtually non-porous structure, RO membranes allow almost exclusive permeation of water molecules, and therefore, are extensively used for water purification and salt concentration. In biorefineries, RO is used for concentration of bioproducts, *i.e.* dewatering of processing solution, prior to the finishing step. The driving force for mass transfer through RO membranes is pressure gradient and the transport mechanism is governed by the solution-diffusion mechanism. In the solution-diffusion model, it is hypothesized that the overall transport is a combination of two steps—initial dissolution of the molecules (solute and solvent) onto the membrane surface followed by the diffusive flow of the molecules through the non-porous, permselective layer of the membrane. The solvent flux (J_w) and the solute flux (J_s) through the membrane are mathematically represented as follows:

$$J_w = \frac{D_w C_w}{RTl_m}\bar{V}_w(\Delta P - \Delta\pi) = A_w(\Delta P - \Delta\pi) \tag{6.25}$$

$$J_s = \frac{D_s K_s}{l_m}(C_f - C_p) = A_s(C_f - C_p) \tag{6.26}$$

A_w = permeability of solvent through the membrane = $\dfrac{D_w C_w \bar{V}_w}{RTl_m}$

D_w = diffusivity of solvent through membrane; C_w = concentration of solvent

$\bar{V}_w$ = partial molar volume of water in solution; l_m = thickness of the permselective barrier of the membrane

R = universal gas constant, T = absolute temperature

ΔP = transmembrane pressure drop; $\Delta\pi$ = osmotic pressure gradient

A_s = permeability of solute through the membrane = $\dfrac{D_s K_s}{l_m}$;

D_s = diffusivity of the solute in membrane

$$K_s = \text{partition coefficient} = \frac{C_m}{C_f} = \frac{\text{concentration of solute in membrane}}{\text{concentration of solute in feed}}$$

C_p = concentration of solute in permeate

6.2.4.3 *Electrodialysis (ED) and Electrodeionization (EDI)*

ED is a membrane separation technique based on electrical potential such that the cations and anions are transported from one stream to another through charged membranes. In ED, cation exchange (CEX) and anion exchange (AEX) membranes are alternatively arranged between two electrodes to form diluate and concentrate compartments parallel to each other as demonstrated in Figure 6.3.

Target ionic species, present in the process solution along with the non-ionic components, is fed as the diluate stream. Under an applied electric field, ions move out (removed) from the diluate compartment and move (accumulate) into the concentrate compartment. This leaves the non-ionic components of the process solution in the diluate stream. Overall, ionic species are recovered from process solution into a purified solution, *e.g.* organic acids are recovered from fermentation broth. Efficacy of ion

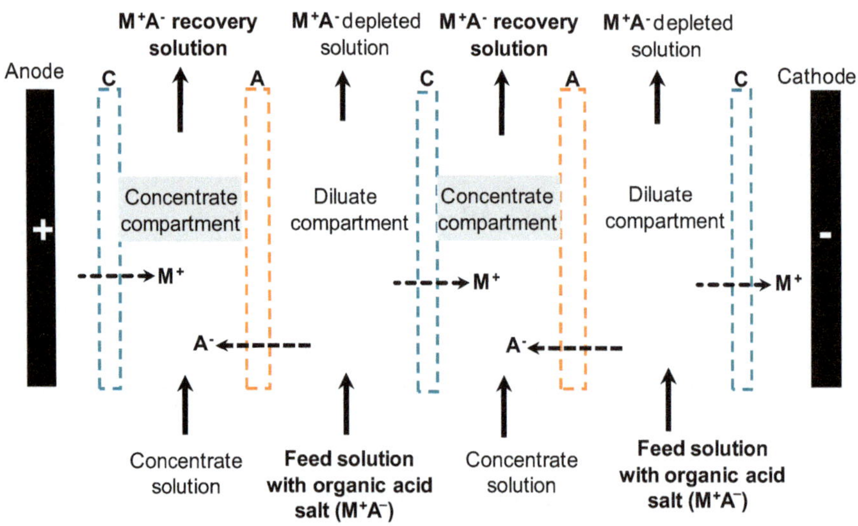

Figure 6.3 Schematic of electrodialysis (ED) for separation of a cationic salt of organic acid ($M + A^-$) from fermentation broth. Both cations and anions transfer from the diluate compartments into the concentrate compartments. The end result is recovery of the organic acid salt from fermentation broth. "C" and "A" denote cation exchange and anion exchange membranes, respectively.

transport across the IEX membranes in ED is quantified by calculating current efficiency (η) as follows:

$$\eta = \frac{zQF\left(C_{\mathrm{D_{in}}} - C_{\mathrm{D_{out}}}\right)}{nI} \tag{6.27}$$

z = charge of the transported ion; Q = flow rate; F = Faraday constant $= 96485 \frac{\mathrm{Amp} - s}{\mathrm{mole}}$;

n = number of cell pairs; I = current; $C_{\mathrm{D_{in}}}$ = concentration of the ions at the diluate inlet; $C_{\mathrm{D_{out}}}$ = concentration of the ions at the diluate outlet

Higher current efficiency, typically >70%, indicates a favorable process. Lower current efficiency indicates the loss of electrical energy in activities other than the ion transport. Water splitting, a phenomenon where water is dissociated into protons and hydroxyl ions, is a major cause of lower current efficiency in ED. Particularly for dilute solutions, when electrical conductivity within the ED cells remains low due to the ion-deprived condition, water splitting occurs to maintain sufficient electrical conductivity. Uncontrolled water splitting from the entire diluate compartment of ED incurs substantial electrical energy loss. Electrodeionization (EDI), a modified form of ED, is designed to facilitate ion transport for dilute solutions. In EDI, conductive IEX resins are inserted into the diluate compartment as demonstrated in Figure 6.4.

Therefore, ion transport occurs through a highly conductive bulk solution–resin–membrane pathway for EDI rather than the less conductive bulk solution–membrane pathway for ED. As a result, electrical conductivity and current efficiency increase significantly. Besides, water splitting on the resin beads' surfaces facilitates continuous, *in situ* regeneration of the IEX properties of the beads. ED and EDI are gaining significant importance in bioproducts industries, particularly for organic acid purification, because of their ability to reduce the number of unit operations and ability to provide economically favorable pathways. They also provide an opportunity for the *in situ* recovery of organic acids during fermentation using bipolar membrane electrodialysis (BPED). A detailed example of this process is presented in Chapter 7.

6.2.4.4 Bipolar Membrane Electrodialysis (BPED)

The bipolar membrane (BP) is a composite material consisting of an AEX membrane and a CEX membrane bound together. Therefore, it does not allow permeation of any ionic species through it. However, it allows water splitting on its structure and liberates hydroxyl ions from the anionic face and protons from the cationic face. In BPED, either CEX membranes or AEX membranes from a regular configuration of ED are substituted with BP

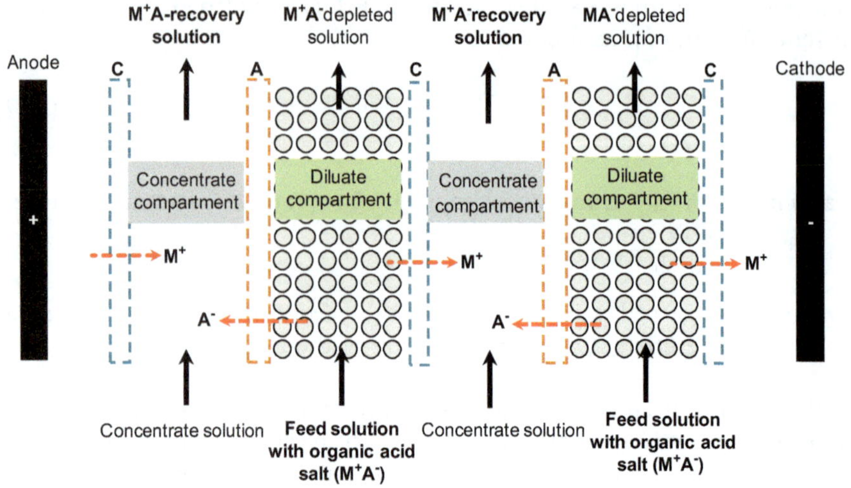

Figure 6.4 Schematic of electrodeionization (EDI), a modified form of ED, for separation of a cationic salt of organic acid ($M+A^-$) from fermentation broth. Ion exchange resin beads are inserted into the diluate compartments for EDI. Both cations and anions transfer from the diluate compartments into the concentrate compartments. The end result is recovery of the organic acid salt from fermentation broth. "C" and "A" denote cation exchange and anion exchange membranes, respectively.

membranes (with proper orientation). For example, in order to recover protonated organic acid from the cation salt of an organic acid formed during fermentation, the CEX membranes are replaced with BP membranes as illustrated in Figure 6.5. When the fermentation broth is fed into the diluate compartments, carboxylate anions transport across AEX membranes, reach the concentrate compartment, and then combine with the protons generated from water splitting to form the organic acid. The cations cannot transport due to the protective barrier of the BP membranes and stay in the diluate compartments. There, they form the corresponding inorganic base by combining with the hydroxyl ions generated from water splitting. The net result is purification and conversion of the cationic salt of the organic acid into protonated organic acid (and formation of the corresponding inorganic base in the diluate compartment).

6.3 Application of Separations Technologies for Biobased Product Recovery

Biobased product discovery and development is in a nascent state of research. As discussed elsewhere in this book, the list of value-added biobased

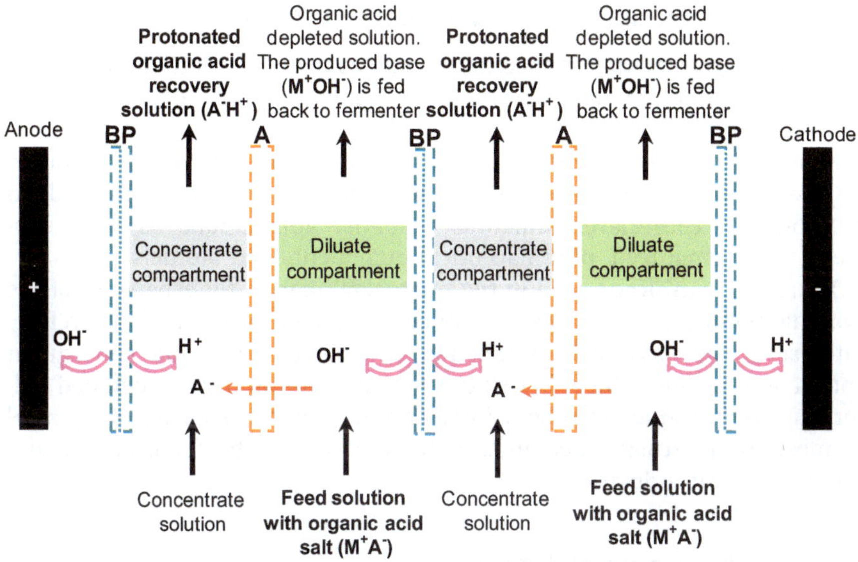

Figure 6.5 Schematic of bipolar membrane electrodialysis (BPED) for separation of a cationic salt of organic acid ($M+A^-$) from fermentation broth and recovery as a protonated organic acid. In BPED, cation exchange membranes from an ED configuration are replaced with bipolar membranes (BP). Anions transfer from the diluate compartments into the concentrate compartments, where they combine with protons generated due to water splitting by the BP membranes. In the diluate compartments, cations combine with the hydroxyl ions formed due to water splitting. The end result is the recovery of protonated organic acid from the organic acid salt present in the fermentation broth and formation of the corresponding inorganic base. "A" denotes the anion exchange membrane.

products is quite large and versatile, and there is frequent addition of new candidates to the list. We highlight the biobased products that are already recognized as valuable building blocks and are already commercialized or close to commercialization. We specifically give importance to the value-added oxygenated biobased products enlisted by the US Department of Energy[4] and another prominent source.[15] The biobased products are classified into four broader categories—algal biomass-derived products, organic acids, furan derivatives and sugar alcohols. Ethanol, although a valuable oxygenated species, is more important as a biofuel rather than biobased product, and hence is not included here. Algal biomass has attracted tremendous attention as an advanced 3rd generation biofuel feedstock. Keeping that in mind, algal biomass-derived products are discussed in detail including types of products, potential applications and separation technologies involved. The focus for the other three categories of bioproducts is predominantly separations technologies.

6.3.1 Algal Biomass-derived Valuable Products

Algae are highly diversified species ranging from single cell to multi-cellular organisms and are found in a wide range of environmental conditions. Due to the enormous biodiversity of microalgae and recent developments in genetic and metabolic engineering, this group of organisms has garnered worldwide attention as one of the most promising sources for value-added bioproducts. They have numerous potential applications as food supplements, animal feed, pharmaceutical ingredients and biofuels.

There is a significant need in the microalgal market for expansion of the existing products as well as introduction of new products.[16] Therefore, interest in the large-scale production of algal biomass and subsequent conversion to bioproducts is rising. With the development of detailed culture and postproduction processing techniques, microalgae are well poised to meet the increasing demands of bioproducts in the food, energy and pharmaceutical industries.[17]

6.3.1.1 *Production of Algal Biomass*

Production of microalgal biomass can be carried out in an open raceway pond (open system) or closed photo bioreactors (PBRs). Open-culture systems are almost always located outdoors and rely on natural light for illumination, whereas PBRs are dependent on supplied CO_2 and light. Each of the two systems has pros and cons.

Open pond system: due to the lack of control involved with open systems, the pond becomes a function of the local climate, thus the location significantly contributes to the success of the cultivation.[18] Open ponds are limited by key growth parameters including light intensity, temperature, pH and dissolved oxygen concentration. Contamination by other organisms is another issue involved with open ponds.

Photobioreactor system: nowadays, designing photobioreactors to cultivate photosynthetic microalgae is extensively investigated. Photobioreactors are associated with improved productivity due to the better control of various parameters. Tubular reactors are considered to be more appropriate for outdoor cultivation.[19] The large illumination surface of the reactor, created by translucent tubing, along with an optimum supply of CO_2 and continuous sparging or shaking of the culture are the main factors behind its outdoor suitability.

6.3.1.2 *Recovery of Algal Biomass*

Recovery of algal biomass (harvesting) from the broth contributes 20–30% to the total production cost of biomass.[20] Harvesting of biomass involves various solid–liquid separation steps, such as flocculation, centrifugation, filtration or gravity sedimentation. No single harvesting method may be suitable to every case. Recovery of biomass is a major problem because of the small size (3–30 μm diameter) of the algal cells. Culture broths are generally relatively dilute (<0.5 kg dry biomass per m^3 broth in some commercial production systems), and hence, large volumes need to be handled to recover the biomass. Gravity sedimented sludge is generally more dilute than centrifugally recovered biomass. A further consideration in selecting a suitable harvesting method is the acceptable level of moisture in the product.[21] After harvesting, dehydration or drying of the biomass is done to extend the shelf-life of the biomass, especially if the biomass is the final product. Drying methods that have been used for microalgae include spray drying, drum drying, freeze-drying and solar drying. Here, we are primarily focusing on separation techniques implemented to recover valuable products from microalgae.

Recovery of valuable products from algal biomass consists of five major steps starting from algae cultivation to utilization of residual biomass as depicted in Figure 6.6. At each and every step, certain decisions need to be taken to improve the overall process economics. Figure 6.6 exhibits some of the possible decision making scenarios that depend on the products of

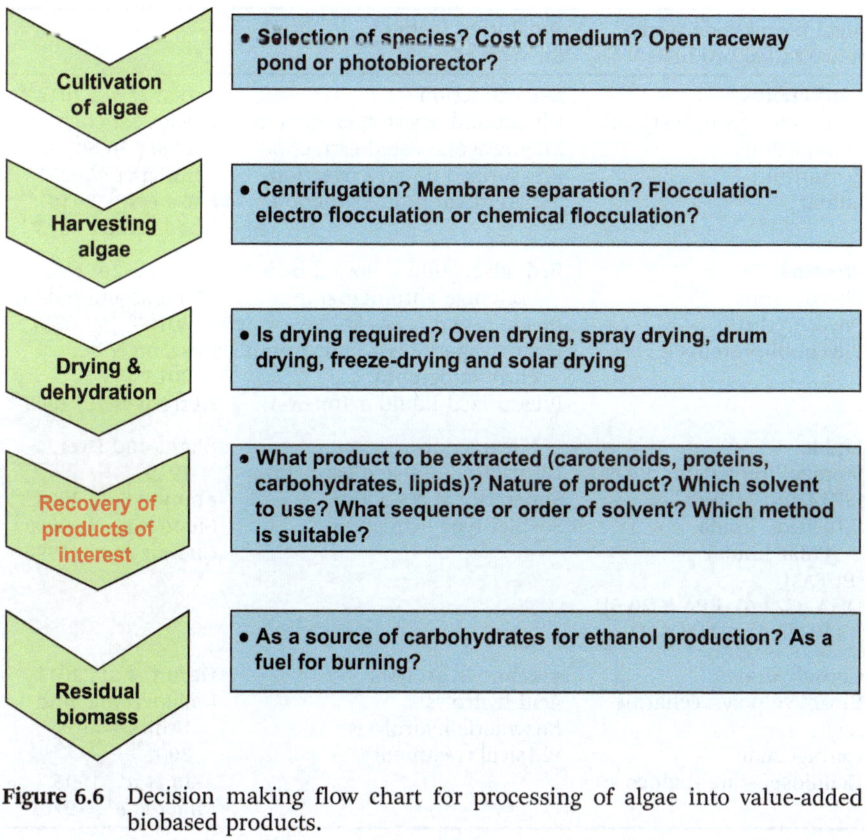

Figure 6.6 Decision making flow chart for processing of algae into value-added biobased products.

interest, economic feasibility, environmental conditions and availability of resources.

6.3.1.3 Products from Microalgae

Various value-added products can be obtained from different species of microalgae. To extract these value-added products, several separations techniques are widely used. Table 6.1 summarizes the algal biomass-derived product portfolio along with the separation techniques used for the recovery of the products. Detailed descriptions of the products are given next.

6.3.1.4 Carotenoids

Carotenoids are strong anti-oxidants and are useful for pharmaceutical, nutraceutical and pigmentation applications. There are several carotenoids, such as β-carotene (hydrocarbon), α-carotene, astaxanthin, lutein,

Table 6.1 Algal biomass-derived value-added products portfolio along with the separations technologies used for recovery of the products from processing solution.

Algal biomass-derived value-added products	Separation technologies used for recovery	Ref.
Carotenoids β-Carotene (hydrocarbon) Astaxanthin Zeaxanthin Lutein	L–L extraction Ultrasound assisted extraction Microwave assisted extraction Pressurized liquid extraction Supercritical fluid extraction	Sarkar *et al.*, 2012[22] Macias-Sanchez *et al.*, 2009[23] Splinter *et al.*, 2013[24] Plaza *et al.*, 2010[25] Kitada *et al.*, 2013[26]
Proteins Phycocyanin Phycoerythrin Phycobilliproteins	Bed adsorption followed by ion exchange chromatography Precipitation Hydroxyapatite (HA) adsorption chromatography Pressurized liquid extraction	Ce *et al.*, 2013[27] Chakdar and Pabbi, 2012[28] Rossano *et al.*, 2013[29] Herrero *et al.*, 2005[30]
Lipids Storage lipids (Non polar) (SFAs and MUFAs) Structural lipids (Polar lipids) (PUFAs) DHA (C22:6), EPA (C20:5), ARA(C20:4), GAL(C18:3)	L–L extraction Ultrasound assisted extraction Supercritical extraction Soxhlet type extraction	Blight and Dyer, 1959[31] Araujo *et al.*, 2011[32] Mendes *et al.*, 2003[33] Cheung *et al.*, 1998[34]
Carbohydrates Bioactive polysaccharide Agar Carrageenan Cellulose, starch, glucose	Alkaline hydrolysis Acid hydrolysis Enzymatic hydrolysis Physical treatment	Harun *et al.*, 2011[35] Kaliaperumal and Uthirasivan, 2001[36] Akin *et al.*, 1995[37] Millet *et al.*, 1976[38]

zeaxanthin and fucoxanthin, which play a major role in the human diet. The common feature of carotenoids is a linear polyisoprenoid structure; a long conjugated chain of double bonds and a near bilateral symmetry around the central double bond.[39]

6.3.1.4.1 Recovery of Carotenoids from Algae. L–L extraction: based on the equilibrium extraction capacity and economic viability of the L–L extraction process, acetone, petroleum ether, benzene, hexane, diethyl ether, chloroform, dichloromethane, ethanol and methanol are identified as suitable solvents for carotenoid recovery.[40] Ethanol and hexane, having GRAS (generally recognized as safe) status, are widely used by food industries. For selective extraction of free astaxanthin from red encysted *Haematococcus* sp., an alternative procedure was designed using dodecane and methanol.[41] It consisted of dodecane-mediated extraction of the crude mixture followed by secondary extraction with methanol.

Conventional L–L extraction of carotenoids from algae requires large amounts of solvents, several extraction steps and long extraction times. Thus, several advanced extraction methods, such as ultrasound and microwave-assisted extraction, extraction under pressurized conditions and supercritical fluid extraction are proposed as alternatives.

Ultrasound assisted extraction (UAE): UAE is a promising alternative to the conventional L–L extraction due to its high efficiency, low energy requirements and low solvent consumption. This technique has been used systematically in the extraction of substances with low molecular weights.[42] UAE is divided into two categories—(i) low intensity high frequency, and (ii) high intensity low frequency. High frequency ultrasonic waves may alter the physical or chemical structure of a material. Ethanol, hexane or acetone can be used as solvents during UAE.[43] A high yield of lutein was achieved in the UAE of *Chlorella vulgaris* compared with pressurized liquid extraction. The UAE yield was 3.93 mg g^{-1} from 2 h extraction time at room temperature using 50 mL 90 wt% ethanol with 0.5 g Chlorella powder, whereas for pressurized liquid extraction, the yield was 3.78 mg g^{-1} from 30 min extraction time at 160 °C and 10.3 MPa pressure using 33 mL ethanol. The extraction time in the earlier case of UAE, however, was four-fold higher compared to pressurized liquid extraction. The carotenoid yield from UAE with methanol was 14 mg g^{-1} from *Dunaliella salina* and it was slightly lower than supercritical CO_2 extraction (yield 15 mg g^{-1}).[23] The carotenoid selectivity in UAE was lower than that with pressurized liquid extraction and supercritical CO_2 extraction.

High pressure liquid extraction (HPLE): HPLE is conducted at a temperature and pressure high enough to maintain the solvent in the liquid state during the whole extraction procedure. The combined use of high pressure and high temperature enable faster extraction processes that require smaller volumes of solvents (*e.g.*, 20 min extraction time using 10–50 mL of solvent in HPLE *vs.* 10–48 h extraction time using up to 300 mL solvent in a traditional extraction).[44] The increased temperature enables higher analyte recovery by increasing both solubility and mass transfer rate. In addition, a high temperature decreases the viscosity and the surface tension of the solvents, resulting in increased accessibility, thus improving the extraction efficiency. A drawback of HPLE is its requirement of special instrumentation to maintain relatively high pressure and temperature. HPLE has been used to extract carotenoids from brown macroalgae, such as *Eisenia bicyclis*,[45] *Cystoseira abies-marina*, and *Himanthalia* elongate.[25] It is reported that ethanol at high temperature provides a high recovery of fucoxanthin and other oxygenated carotenoids. Depending on the selected solvent, HPLE could be an ideal technique for the separation of bioactive phenolic compounds as well.

Microwave-assisted extraction: microwave-assisted extraction may be microwave-assisted solvent extraction (MASE) or microwave solvent-free extraction (MSFE). In MASE, there are certain benefits over conventional extraction—reduced extraction time, reduced solvent use and improved extraction yield. However, recent concerns about the environmental hazards of organic solvents provided the impetus to develop solvent-free extraction, MSFE, which is based on the hydro-diffusion phenomenon. For example, astaxanthin was extracted (yield 2.1 wt%) from *Haematococcus pluvialis* using safflower oil as a solvent in the presence of microwaves (300 W, 5 min, 100 g biomass and 500 ml oil).[24]

Supercritical fluid extraction (SFE): SFE using supercritical CO_2 is gaining significant attention for carotenoids recovery from algae. A disadvantage of using supercritical CO_2 is its low polarity. However, the lower polarity of CO_2 can be overcome by using polar modifiers or co-solvents, which alter the polarity of supercritical fluid and enhance its extraction capacity. SFE of astaxanthin from *Haematococcus* was done efficiently using CO_2 flow rates of 2–4 mL min^{-1} at 20–55 MPa and 313–353 K, and with ethanol as an entrainer.[46] SFE of β-carotene with ethane or ethylene was successfully demonstrated, since β-carotene has higher solubility in these hydrocarbons than CO_2. In *Chlorella vulgaris*, the extracts have pigment components, such as lutein, β-carotene, chlorophyll a and b. It was observed that high pressure and high temperature promote the supercritical CO_2 extraction of lutein from *Chlorella vulgaris* at 50 MPa and 80 °C.[26] This demonstrates that yield increases with increasing pressure and temperature, if the carotenoid is thermolabile enough. The optimal conditions for astaxanthin extraction from *Haematococcus pluvialis* were 70 °C, 50 MPa, and 4 h of extraction time.[47] Bioactive compounds, such as diolefins, have been extracted from *Botrycoccus braunii* cells by SFE.[33] This organism can store high amounts of long-chain hydrocarbons (*i.e.*, 25–31 carbon atoms) that can be used as substitutes of paraffinic and natural waxes.

6.3.1.5 Proteins

Recently, algal proteins have received much attention because of their potential bioactive and functional properties. Phycobiliproteins are one of the most important groups of proteins from seaweed. These water-soluble proteins possess a tetrapyrrolic ring covalently attached to their structure. This pigment can be either phycocyanobilin (in blue-green algae) or phycoerythrobilin (in red algae), and it is partially responsible for the functional properties associated with these proteins, mainly hepatoprotective, anti-inflammatory, and antioxidant activities.[48] Some genus of algae, like *Spirulina*, *Scenedesmus*, *Anabaena*, have a protein content of up to 50% of dry biomass. Non-pigment proteins from microalgae are actually not studied or are poorly investigated, although some of them are used in animal feed and food. Guil-Guerrero *et al.*[49] reported on the functional properties (water and fat adsorption capacity, and emulsification capacity) of the biomass of some microalgae including *Porphyridium cruentum*, *Nannochloropsis* spp. and *Phaeodactylum tricornutum*. As industrial demands for new emulsifying agents increase, the price of commercial ingredients, such as caseinate or soybean proteins also increase. Microalgae, as a rich source of proteins, can help stabilize the market.

6.3.1.5.1 Recovery of Proteins. Recovery of proteins requires a pretreatment step to disrupt the microalgal cells prior to extraction. Cell disruption enables easy access to the internal components and facilitates the recovery process. There are several disruption techniques to break the cell wall of microalgae to recover the intracellular proteins of the cell. These include (i) bead milling, (ii) ultrasonication,[50] (iii) microwave radiation,[51] (iv) enzymatic treatment,[52] (v) cell homogenization,[53] and (vi) high pressure cell disruption.[54] After cell lysis, the microalgae suspension is centrifuged followed by separation of protein from supernatant by either of these two methods—precipitation and ultrafiltration.

Precipitation: precipitation of protein is conducted at its isoelectric point, where solubility of the protein is minimal. After precipitation, the protein is separated from solution using centrifugation. Solvents, such as TCA (trichloroacetic acid), are sometimes used to facilitate precipitation.

Ultrafiltration (UF): UF is used to concentrate the supernatant of protein-rich algal process solutions. The operation can be performed using a tangential-flow filtration unit equipped with an UF membrane. Typically, 300 KDa molecular weight cut-off (MWCO) polyethersulfone (PES) UF membrane is used to recover algal proteins.

6.3.1.6 Carbohydrates

Macroalgae are regarded as rich sources of polysaccharides. Types of polysaccharides vary depending on the group of macroalgae. Different carbohydrates including agar, carrageenan, or alginates are extracted from macroalgae, and these carbohydrates are widely used in food and pharmaceutical industries as functional ingredients, such as stabilizers. For example, the alga *Chondrus crispus* is traditionally employed for the extraction of carrageenan (also known as Irish moss), a highly sulfated polysaccharide.

Polysaccharides extracted from *Ulva pertusa* are a group of hetero-polysaccharides, mainly composed of rhamnose, xylose, glucose, glucuronic acid, iduronic acid, and sulfate, with smaller amounts of mannoses, arabinose, and galactose. Sulfated polysaccharides are antioxidants, which can delay or prevent oxidative cellular oxidizable substrates.[55] Besides these, algal carbohydrates can be used for biofuel production. Production of biofuels from algal polysaccharides involves extensive processing that varies depending on the type of biomass. In general, the process involves several upstream and downstream operations including pre-treatment of the biomass, saccharification, fermentation and product recovery. Biomass pre-treatment is essential for the extraction of algal carbohydrates, which remain intact in the cell wall and are released by hydrolysis methods, such as physical pretreatment, chemical pretreatment and biological pretreatment.

6.3.1.7 Lipids

Lipid-accumulating microalgae have the potential to contribute extensively to large-scale biodiesel production without competing for arable land and food. High lipid content and high biomass yield are primary requisites for the commercial production of biodisel. The lipid content of algae can be manipuleted or enhanced under stress conditions, albeit at the expense of a drastic loss of biomass yield. This problem can be solved to some extent by using a biphasic culture strategy.

Lipids produced by microalgae can be grouped into two categories—storage lipids (non-polar lipids) and structural lipids (polar lipids). Storage lipids are mainly in the form of Triacylglyceride (TAG), made of predominately saturated fatty acids and some unsaturated fatty acids, which can be *trans*-esterified to produce biodiesel. Structural lipids typically have a high content of polyunsaturated fatty acids (PUFAs), which are essential nutrients for aquatic animals and humans. The main polar lipids found in these substrates include monogalactosyl diacylglycerols (MGDG), digalactosyl diacylglycerols (DGDG), and phosphatidylglycerols (PG). These polar lipids possess several functional activities, such as anti-inflammatory activities. Lipids from microalgae can be extracted by several methods such as expeller/press, L–L extraction,[31] high pressure liquid extraction, supercritical extraction,[32] and ultrasound-assisted extraction.[33]

6.3.2 Organic Acids

Organic acids, an influential group of biobased products, are widely used in the chemical, pharmaceutical, food, beverage, and cosmetics industries due to their ability to replace several petroleum-derived chemicals. The fermentation broth of organic acids consists of a complex mixture of microbes, fermentation medium, acids and other byproducts. Therefore, downstream processing (DSP) involving separation, purification and concentration of organic acids from the broth are required. Typically, DSP of organic acid contributes around 30–40% of the total production cost.[56] In this section, different techniques for the recovery of organic acids from fermentation broth are discussed focusing on the challenges and opportunities.

DSP for the recovery of organic acids from fermentation broth could be represented by a typical flow chart as demonstrated in Figure 6.7. Cell removal from the fermentation broth by microfiltration/centrifugation is usually the first step (pretreatment) towards separation. Cell removal is followed by primary recovery or purification, which involves separation of the organic acid from bulk aqueous solution containing major impurities. Thereafter, removal of the bulk solvent and the remaining impurities is achieved to get the desired concentration of organic acid in the concentration step. Finally, if needed, upgrading and formulation of the purified organic acid is conducted to further improve the quality.[57]

Due to the variation in physicochemical properties of different organic acids, a single recovery process cannot be implemented for all cases. Therefore, knowledge of different separations techniques along with the knowledge of their respective benefits and challenges are needed before its implementation at industrial scale. Separations techniques for the energy efficient recovery of organic acids during fermentation can be divided into the following categories—conventional separation processes and membrane separation processes. Conventional processes include precipitation and acidification, L–L extraction, adsorption and ion-exchange, and

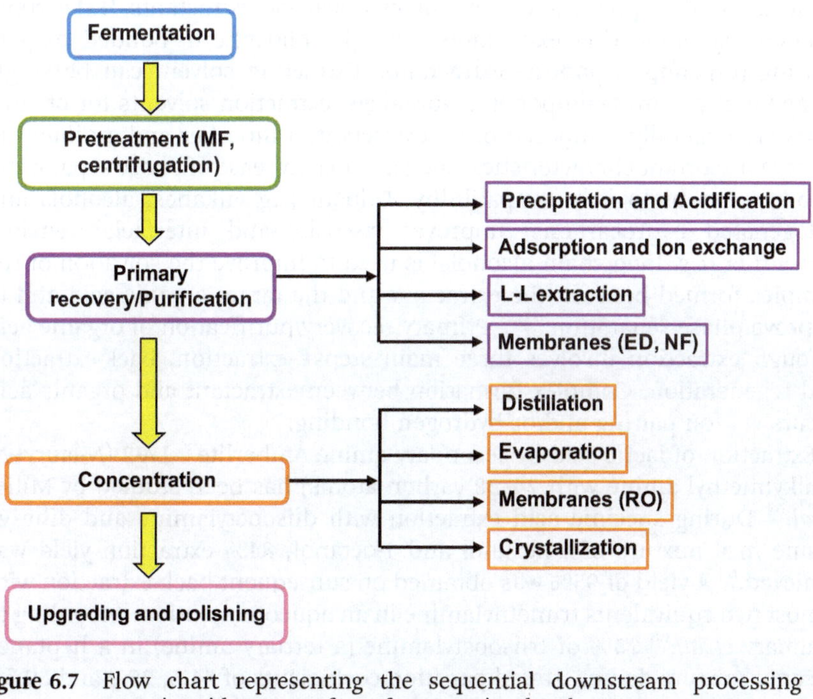

Figure 6.7 Flow chart representing the sequential downstream processing of organic acid recovery from fermentation broth.

crystallization. Membrane processes include reverse osmosis (RO), nanofiltration (NF) and electrodialysis (ED).

Precipitation and acidification: in this technique, fermentation broth is treated with calcium hydroxide or calcium oxide to precipitate the cationic salt of the organic acid. Then, the precipitate is acidified with sulfuric acid, which leads to the separation of the organic acid in the filtrate along with the production of $CaSO_4$ sludge. Precipitation is usually followed by concentration steps, such as adsorption and/or crystallization. Calcium salts of fumaric, citric and lactic acids have low solubility in water, and hence, they were successfully separated by precipitation and acidification.[58] Succinic acid was recovered with 92% purity and 52% yield with a calcium precipitation method.[59]

However, this approach of separating organic acid is often economically challenging, because of its low yield and $CaSO_4$ sludge generation (solid handling and disposal). Also, since short chain organic acids have similar characteristics in solution at room temperature, it is very difficult to remove low molecular weight organic acids through precipitation.[60]

L–L extraction: for the primary recovery of organic acids, extraction is probably the most studied separation technology. The extracting solvent (extractant) is involved in transfer of the carboxylic acid or carboxylate to the solvent phase. Based upon the chemical properties of the target organic acid and the mechanism involved, separation of organic acid by extraction can be achieved by any of the following four extractants:[57,61] (i) amine-based extractants, (ii) aliphatic and aromatic hydrocarbon extractants, (iii) carbon-bonded oxygen-bearing extractants, and (iv) phosphorus-bonded oxygen-bearing (organophosphorus) extractants. Extraction solvent can be single component or multi-component. Nowadays, extraction solvents for organic acids are generally composed of an extractant, a modifier and a diluent to achieve important characteristics, such as capacity, ease of phase separation, stability and biological compatibility. Diluent (*e.g.* alkanes, alcohols and halogenated hydrocarbons) improves viscosity and interfacial tension. A modifier (*e.g.* long chain alcohols) is used to improve the solvation of the complex formed between the extractant and the target organic acid and to improve phase separation.[62,63] Primary recovery/purification of organic acid through extraction involves three main steps—extraction, back-extraction and regeneration. Complex formation between extractant and organic acid occurs *via* ion pairing and/or hydrogen bonding.

Extraction of lactic acid by secondary amine Amberlite® LA-2 (*N*-lauryl-*N*-trialkylmethyl amine with 24–28 carbon atoms) has been studied by Miller *et al.*[64] During succinic acid extraction with diisooctylamine and dihexylamine in a mixture of 1-hexanol and 1-octanol, 84% extraction yield was achieved.[65] A yield of 95% was obtained on subsequent back-extraction with almost two equivalents trimethylamine in an aqueous solution. According to Malmary *et al.*,[66] 25% of triisooctylamine (a tertiary amine) in a heptane–hexanol mixture demonstrated partition coefficients of 41.5, 25.9 and 59 for citric, lactic and malic acids respectively. A pilot scale extraction of citric acid was performed using Alamine® 336 (a tertiary amine) in kerosene as an extractant with a yield of 97%.[67] Large-scale citric acid extraction by tertiary amine extractant was also reported.[61,68] 90% recovery of citric acid was achieved by a modified extraction procedure.[69] Lactic acid was also purified commercially by using countercurrent extraction with diisopropyl ether.[70]

Adsorption and ion-exchange: adsorbent selectively adsorbs organic acid leaving behind the unwanted impurities. The adsorbed organic acid is recovered by elution with a bulk solvent (extraction, changing pH, *etc.*). Adsorption of organic acids using a packed-bed column is an important technique for industrial separations. For industrial applications, a minimum capacity of 0.05 g g^{-1} resin has been suggested by Davison *et al.*[71] Adsorption not only provides selective recovery of the target molecules,[72] but simplifies the regeneration step as well. The energy demand in the regeneration step of adsorption is significantly lower than that of L–L extraction, because of the confinement of the solid adsorbents in the column. However, adsorption is associated with a higher pressure drop and involvement of expensive resin beads.

Based on the electronic properties and interaction with organic acids, adsorbents can be classified in the following two broad categories—ionic materials (also known as ion-exchangers) and non-ionic materials.

Adsorption with ionic materials: ion-exchangers are used for charged organic acids present in aqueous solution, where electrostatic interactions between adsorbent and organic acid prevail. Materials functionalized with weak base groups such as pyridine, imidazole and primary, secondary and tertiary amines, constitute weak anion exchangers, while materials functionalized with quaternary ammonium compounds come under strong anion exchangers.

Commercial anion exchange resin adsorbents such as Cellufine™ A-200, DIAION™ WA21J, DOWEX™ SBR have satisfactory adsorption capacities for low molecular weight organic acids such as acetic and formic acid.[59] Kawabata *et al.* studied the recovery of lactic acid, malic acid and citric acid with a capacity of 0.20, 0.54 and 0.83 g g^{-1} resin, respectively, by using commercially available adsorbents.[73] Similarly, Davison *et al.*[71] has suggested a weak base pyridine polymer for the uptake of succinic acid from aqueous solution, which was found to provide adequate capacity and stability. Lactic acid adsorption by a quaternary anion exchange resin was studied by Cao *et al.*[74] and it was suggested that 70% and 80% recovery can be made after desorption, *e.g.* with sulfuric acid, ammonia, methanol and pure water. Separation of acetic, lactic, and formic acid using anion exchange resins as adsorbents has also been studied by Yang *et al.*[58] Semi-continuous simulated moving beds were also applied for the purification of lactic[75] and citric acids.[76]

Adsorption with non-ionic materials: non-ionic adsorbents are used to recover various organic acids based on hydrophobic interactions between the adsorbent and the organic acid.[57] Succinic acid was recovered using polybutylene terephthalate (PBT)–polyether copolymer as an adsorbent,[77] however, adsorption capacity was not satisfactory. Major limiting factors in using non-ionic materials for the recovery of undissociated acids are low capacities and lesser stability of the materials.

Basic challenges in using adsorption and ion exchange as primary recovery techniques in industries are the shorter lifetime of adsorbents due to fouling and low capacity.

Membrane separations: membrane technologies have attracted significant attention from biochemical industries, particularly for the downstream processing of organic acids, due to their high efficiency of separation and low energy consumption.

Electrodialysis (ED) is specifically considered to be one of the most attractive tools for organic acid separation. During fermentation, due to a slightly basic pH of the medium, organic acids are often formed as cationic salts. ED is used for the recovery and concentration of these cationic salts of organic acids. Recovered organic acids by ED are more concentrated and purified than NF or RO.[78–80] ED was utilized to concentrate succinate up to 55 g L^{-1}.[78] After clarifying the fermentation medium and a 50% increase in electrical current density, 80% total succinate removal from the feed stream was achieved. Meynial-Salles *et al.* implemented ED to recover succinate from continuous fermentation on a pilot scale.[79] Concentration of recovered succinate was 80 g L^{-1}. Citrate was concentrated by a factor of two using ED.[80] Bipolar membrane electrodialysis (BPED) was able to recover organic acid salt as a protonated organic acid.[81,82] A fermentation solution containing sodium acetate was recovered in acid form using BPED.[83] BPED was examined for lactic acid[82,84,85] and gluconic acid[86] recovery from fermentation broth and concluded to be economically favorable. Since the costs of the bipolar membranes are higher than IEX membranes, prior concentration using ED was proposed.[81] Recovery of sodium gluconate and sodium citrate was demonstrated using a coupled ED-BPED approach.[87]

Basic challenges towards industrial implications of ED are improving antifouling characteristics and increasing selectivity for co-ions.[82] Also, the cost of the membrane is another major limiting factor for industrial applications. Besides, monovalent organic acid salt separation using ED was found to be economically more favorable than multivalent organic acid salt separation.[68] This can be attributed to the higher electrical energy demand and lower membrane flux for multivalent organic acid salts.

RO and NF membranes are employed to concentrate organic acids from aqueous solutions using their ultrathin permselective skin layer on a microporous polymeric support. NF membranes can be used for separation of organic acids in continuous fermentation processes, while selectively rejecting other large macromolecular components.[88] RO is primarily used for dewatering, *i.e.* concentrating the organic acid.

Another emerging membrane process for organic acid separation from fermentation broth is a hybrid technique of membrane-assisted extraction, where an organic solvent is used on the permeation side of the membrane.

A membrane matrix provides a higher interfacial surface area, while the solvent provides an additional driving force for mass transfer. Some organic extractants may affect microbial activity. The membrane matrix acts as a barrier to eliminate direct contact between the fermentation broth and the extractant. The disadvantage of this technique is the requirement for energy intensive distillation for further processing as previously mentioned for L–L extraction. Ionic liquids are an emerging group of solvents that have potential applications in many industrial extractions, including membrane extraction. Ionic liquids are versatile, non-toxic, can be recycled, and are non-volatile and non-flammable. Therefore, they are termed as "designer solvents" and "green solvents".[89] Though the present production costs of ionic liquids is high, researchers are working towards synthesizing low-cost ionic liquids.

Variations in the physical properties and type of fermentations for different organic acids do not allow a single method to be applicable for recovery. Therefore, a combination of several methods should be applied for the removal of major impurities in a primary recovery/purification step followed by removal of water and minor impurities in a concentration step.

6.3.3 Furan Derivatives

Furan is a five-membered, oxygen-containing heterocyclic ring compound. Furan derivatives can be obtained from lignocellulosic biomass. They have high thermal stability and the capability to improve the mechanical properties of various materials. Major commercially important furan derivatives are discussed further.

6.3.3.1 Furfural

Furfural or furanaldehyde is an aldehyde derivative of furan. It has numerous applications as a starting material for resins used in metal casting or as a solvent precursor. Dehydration of xylose using a modified acidic zirconia catalyst provided a 95% conversion to furfural with 45% selectivity.[90] However, 91% conversion with 82% selectivity was obtained when dehydration was done with mesoporous silica functionalized with sulfonic acid groups.[91] During dilute acid pretreatment of lignocellulosic biomass, xylose is dehydrated under thermally elevated conditions to form the by-product furfural.

Separation of furfural from lignocellulosic biomass hydrolysate liquor by conventional methods, such as distillation, is energy-intensive. Pervaporation with PDMS (polydimethylsiloxane) membrane is an energy-efficient, "clean technology" alternative to distillation. Permeate furfural concentration and furfural flux of 62.4 wt% and 3222 g m^{-2} h^{-1}, respectively, were obtained with pervaporation.[92] This approach provides higher selectivity, while consuming 70% less energy compared to distillation.

6.3.3.2 *Hydroxymethylfurfural (HMF)*

HMF can be produced by acid-catalyzed dehydration of six-carbon sugars, such as fructose. Lignocellulosic biomass can also be used as a raw material. Being highly reactive, HMF functions as an intermediate to other useful chemicals, such as levulinic acid, formic acid and furan-2,5-dicarboxylic acid (FDCA).

Different strategies have been suggested for effective recovery of HMF from aqueous solution, such as use of anhydrous organic solvents, sub- or supercritical media, biphasic media in a reactive aqueous phase and a 5-HMF extractive solvent immiscible in water and ionic liquids.[93] 95% yield of HMF was obtained using ionic liquid media after dehydration of glucose in methyl imidazolium chloride.[94] However, separation of HMF from the solvent may require continuous extraction for several hours.[95] HMF yields of 90% using choline chloride/citric acid solution as the ionic liquid medium and HMF recovery of 70% using a biphasic ionic liquid/ethyl acetate system, in 20 minutes have been reported.[96] Dimethylsulfoxide (DMSO) was observed to dehydrate fructose to HMF with >90% yield, eliminating HMF decomposition to levulinic and formic acids. However, recovery of HMF for further use becomes difficult due to DMSO.[97]

6.3.3.3 *Furan-2,5-Dicarboxylic Acid (FDCA)*

FDCA is an oxidized furan derivative and can be used as a substitute to a wide variety of petrochemicals, such as adipic acid and terephthalic acid,[98] and biobased intermediates, such as levulinic acid and succinic acid. The major technical barrier for commercializing FDCA is its selective and cost-effective production. FDCA with 99% purity can be obtained from HMF oxidation under strongly alkaline conditions using a Pt/Pb catalyst for two hours.[99] 72% yield of FDCA was observed from fructose with 99% selectivity using a cobalt catalyst (Co tris-acetylacetonate) in a silica sol–gel system.[100] Being an organic acid, FDCA could also potentially be recovered using ED, adsorption and L–L extraction.

6.3.4 Sugar Alcohols

Sugar alcohols are naturally occurring polyols. The structural difference of sugar alcohol from the corresponding parent sugar is the conversion of the terminal aldehyde group in sugar into an alcohol group in sugar alcohol. For example, the corresponding sugar alcohol of xylose [$C_5H_{10}O_5$, CH_2OH–$(CHOH)_3$–CHO] is xylitol [$C_5H_{12}O_5$, CH_2OH–$(CHOH)_3$–CH_2OH]. Therefore, sugar alcohols can easily be derived from various sugars by reduction (hydrogenation). Like sugars, sugar alcohols also occur naturally, are isolated in pure, crystalline form and have wide applications as sweeteners in food and beverage industries. Additionally, sugar alcohols have lower energy content than the sugars. Furthermore, sugar alcohols are intermediates for various useful drop-in hydrocarbon products. For example, conversion of sorbitol to hexane is reported with 50% selectivity using Pt/Al_2O_3 catalyst.[101] For the reasons mentioned, and for the recent interest in obtaining biobased products, sugar alcohols are synthesized by a biochemical route of fermentation of sugars obtained from renewable resources. Among all possible sugar alcohols, xylitol, sorbitol and glycerol have exhibited tremendous potential as biobased value-added products and have been enlisted in the "top value-added bioproducts" list [4] (arabinitol is not included in another modified list).[15] Glycerol, a three-carbon polyol, is sometimes not categorized as a sugar alcohol as it is derived mostly from non-sugar sources, such as lipids (triglycerides). However, due to the structural similarity of glycerol with sugar alcohols, we prefer to discuss glycerol in this section.

6.3.4.1 *Xylitol*

Xylitol is a five-carbon sugar alcohol and has similar sweetness to "table sugar", sucrose. It possesses anti-caries properties,[102] non-carcinogenic properties[103] and has lower dietary energy content than sucrose. Hence, it is emerging as a potential alternative to sucrose in food and beverage industries. Apart from that, because of its favorable thermodynamic properties in solution, it has found potential application as a coating material in pharmaceutical products.[103]

Xylitol could be produced either by a biochemical pathway of fermentation of xylose using suitable microorganisms or by catalytic dehydrogenation of xylose. Xylitol production by fermentation of yeast species, such as *Pachysolen tannophilus*, *Debaryomyces hansenii* and *Candida guillermondii*, is reported.[104] The benefit of the biochemical pathway is the ability to use pretreated hemicellulosic biomass hydrolysate liquor as the primary source of xylose, rather than purified xylose. Engineered *Saccaromyces cerevisiae* and various candida yeasts have been examined for high yield xylitol production.[15] On the other hand, a xylose solution comprising 75% xylose and 25% other sugars is hydrogenated using a raney-nickel catalyst at a temperature of 135 °C and at a hydrogen pressure of 40 atmospheres.[104] Separations schemes for these two approaches vary due to the variability in the contents of the process solution.

For the fermentation approach, typical primary treatments, such as centrifugation, microfiltration and IEX adsorption, are conducted to remove the cellular components, salts and nutrients. Coloring components and other trace amount of organics are removed by adsorption with activated carbon. The resulting solution that contains primarily xylitol and unreacted xylose is then subjected to secondary treatments for removal and recovery of xylitol. Secondary treatment involves preferential crystallization of xylitol (discussed later) over xylose using suitable solvents and by manipulating the operating conditions.

For the hydrogenation pathway, lignocellulosic biomass is first hydrolyzed to obtain xylose. Purification of the hydrolyzed liquor consists of two major steps—(i) removal of the salt and organic impurities by the ion-exclusion technique, and (ii) color removal by an IEX system comprising a strong cation exchanger followed by a weak anion exchanger, and then subsequent passage through an activated carbon bed. Xylose is then hydrogenated to form xylitol. Xylitol is purified by chromatographic fractionation using CEX resin beads. Alkaline earth metals (Ca^{+2}, Ba^{+2}, Sr^{+2}) containing CEX resin beads are reported to be efficient in fractionating a mixture of polyols and recovering pure xylitol. Trivalent metal salts, such as Fe^{+3} and Al^{+3}, containing CEX resin beads were also found to be efficient in recovering pure xylitol solution from a mixture of polyols.[105] The concentrated solution of xylitol, thus obtained, is then subjected to crystallization.

Crystallization is important for xylitol production as it dictates the yield and quality of the final product. An ethanol–water mixture is a useful solvent for crystallization of xylitol. Sequential crystallization in the form of preliminary crystallization from supersaturated solution followed by final crystallization in an ethanol–water mixture is found to be more efficient than one-step crystallization.[101] For most crystallization operations, the evaporator–crystallizer is implemented using a rotavapor under vacuum. Crystallization is typically followed by centrifugation, vacuum filtration and drying to obtain anhydrous crystals of xylitol.[103] Crystallization under optimal conditions is able to produce xylitol crystals with more than 95% purity. A major concern for crystallization is high energy consumption. Energy consumption to reduce the temperature enough to maintain supersaturation conditions and to get a xylitol solution concentrated enough for the onset of crystallization is high. For example, 730 g L^{-1} xylitol concentration and $-5\,^{\circ}C$ temperature were required to obtain a crystallization yield of 0.56. In other conditions, low yields of 20–30% were obtained, although the degree of purification was close to 1.[103] Another concern for xylitol synthesis is the formation of other sugar alcohols, such as arabinitol, mannitol and sorbitol. Genetic modification of the microorganism increased the preference of xylose conversion over other sugars, such as arabinose.[106]

6.3.4.2 Sorbitol

Sorbitol is a six-carbon ($C_6H_{14}O_6$) sugar alcohol that is used in food and beverage industries as a sweetener with low dietary energy content and in pharmaceutical industries due to its laxative properties. Production of sorbitol by a biochemical pathway is dominated by *Zymomonas mobilis*. Conversion of sucrose or a mixture of glucose and fructose to sorbitol and gluconic acid by *Zymomonas mobilis* is well established.[107] Isolation of the enzyme, glucose fructose oxidoreductase (GFOR), from the same microorganism has enabled the direct enzymatic pathway for the synthesis of sorbitol from a mixture of glucose and fructose.[108]

Chun and Rogers[109] have demonstrated the production of sorbitol and gluconic acid using both free and immobilized (alginate beads) *Zymomonas mobilis*. Sorbitol and gluconic acid were separated using strongly basic anion exchange resin beads (both Cl^- and OH^- form) in a fixed-bed column. The products' separation was close to 100%. Although this separations strategy was cost effective at a smaller scale, the effectiveness for commercial scale separations was questionable.

An integrated system consisting of a hollow fiber membrane bioreactor and electrodialysis as a separations unit was designed to recover sorbitol from gluconic acid in the product mixture.[110] *Zymomonas mobilis* cells were first entrapped within alginate beads followed by confinement of the beads in the membrane bioreactor. Sorbitol and gluconic acid produced in the membrane bioreactor were separated in the ED cell, because the charged molecules of gluconic acid were removed within the ED cell under an applied electrical potential. Uncharged sorbitol molecules stay in the process stream. This integrated approach enables removal of the organic acid, thereby avoiding the strong acidic pH of the medium—a condition unsuitable for functioning of *Zymomonas mobilis* cells.

Researchers from Argonne National Laboratory, US, have further developed the integrated reaction–separation approach by designing a separative bioreactor (see Chapter 7). It is based on an electrodeionization (EDI) platform and performs the biochemical synthesis of gluconic acid and sorbitol from a mixture of glucose and fructose.[111] In this approach, GFOR enzyme was immobilized onto the porous structure of the proprietary resin wafers[112] in the diluate compartments of the EDI cell. GFOR oxidizes glucose to gluconic acid and reduces fructose to sorbitol in the diluate compartments. Due to the applied electrical potential, charged molecules of gluconic acid undergo *trans*-membrane transport to the adjacent concentrate compartment, whereas the uncharged sorbitol molecules remain in the diluate side. This technique enables simultaneous, *in situ* production and purification of two important biobased products in a single unit operation. The integrated separative bioreactor concept provides an economically favorable pathway for the production of valuable bioproducts.

6.3.4.3 Glycerol

Glycerol or glycerine ($CH_2OH–CHOH–CH_2OH$) is identified as a building block biobased chemical and is listed in the top 12 value-added chemicals from biomass by the US Department of Energy.[4] It is extensively used in food, beverage, pharmaceutical and personal care products, and as an antifreezing chemical. Glycerol can be produced either by microbial fermentation or by chemical synthesis from petrochemical feedstocks or can be recovered as a by-product of soap manufacture from fats. It is also produced as a co-product during biodiesel production by transesterification and ethanol production by fermentation. With the growing production of biodiesel, a significant surge in crude glycerol availability is imminent. Since glycerol has a high market demand, a suitable separation technology for recovering pure glycerol is critical for the economic viability of biofuel production.

Glycerol purification from a biodiesel processing stream involves initial separation of the crude glycerol layer (bottom) and the biodiesel layer (top) by settling based on density difference. The crude glycerol layer is then subjected to three major purification steps—neutralization, methanol removal and glycerol recovery. Strong acid neutralization is conducted to remove catalyst and soaps. The reaction of acid with soap produces free fatty acids and its reaction with base catalyst forms salt and water. This results in a three-layer mixture—a top layer of free fatty acids (and some salts), a glycerol-rich middle layer and a bottom layer of inorganic salts. Individual phases are separated by decantation. The middle layer containing glycerol and excess methanol is then subjected to vacuum evaporation to remove the volatile methanol.

Ion-exchange resin based adsorption is also used to remove the impurities, such as inorganic salts and free ions, from crude glycerol. However, crude glycerol purification with IEX resin has some drawbacks, such as fouling by fatty acids, oils and soaps, delicate regeneration of the beds and a large quantity of waste water produced, which need further improvement.

Separation of glycerol from a biodiesel stream can also be done by membrane filtration. Ultrafiltration is used to separate glycerol from canola oil and methanol.[113] Although conventional methods, such as combinations of chemical (neutralization) and physical treatments (vacuum distillation, evaporation), produce good quality purified glycerol, they are not economically favorable. Membrane technology is considered as an emerging technology for glycerol purification, because it provides a cost effective alternative.

Separation of glycerol from fermentation broth is conducted by vacuum distillation after removal of ethanol by distillation and precipitation of salts (sulfite, sulfates and phosphates).[114] The cell debris and other dissolved organic substances in the fermentation broth form a solid aggregate that traps glycerol and hinders distillation (recovery of $\sim$50% of glycerol). This can be resolved by removing the debris from the broth followed by adding an inorganic inert material to the broth, which enables the recovery of approximately 90% of glycerol by distillation. This technique is patented as carrier-distillation.[115]

A packed-bed reactor was designed for enzymatic biodiesel production, which also simultaneously separates glycerol. Glycerol tends to adsorb on enzyme immobilized carriers and forms a hydrophilic layer. However, a major drawback associated with this technique is due to the inhibition from the reactant oils. Also, the change in flow rate affects the adsorption of glycerol on the solid surface.[116]

6.4 Summary

In order to transform the biorefinery concept into a commercial success, economically viable formation of biobased products is absolutely necessary. Separations are an integral part of that as they contribute significantly to overall production costs. Development and implementation of highly-efficient, cost-effective, environmentally-friendly and energetically-favorable separations technologies is the key for successful commercialization of biobased products. We presented several important separations technologies that are relevant to biobased product formation. The fundamental principles of these technologies, their limitations and potential solutions to overcome the limitations are discussed. Applications of separations technologies for certain value-added biobased products, such as algal biomass-derived products, organic acids, furans and sugar alcohols, are illustrated. Considering the complexities of biological solutions and variations in their contents, it is pretty evident that no single separation technology can be highlighted as superior to others. A comprehensive separations strategy consisting of various separations technologies is needed for bioproduct formation. Although conventional separations technologies, such as distillation and evaporation, are still extensively used in industry, modern, energy efficient technologies, such as membrane separations and supercritical fluid extraction, are emerging as potential alternatives. Overall, the development of innovative separation technologies and judicious refinement of the existing technologies can pave the foundation for successful biorefineries.

Acknowledgements

We acknowledge the Indian Institute of Technology Roorkee (IIT Roorkee) for providing the academic and research facilities needed for this study. We also acknowledge the Ministry of Human Resource and Development, Government of India, for financial assistance to Ms Anju Kumari and the University Grant Commission, India, for financial assistance to Mr Bhanendra Singh.

References

1. Office of the Biomass Program, Energy Efficiency and Renewable Energy, U.S. Department of Energy, *Biomass Multi Year Program Plan*, 2011.
2. Energy Independence – Security Act of 2007, *U.S. Government Printing Office Public Law*, 2007, **110** (140), 19, http://www.gpo.gov/fdsys/pkg/PLAW-110publ140/content-detail.html.
3. The White House, *National Bioeconomy Blueprint*, 2012, http://www.whitehouse.gov/sites/default/files/microsites/ostp/national_bioeconomy_blueprint_april_2012.pdf.
4. T. Werpy and G. Petersen, *Top Value Added Chemicals from Biomass. Volume I – Results of Screening for Potential Candidates from Sugars and Synthesis Gas*, U. S. DOE, 2004, http://www.nrel.gov/docs/fy04osti/35523.pdf.
5. B. E. Dale, *Chem. Technol. Biotechnol.*, 2003, **78**, 1093.
6. E. J. Henley, J. D. Seader and D. K. Roper, *Separation Process Principles – International Student Version*, 3rd edn, John Wiley and Sons, 2011.

7. C. J. Geankoplis, *Transport Processes and Separation Process Principles (Includes Unit Operations)*, 4th edn, Pearson Prentice Hall, 2011.
8. R. E. Treybal, *Mass-Transfer Operations*, McGraw-Hill, 1980.
9. B. K. Dutta, *Principles of Mass Transfer And Separation Processes*, 1st edn, PHI Learning Pvt. Ltd., 2007.
10. A. Aden, M. Ruth, K. Ibsen, J. Jechura, K. Neeves, J. Sheehan and B. Wallace, *Technical Report of National Renewable Energy Laboratory (NREL/TP-510-32438)*, 2002.
11. R. Ghosh, *Principles of Bioseparations Engineering*, 1st edn, World Scientific Publishing Co., 2006.
12. W. Ho and K. K. Sirkar, *Membrane Handbook*, Springer 1992.
13. J. Mulder, *Basic Principles of Membrane Technology*, Springer, 1996.
14. M. Cheryan, *Ultrafiltration and Microfiltration Handbook*, CRC Press, 1998.
15. J. J. Bozell and G. R. Petersen, *Green Chem.*, 2010, **12**, 539.
16. E. E. M. Luiten, I. Akkerman, A. Koulman, P. Kamermans, H. Reith, M. J. Barbosa, D. Sipkema and T. H. Wijffels, *Biomol. Eng.*, 2003, **20**, 429.
17. R. Harun, M. Singh, G. M. Forde and M. K. Danquah, *Renewable Sustainable Energy Rev.*, 2010, **14**, 1037.
18. J. Masojidek and G. Torzillo, Mass cultivation of fresh water microalgae, in *Encyclopedia of Ecology*, Academic press, 2008, 2226.
19. W. Ugwu, H. Aoyagi and H. Uchiyama, *Bioresour. Technol.*, 2008, **99**, 21.
20. C. Gudin and C. Therpenier, *Adv. Biotechnol. Proc.*, 1986, **6**, 73.
21. F. H. Mohn, *Arch. Hydrobiol., Beih. Ergeb. Limnol.*, 1978, **1**, 228.
22. C. R. Sarkar, L. Das, B. Bhagawati and B. C. Goswami, *Asian J. Plant Sci. Res.*, 2012, **2**, 546.
23. M. D. Macias-Sanchez, C. Mantell, M. Rodriguez, O. E. Martinez, L. M. Lubián and O. Montero, *Talanta*, 2009, 77(3), 948.
24. S. Splinter, J. J. R. Pare and S. Kadali, *US pat.* 2013/0338234 A1, 2013.
25. M. Plaza, S. Santoyo, L. Jaime, G. Garcia-Blairsy, M. Herrero and F. J. Señorans, *J. Pharm. Biomed. Anal.*, 2010, **51**, 450.
26. K. Kitada, S. Machmudah, M. Sasaki, M. Goto, Y. Nakashima, S. Kumamoto and T. Hasegawa, *J. Chem. Technol. Biotechnol.*, 2009, **84**, 657.
27. W. G. Ce, S. H. Bao, T. C. Xiao and F. Kuli, *Acta Bot. Sin.*, 2002, **44**(5), 541–546.
28. H. Chakdar and S. Pabbi, *Phykos*, 2012, **42**(1), 25.
29. U. N. Rossano, A. D. Ambrosio and G. M. Liuzzi, *J. Biotechnol.*, 2003, **101**, 289.
30. M. Herrero, C. Simo, E. Ibáñez and A. Cifuentes, *J. Agric. Food Chem.*, 2005, **26**, 4215.
31. E. G. Blight and W. J. Dyer, *Can. J. Biotechnol. Physiol.*, 1959, **37**, 911.
32. G. S. Araujo, L. J. B. L. Matos, L. R. B. Gonçalves, F. A. N. Fernandes and W. R. L. Farias, *Bioresour. Technol.*, 2011, **102**(8), 5248.

33. R. L. Mendes, B. P. Nobre, M. T. Cardoso, A. P. Pereira and A. F. Palabra, *Inorg. Chim. Acta*, 2003, **356**, 328.
34. P. C. K. Cheung, A. Y. H. Leung and P. O. Ang Jr., *J. Agric. Food Chem.*, 1998, **46**, 4228.
35. R. Harun, W. Jason, T. Cherrington and M. K. Danquah, *Appl. Energy*, 2011, **88**, 3464.
36. N. Kaliaperumal and P. Uthirasivan, *Seaweed Res. Util.*, 2001, **23**(1&2), 55.
37. D. E. Akin, L. L. Rigsby, A. Sethuraman, W. H. Morrison, G. R. Gamble and K. E. Eriksson, *Appl. Environ. Microbiol.*, 1995, **61**, 1591.
38. M. A. Millet, A. J. Baker and L. D. Scatter, *Biotechnol. Bioeng. Symp.*, 1976, **6**, 125.
39. G. Britton, *FASEB J.*, 1995, **9**, 1551.
40. M. C. Cerón, I. Campos, J. F. Sánchez, F. G. Acién, E. Molina and J. M. Fernandez-Sevilla, *J. Agric. Food Chem.*, 2008, **56**, 11761.
41. C. D. Kang and S. J. Sim, *Biotechnol. Prog.*, 2007, **23**, 866.
42. P. M. Arvela, I. Hachemi and D. Y. Muzrin, *J. Chem. Technol. Biotechnol.*, 2014, **89**(11), 1607.
43. M. Plaza, B. Avalo, A. Cifuentes and E. Ibáñez, *Food Sci. Technol.*, 2012, **16**, 245.
44. M. Herrero, L. Jaime, P. J. Martín-Álvarez, A. Cifuentes and E. Ibáñez, *J. Agric. Food Chem.*, 2006, **54**, 5597.
45. Y. F. Shang, S. M. Kim, W. J. Lee and B. H. Um, *J. Biosci. Bioeng.*, 2011, **111**(2), 237.
46. S. Machmudah, A. Shotipruk, M. Goto, M. Sasaki and T. Hirose, *Ind. Eng. Chem. Res.*, 2006, **45**, 3652.
47. P. Thana, S. Machmudah, M. Goto, M. Sasaki, P. Pavasant and Shotipruk, *Bioresour. Technol.*, 2008, **99**, 3110.
48. C. H. Romay, R. Gonzalez, N. Ledón, D. Remirez and V. Rimbau, *Curr. Protein Pept. Sci.*, 2003, **4**, 207.
49. J. L. Guil-Guerrero, R. Navarro-Juarez, J. C. Lopez-Martinez, P. Campra-Madrid and M. M. Rebolloso-Fuentes, *J. Food Eng.*, 2004, **65**(4), 511.
50. T. Furuki, S. Maeda, S. Imajo, T. Hiroi, T. Amaya and T. Hirokawa, *J. Appl. Phycol.*, 2003, **15**, 319.
51. H. Zheng, J. Yin, Z. Gao, H. Huang, H. Ji and C. Dou, *Appl. Biochem. Biotechnol.*, 2011, **164**, 1215.
52. J. Fleurence, *J. Appl. Phycol.*, 1999, **11**, 313.
53. J. Doucha and K. Livansky, *Appl. Microbiol. Biotechnol.*, 2008, **81**, 431.
54. S. Jubeau, L. Marchal, J. Pruvost, P. Jaouen, J. Legrand and J. Fleurence, *J. Appl. Phycol.*, 2012, **25**, 983.
55. J. Wang, Q. Zhang, Z. Zhang and Z. Li, *Int. J. Biol. Macromol.*, 2008, **42**, 127.
56. A. J. J. Straathof, The proportion of downstream costs in fermentative production processes, in *Comprehensive Biotechnology*, 2nd edn, Elsevier, 2011, p. 811.
57. C. S. López-Garzón and A. J. J. Straathof, *Biotechnol. Adv.*, 2014, **32**, 873.

58. S. T. Yang, H. Huang, A. Tay, W. Qin, L. D. Guzman and E. C. S. Nicolas, Extractive Fermentation for the Production of Carboxylic Acids, in *Bioprocessing for Value-Added Products from Renewable Resources*, Elsevier, 2007, p. 421.
59. Q. Li, D. Wang, Y. Wu, W. L. Li, Y. J. Zhang, J. M. Xing and Z. G. Su, *Sep. Purif. Technol.*, 2010, **72**, 294.
60. A. Evren TUGTAS, *Fen Bilimleri Derg.*, 2011, **23**(2), 70.
61. A. S. Kertes and C. J. King, *Biotechnol. Bioeng.*, 1986, **28**, 269.
62. V. Bízek, J. Horáček and M. Koušová, *Chem. Eng. Sci.*, 1993, **48**, 1447.
63. M. Marinova, J. Albet, J. Molinier and G. Kyuchoukov, *Ind. Eng. Chem. Res.*, 2005, **44**, 6534.
64. R. W. Miller, M. C. M. Cockrem, J. J. de Pablo and E. N. Lightfoot, *Ind. Eng. Chem. Res.*, 1996, **35**, 1156.
65. T. Kurzrock and D. Weuster-Botz, *Bioprocess Biosyst. Eng.*, 2011, **34**, 779.
66. G. Malmary, J. Albet, A. Putranto, H. Hanine and J. Molinier, *J. Chem. Eng. Data*, 1998, **43**, 849.
67. R. Wennersten, *J. Chem. Technol. Biotechnol.*, 1983, **33**, 85.
68. M. Moresi and F. Sappino, *J. Membr. Sci.*, 2000, **164**, 129.
69. A. M. Baniel and D. Gonen, *US Pat.* US4994609, 1991.
70. H. Benninga, *A History of Lactic Acid Making*, Kluwer Academic Publishers, 1990.
71. B. H. Davison, N. P. Nghiem and G. L. Richardson, *Appl. Biochem. Biotechnol.*, 2004, **113**, 653.
72. M. Kalyanpur, *Mol. Biotechnol.*, 2002, **22**, 87.
73. N. Kawabata, J. I. Yoshida and Y. Tanigawa, *Ind. Eng. Chem. Prod. Res. Dev.*, 1981, **20**, 386.
74. X. J. Cao, H. S. Yun and Y. M. Koo, *Biochem. Eng. J.*, 2002, **11**, 189.
75. H. J. Lee, Y. Xie, Y. M. Koo and N. H. L. Wang, *Biotechnol. Prog.*, 2004, **20**, 179.
76. J. Wu, Q. Peng, W. Arlt and M. Minceva, *J. Chromatogr. A*, 2009, **1216**, 8793.
77. A. J. Hepburn and A. J. Daugulis, *J. Chem. Technol. Biotechnol.*, 2012, **87**, 42.
78. D. A. Glassner and R. Datta, *US Pat.* US5143834, 1992.
79. I. Meynial-Salles, S. Dorotyn and P. Soucaille, *Biotechnol. Bioeng.*, 2008, **99**, 129.
80. L. P. Ling, H. F. Leow and M. R. Sarmidi, *J. Membr. Sci.*, 2002, **199**, 59.
81. M. Bailly, *Desalination*, 2002, **144**, 157.
82. C. H. Huang, T. W. Xu, Y. P. Zhang, Y. H. Xue and G. W. Chen, *J. Membr. Sci.*, 2007, **288**, 1.
83. S. P. R. Katikaneni, *Ind. Eng. Chem. Res.*, 2002, **41**, 2745.
84. P. Börgardts, W. Krischke, W. Trösch and H. Brunner, *Bioprocess Eng.*, 1998, **19**, 321.
85. R. Datta, S. P. Tsai, P. Bonsignore, S. H. Moon and J. R. Frank, *FEMS Microbiol. Rev.*, 1995, **16**, 221.
86. Y. M. Wang, C. H. Huang and T. W. Xu, *J. Membr. Sci.*, 2011, **374**, 150.

87. S. Novalic, T. Kongbangkerd and K. D. Kulbe, *J. Membr. Sci.*, 2000, **166**, 99.
88. S. H. Kang and Y. K. Chang, *J. Membr. Sci.*, 2005, **246**(1), 49.
89. R. D. Noble and D. L. Gin, *J. Membr. Sci.*, 2011, **369**, 1.
90. A. S. Dias, S. Lima, M. Pillinger and A. A. Valente, *Catal. Lett.*, 2007, **114**, 151.
91. A. S. Dias, M. Pillinger and A. A. Valente, *J. Catal.*, 2005, **229**, 414.
92. F. Qin, S. Li, P. Qin, M. N. Karim and T. Tan, *Green Chem.*, 2014, **16**, 1262.
93. A. J. Sanborn and P. D. Bloom, *US Pat.* US 7393963 B2 20080701, 2008.
94. C. H. C. Zhou, J. N. Beltramini, Y. X. Fan and G. Q. M. Lu, *Chem. Soc. Rev.*, 2008, **37**, 527.
95. C. Moreau, A. Finiels and L. Vanoye, *J. Mol. Catal. A*, 2006, **253**, 165.
96. S. Q. Hu, Z. F. Zhang, Y. X. Zhou, B. X. Han, H. L. Fan, W. J. Li, J. L. Song and Y. Xie, *Green Chem.*, 2008, **10**, 1280.
97. K. Seri, Y. Inoue and H. Ishida, *Chem. Lett.*, , 2000, 22.
98. C. Moreau, M. N. Belgacem and A. Gandini, *Top. Catal.*, 2004, **27**, 11.
99. P. Verdeguer, N. Merat and A. Gaset, *J. Mol. Catal.*, 1993, **85**, 327.
100. M. L. Ribeiro and U. Schuchardt, *Catal. Commun.*, 2003, **4**, 83.
101. G. W. Huber and J. A. Dumesic, *Catal. Today*, 2006, **111**, 119.
102. V. V. Vyglazov, *Russ. J. Appl. Chem.*, 2004, 77(1), 26.
103. D. D. Faveri, P. Perego, A. Converti and M. D. Borghi, *Chem. Eng. J.*, 2002, **90**, 291.
104. A. Converti, P. Perego and J. M. Domınguez, *Appl. Biochem. Biotechnol.*, 1999, **82**, 141.
105. A. J. Melaja, N. Kantvik, L. Hamalainen and C. Rajakallio, *US Pat.* US4008285, 1977.
106. N. U. Nair and H. M. Zhao, *Chem. Biol. Chem.*, 2008, **9**, 1213.
107. K. D. Barrow, J. G. Collins, D. A. Leigh, P. L. Rogers and R. G. Warr, *Appl. Microbiol. Biotechnol.*, 1984, **20**, 225.
108. M. Zachariou and R. K. Scopes, *J. Biotechnol.*, 1986, **167**, 863.
109. U. H. Chun and P. L. Rogers, *Appl. Microbiol. Biotechnol.*, 1988, **29**, 19.
110. H. C. Ferraz, T. L. M. Alves and C. P. Borges, *J. Membr. Sci.*, 2001, **191**, 43.
111. M. B. Arora, J. A. Hestekin, S. W. Snyder, E. J. St. Martin, Y. J. Lin, M. I. Donnelly and C. S. Millard, *Sep. Sci. Technol.*, 2007, **42**, 2519.
112. Y. J. Lin, M. P. Henry and S. W. Snyder, *U.S. Pat.* US7452920 B2, 2008.
113. J. Saleh, A. Y. Trembley and M. A. Dube, *Fuel*, 2010, **89**, 2260.
114. L. A. Underkofler and R. J. Hickey, *Industrial Fermentations*, Chemical Publishing, 1954, 252.
115. J. Zhuge and X. Z. Liu, *Chinese Pat.* CN 1047532A, 1990.
116. S. Hama, S. Tamalampudi, A. Yoshida, N. Tamadani, N. Kuratani, H. Noda, H. Fukuda and A. Kondo, *Bioresour. Technol.*, 2011, **102**(22), 10419.

CHAPTER 7

Lignin as Feedstock for Fibers and Chemicals

STEVEN W. PERETTI,*[a] RYAN BARTON[a] AND REGIS TEIXEIRA MENDONCA[b]

[a] Department of Chemical and Biomolecular Engineering, North Carolina State University, Box 7905, Raleigh, NC 27695-7905, USA; [b] Centro de Biotecnología, Barrio Universitario s/n, Universidad de Concepción, Concepción 4070386, Chile
*Email: peretti@ncsu.edu

7.1 Introduction

The US Energy Information Administration recently made projections regarding the world levels of biofuel production: by 2040, biofuel production is expected to rise between 55% and 100% relative to 2014 levels.[1] Lignocellulose will be a primary feedstock for the second and third generation biorefineries generating those fuels. First generation cellulosic biorefineries have been focused on ethanol production. They are designed to separate cellulose from hemicellulose and lignin, to hydrolyze the cellulose to glucose, and to use the sugar as a substrate for ethanol fermentation. Lignin, because of its relative resistance to degradation, is typically relegated to use as boiler fuel to satisfy process energy requirements. Second generation biorefineries will need to capture additional value from the traditionally low-value, high volume lignin byproduct. In a recent review of lignin valorization,[2] the challenge is defined clearly: *"Lignin…is underutilized in these first generation cellulosic projects, with about 60% more lignin generated than is needed to meet*

RSC Green Chemistry No. 43
Commercializing Biobased Products: Opportunities, Challenges, Benefits, and Risks
Edited by Seth W. Snyder

Published by the Royal Society of Chemistry, www.rsc.org

internal energy use by its combustion.[3,4] *Therefore, new processes are needed that generate value-added products from lignin.*[5]"

Figure 7.1 illustrates the recent development in thinking about processing lignocellulosic biomass, one that embodies the highest value use of available carbon through the modification of natural polymers (cellulose, hemicellulose, lignin) into valorized materials and chemical intermediates. Material processing in a first generation cellulosic biorefinery is indicated by the black lines: pretreatment to separate lignin from celluloses, hydrolysis to generate sugars, fermentation of sugars to ethanol, and combustion of the lignin. The red lines represent areas of research efforts geared towards expanding the number of products generated from lignocellulosic biomass. The dashed green lines indicate feedback interactions established to support this expanded product line.

The structure of lignin is highly dependent on the time, temperature, and chemical exposure imposed by the pretreatment technology utilized to separate it from cellulose. Since traditionally the objective was to generate the "cleanest" possible cellulose, until recently little concern was paid to the effects of pretreatment on lignin structure and composition. In the pretreatment arena, significant recent development of genetic engineering techniques that effectively manipulate lignin synthesis and composition (pathway engineering) impact the type and severity of pretreatment technology required to generate "clean" cellulose and lignin. Developments in plant cloning and gene expression applications to lignin would result in improved economic viability of lignocellulosic biorefineries,[6,7] but it is not our intention to summarize that material. Instead, this review will focus on

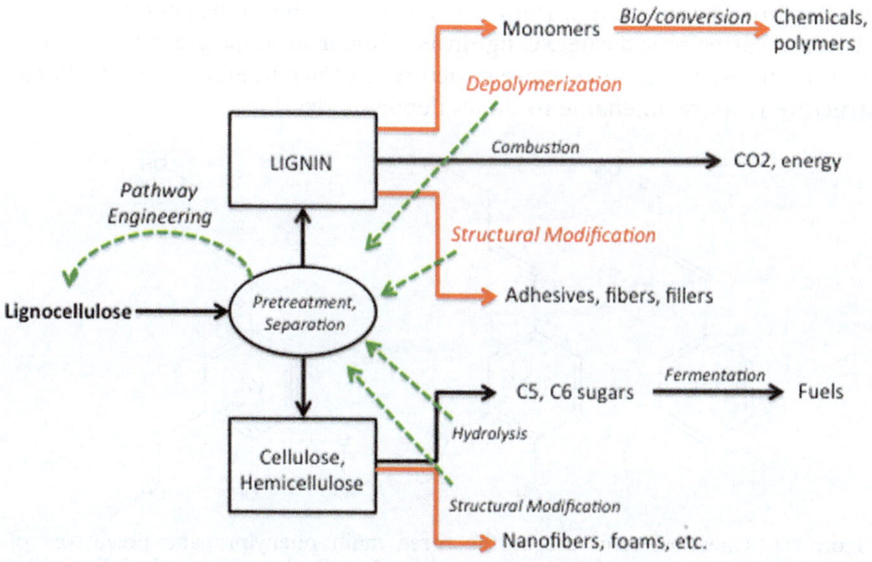

Figure 7.1 Processing lignocellulose for maximum value.

the lignin structures that result from various pretreatment technologies, and efforts to valorize that lignin through material property modification or depolymerization.

7.2 Lignin Fundamentals

7.2.1 Lignin Structure and Chemistry

Lignin (a cross-linked aromatic macromolecular network), cellulose, and hemicellulose comprise the main structural components of lignocellulosic materials (*i.e.* wood, agro-forestry residues and grasses). Lignin corresponds to roughly 20–30% of the plant cell wall, and acts as a cementing agent for the polysaccharides, providing mechanical strength and support for plant growth.[8] Lignin also serves as a barrier against microbial degradation of wood by bacteria, molds and fungi.[8] Lignin is heterogeneous and formed mainly by three cinnamyl alcohols, also known as monolignols: *p*-coumaryl alcohol, coniferyl alcohol and sinapyl alcohol. The main difference among these phenylpropane derivatives is the number of methoxyl groups attached to the phenolic unit. The *p*-hydroxyphenyl or H-unit has no methoxyl group, the guaiacil or G-unit has one methoxyl group at the 3-position of the aromatic ring, and the syringyl or S-unit has two methoxyl groups in the 3- and 5-positions of the aromatic ring (Figure 7.2). Softwood and hardwood lignins are mainly of the G and GS types, respectively, while grasses and other non-wood lignocellulosic plants contain different proportions of H, G and S units in their lignin structure.[4,8] An example of a lignin macromolecule and major intermolecular linkage types is shown in Figure 7.3.

C lignin, a polymer derived from a modified monolignol, caffeyl alcohol, was recently discovered in certain seed coats.[9–11] C lignin has not been found in stem, leaf or root tissues. C lignin is a linear homopolymer that is relatively easily extracted. Its putative structure is shown in Figure 7.4. The linear structure is more amenable to direct depolymerization.

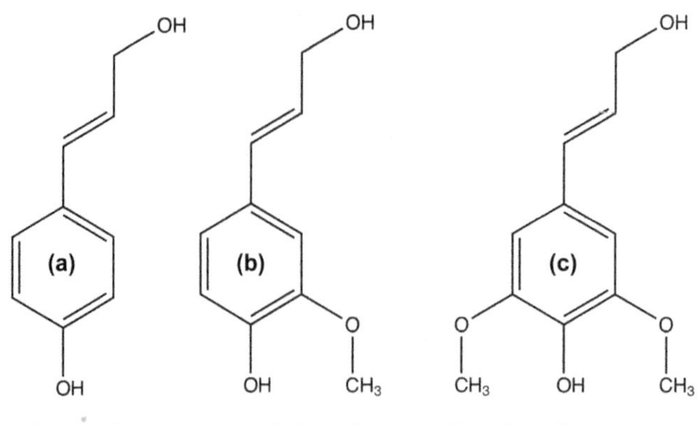

Figure 7.2 Chemical structure of the three main phenylpropane precursors of lignin, where (a) is *p*-coumaryl alcohol, (b) is coniferyl alcohol, and (c) is sinapyl alcohol.

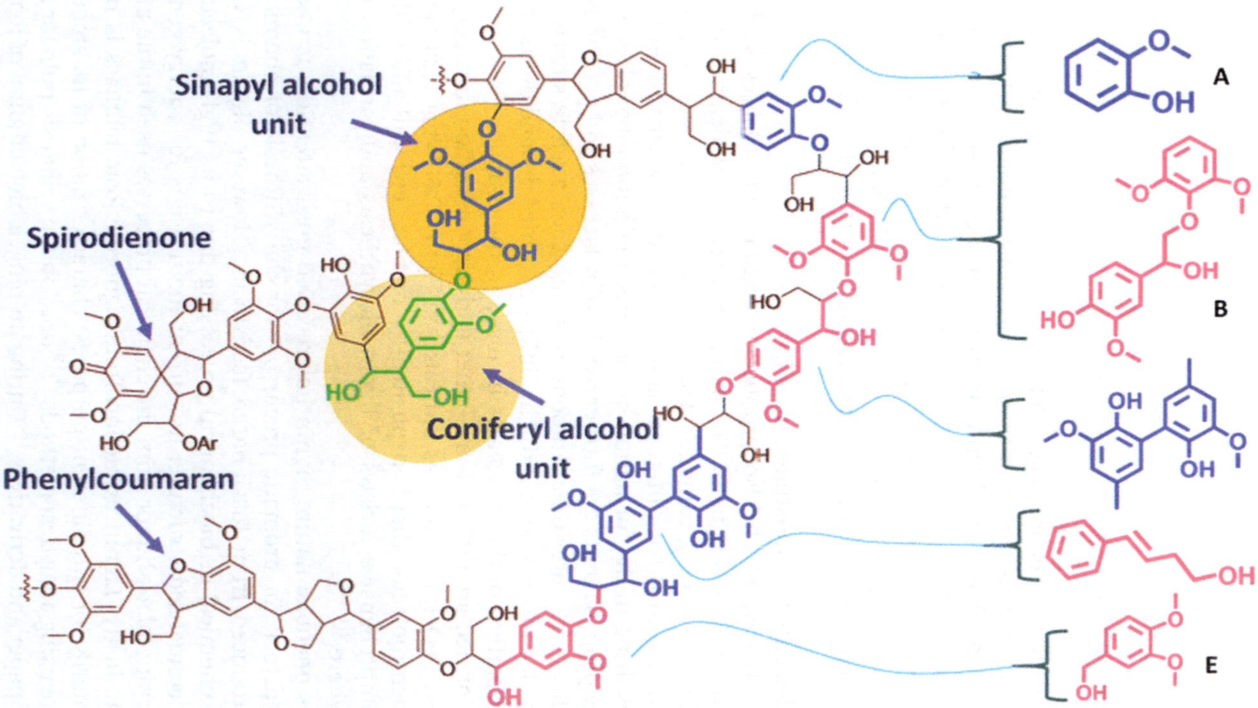

Figure 7.3 Schematic depiction of lignin showing various linkages and lignin model compounds. Model (A) phenol and methoxy functionalities, (B) a β-O-4 linkage, (C) a 5-5′ linkage, (D) a propyl side chain, and (E) a benzylic group (reproduced from ref. 148).

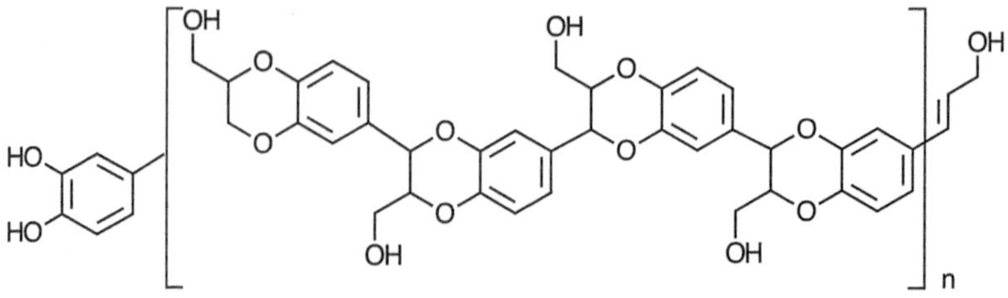

Figure 7.4 Putative chemical structure of C lignin.

As of yet, C lignin has only been detected in certain seed coats, and one preliminary study involving an attempt to genetically modify a plant (Arabidopsis) that would lead to C lignin production in stem tissue resulted in (i) no detectable C lignin in stems, and (ii) dwarf plants that did not grow past the sixth internode.[12] This is an area that will likely be the focus of intensive research and development efforts moving forward, but is not currently available in quantities relevant for generating commercial scale lignin-derived products.

Lignin biosynthesis, illustrated in Figure 7.5, is part of the phenylpropanoid pathway, which also produces secondary metabolites such as tannins, flavonoids, and lignans. The main route starts with deamination of phenylalanine, generating cinnamic acid by a series of enzymatic reactions. It then branches to three different pathways depending on the enzymes involved, resulting in the formation of the three main monolignols shown. These phenylpropane units are further oxidized by laccases and peroxidases, forming radical structures that can bond one with each other in sites where an unpaired electron can be found. The linkages formed depend on the ring-substitution pattern of the molecules involved in the coupling (H, G, or S). The bonds formed are different aryl–ether and carbon–carbon linkages, including α-O-4, β-O-4, β-β′, 5-5′, as indicated in Figure 7.5. What results is a complex macromolecule that is deposited in the plant cell wall and middle lamella. Comprehensive reviews of lignin biosynthesis and formation are found elsewhere.[4,13,14]

Due to its phenolic nature, lignin has received great attention as a sustainable feedstock for aromatic precursors for the synthesis of chemicals currently produced from fossil-based feedstocks. However, lignin is very resistant to chemical or biochemical processing due to its high molecular weight and condensed structure; its current main use is for energy generation by burning black liquor from the chemical processes used in the pulp and paper industry. Lignin extraction from lignocellulosic biomass is performed primarily for two different purposes: characterization of its amount and structure using analytical methods[16,17] or to obtain cellulosic pulp for the synthesis of paper and derivatives.[18] During pulping, a high amount of lignin should be removed from the middle lamella and cell walls to release wood fibers and the fibers should be fibrillated further for paper production or chemical modification of cellulose.[18] Solid lignin can also be converted into

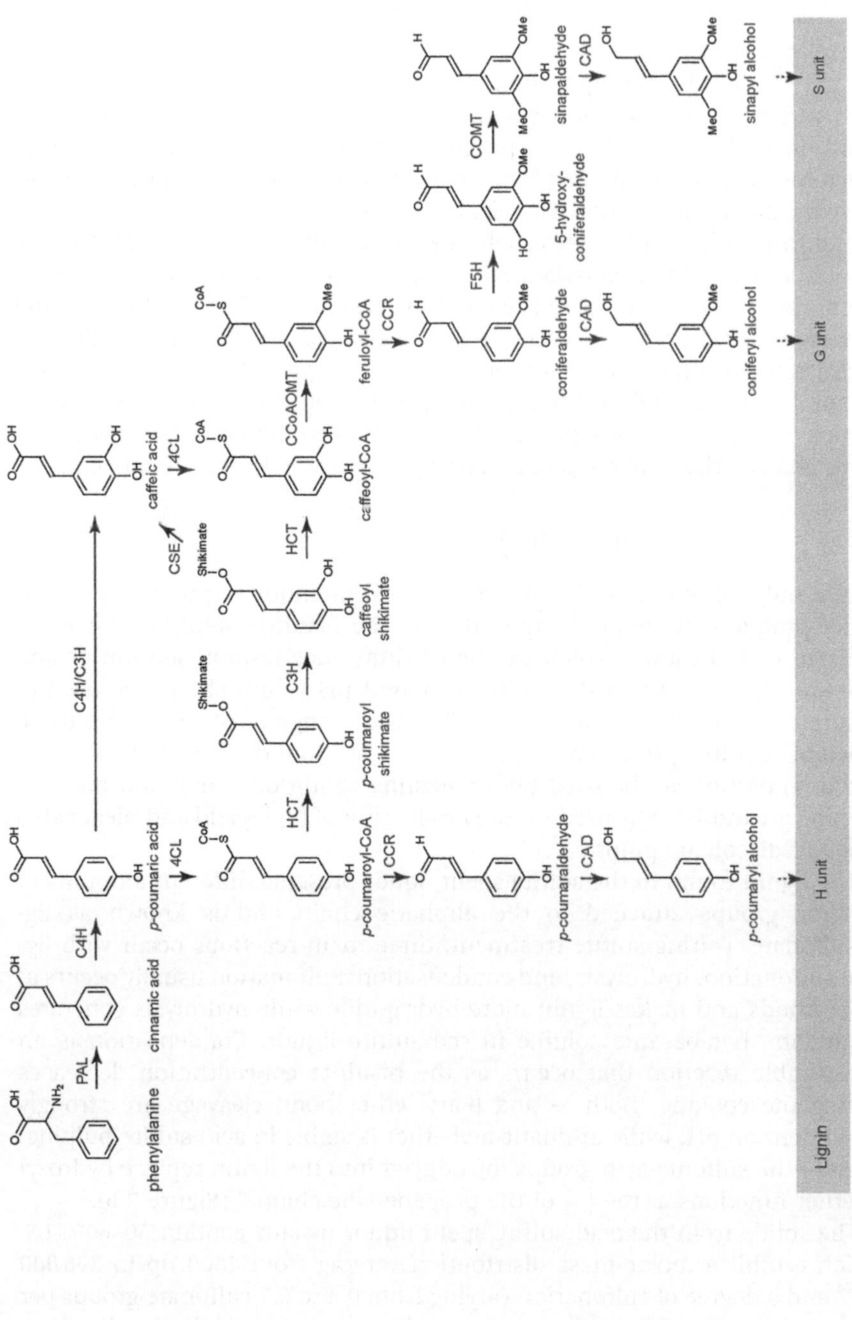

Figure 7.5 Proposed pathway for lignin biosynthesis in plants.[15]

liquids by thermochemical transformations using direct liquefaction under hydrogen pressure and a catalyst, by liquefaction or pyrolysis, for instance.[19]

7.2.2 Lignin from Industrial Processes

Chemical delignification processes have been used for lignin removal from the plant cell wall for pulp or biofuel production, generating residual lignin with varied molar mass and chemical characteristics. These processes depend on their effectiveness to progressively cleave and modify the lignin macromolecule until the resulting fragments become small enough to dissolve in the pulping liquor. The cleavage of aryl–ether linkages in lignin occurs during pulping processes such as Kraft (an alkaline process) and sulfite (which is generally undertaken with pH varying between 2 and 12) treatments, generating lignophenols containing residual sulfur.[20] Delignification processes using organic solvents (organosolv) generate sulfur-free lignins. Pretreatment processes used for fractionation of lignocellulosic biomass such as autohydrolysis and steam explosion, generated more highly condensed lignins with higher molecular weight compared to the other processes.[21–23] Some characteristics of the process and lignin obtained are summarized next.

7.2.2.1 Lignosulfonates (LS)

Sulfite pulping is one of the oldest chemical methods to produce pulp for paper products. Its name is derived from the bisulfite solution that is associated with a cation, which can be calcium, magnesium, sodium or ammonium. Depending on the cation, different pH values should be used to ensure the solubility of the sulfite salt; a wide range of pH levels are used. Calcium requires pH 2, magnesium is used at around pH 4, while sodium and ammonium can be used under alkaline conditions. Acid and bisulfite pulping are industrially used for the production of high-yield and bleachable grade or dissolving pulps.[18,20]

The lignin found in the sulfite spent liquor presents different amounts of sulfonic groups attached to the aliphatic chain and is known as lignosulfonate. During sulfite treatment, three main reactions occur with lignin: sulfonation, hydrolysis, and condensation. Sulfonation usually occurs at ether bonds and makes lignin more hydrophilic while hydrolysis generates fragments that became soluble in the sulfite liquor. Condensation is an undesirable reaction that occurs as the bisulfite concentration decreases during the cooking. Both α- and β-aryl ether bond cleavage are strongly dependent on pH, while aromatic-aryl-ether is stable in acid sulfite pulping. Most of the sulfonic acid groups introduced into the lignin replace hydroxyl or ether functions at the Cα of the propane side chain[18] (Figure 7.6).

The solids from the acid sulfite spent liquor usually contain 50–60% LS, which exhibit a molar mass distribution varying from 4600 up to 398 000 Da[24] and a degree of sulfonation varying from 0.4 to 0.7 sulfonate groups per phenylpropane residue.[25,26] Askvik *et al.*[27] characterized lignosulfonates,

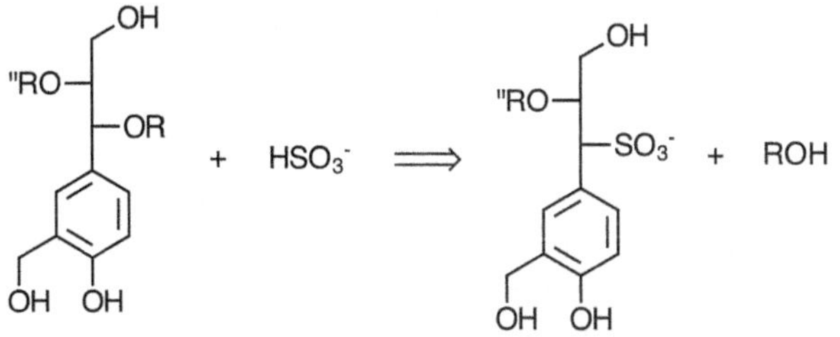

Figure 7.6 Representative sulfonation of Cα in a free phenolic structure of lignin.

obtaining total sulfur content around 7%, methoxyl groups 9–12%, carboxyl 4–6%, and a phenolic hydroxyl content of 2.4%. LS can be considered anionic polymers in which the hydrophobic backbone is rendered hydrophilic by substitution of sulfonate groups. The molecule is highly cross-linked and also contains some carboxylate groups. Lignosulfonates are used in several industrial fields, mainly as dispersants and binders, and their functional properties depend on the molecular weight. LS are also effective emulsion stabilizers and have been used to control viscosity in oil well drilling fluids.[28]

7.2.2.2 *Kraft Lignin*

Kraft pulping is the main industrial process for chemical conversion of wood into cellulosic pulp for paper and paperboard products, and for cellulose derivatives (viscose, rayon, cellulose acetate, carboxymethylcellulose and others). Kraft is an alkaline process that uses sodium hydroxide (NaOH) and sodium sulfide (Na_2S) in the pulping liquor to fragment and solubilize lignin. The process is carried out between 150–180 °C for 1–5 h depending on the final kappa number of the pulp (which represents the residual lignin amount).[20] Bleaching operations are further performed to remove the residual lignin with the objective to improve pulp brightness and economic value. In the Kraft process, the yield of bleachable grade pulp is approximately 45–55% depending on the chemical characteristics of the wood used. Kraft pulp is made up of more than 85% cellulose; most of the hemicelluloses and lignin are dissolved during processing and end up in the black liquor. The black liquor, which contains up to 95% of the original lignin in wood, is then concentrated to increase solid content and burned for energy, and inorganic chemical recovery. However, if a part of the lignin is extracted, the heat load of the recovery boiler will decrease, which may be beneficial for pulp mills if the capacity of the recovery boiler limits the overall production rate of pulp, and can be a step toward a biorefinery generating products from lignin.

Lignin can be recovered from Kraft black liquors by precipitation after acidification of the medium to pH 3–4. To enhance industrial lignin recover, a process called "Lignoboost" was developed to extract lignin from black liquor based on bubbling carbon dioxide to obtain a precipitate that is recovered by filtration, washing and drying.[29] This process can be integrated into a pulp mill and the lignin used to produce several valuable green biobased products.[30–32] Results of the elemental analysis show that the purified lignin contains organic elements up to 97%.[31] Zhu[29] showed that the molar mass of recovered lignin can vary according to the pH and temperature applied during recovery, and can have Mw values ranging from 10 000 to 15 000 Da, and Mn from 4000 to 8000 Da. A variation in the amount of methoxyl groups in the precipitated lignin was also observed and its content decreased with an increase in the molecular weight of the lignin. Phenolic groups are also important for lignin reactivity and presented values between 0.8 and 1.2 OH/aromatic ring. Like the lignosulfonates, Kraft lignins have also been used as dispersants and coagulants.[33]

7.2.2.3 Organosolv Lignin

Organosolv pulping was developed as an alternative to conventional chemical delignification processes. Semi-commercial and pilot plants were built in North America and Europe in the 1990's. Despite good pulp quality and bleachability, the processes failed to achieve economical sustainability for large-scale operations.[34] Organosolv processes use organic solvents such as alcohols, ketones, and organic acids, as delignifying agents. In most of the organosolv processes, a mixture of water and an organic solvent is used for the delignification of the lignocellulosic material to produce a pulp enriched in cellulose. Organosolv delignification can be noncatalyzed or catalyzed with mineral or organic acids. Noncatalyzed processes are used for hardwoods and herbaceous or agricultural residues owing to the higher amount of acetylated xylans in these feedstocks, from which acetyl groups are released as acetic acid, autocatalyzing the reaction. Acid-catalyzed organosolv processes are applied to softwoods with good results regarding delignification and pulp properties.[35]

Recently, organosolv processes have gained new importance as an alternative for fractionation of lignocellulosic biomass in a biorefinery concept, due to the economic necessity for multiple co-products and particularly a high-quality lignin fraction with potential for several industrial applications.[36] The organosolv process known as Alcell®, developed further by Lignol Innovations, Inc., uses ethanol–water solutions for delignification and is one of the most studied organosolv pretreatment processes for biomass fractionation, mainly for cellulosic ethanol production. The obtained cellulosic pulp is treated with cellulases and converted into glucose for bioethanol by fermentation, while lignin, hemicelluloses, acetic acid, furfural and other co-products can be recovered and used for different purposes.[36,37] Lignin is recovered as a fine precipitate by flashing the pulping liquor to atmospheric pressure. In contrast to lignin produced by other chemical processes, the organosolv lignin is sulfur-free and of low molecular weight.[38,39] The organosolv lignin process mainly attacks the α-ether linkages and thus retains relatively larger amounts of β-O-4 linkages.[40] Organosolv lignin is supposed to contain relatively more oxidized functionalities, compared to alkaline lignin. High-quality lignin can be used as a substitute for polymeric materials, such as phenolic powder resins, polyurethane foams, epoxy resins and phenol derivatives, as described later in this review.

7.2.2.4 *Autohydrolysis and Steam-explosion Lignins*

Hydrothermal processes used for the fractionation of lignocellulosic biomass are primarily based on steam or aqueous treatments performed at temperatures between 150–220 °C, which separates the biomass into a liquid stream composed mainly of hemicelluloses, oligomers and a small fraction of water-soluble lignin, and a fibrous pulp composed of cellulose and the remaining lignin.[41–43] Steam explosion was based on the Masonite process[44] developed for the production of fiberboards and later applied as a pretreatment in the biochemical platform for bioethanol production. In the steam-aqueous treatment of lignocellulosics, with or without explosive decompression, hydrolysis is catalyzed by the acids formed from wood during the heating in aqueous media (autohydrolysis), or by mineral acids or SO_2 added to the system.[45,46]

Lignin is depolymerized by acid hydrolysis through the breaking of the aryl–ether bonds and also by cleaving lignin–carbohydrate linkages within the hemicelluloses.[22,47] A small fraction of the depolymerized lignin is soluble in the aqueous media; most of it remains in the pulp. Some of the residual lignin co-condenses with monosaccharides or degradation products such as furanic compounds, forming spherical aggregates commonly referred to as pseudolignin that can migrate to the pulp surface or cell lumen.[48] Lignin can be partially removed from pulp by alkaline extraction with NaOH or organic solvents for further uses.[49,50] Leschinsky *et al.*[50,51] studied the structural characteristics of lignin from *Eucalyptus globulus* after autohydrolysis. They found that extensive lignin degradation occurs during autohydrolysis through homolytic cleavage of the aryl–ether bonds, resulting in low molar mass lignin in both treated wood and hydrolysate. The aryl–ether cleavage is coupled with a strong increase in phenolic hydroxyl groups, a decrease in aliphatic hydroxyl groups and, a reduced amount of β-O-4 bonds.

Li *et al.*[52] studied the characteristics of lignin from softwoods and hardwoods after SO_2-catalyzed steam explosion. Size exclusion chromatography (SEC) revealed that substantial degradation of lignin took place during steam explosion. Isolated lignins had a molecular size of 3000 Da, lower than wood lignin from spruce (6155 Da) or aspen (4580 Da). Methoxyl group concentration in softwood lignins recovered from the steam-explosion pulps and milled wood lignin were between 0.84–0.97/C_9 units, while for hardwoods it was between 0.95–1.48/C_9 units. Phenolic groups varied over a wide range for all samples, from 0.45 to 2.66 mmol g^{-1} lignin.[52]

7.3 Lignin Depolymerization

While there are multiple pathways for the catalytic processing of lignin, the two most favorable are detailed in Figure 7.7. One pathway degrades the

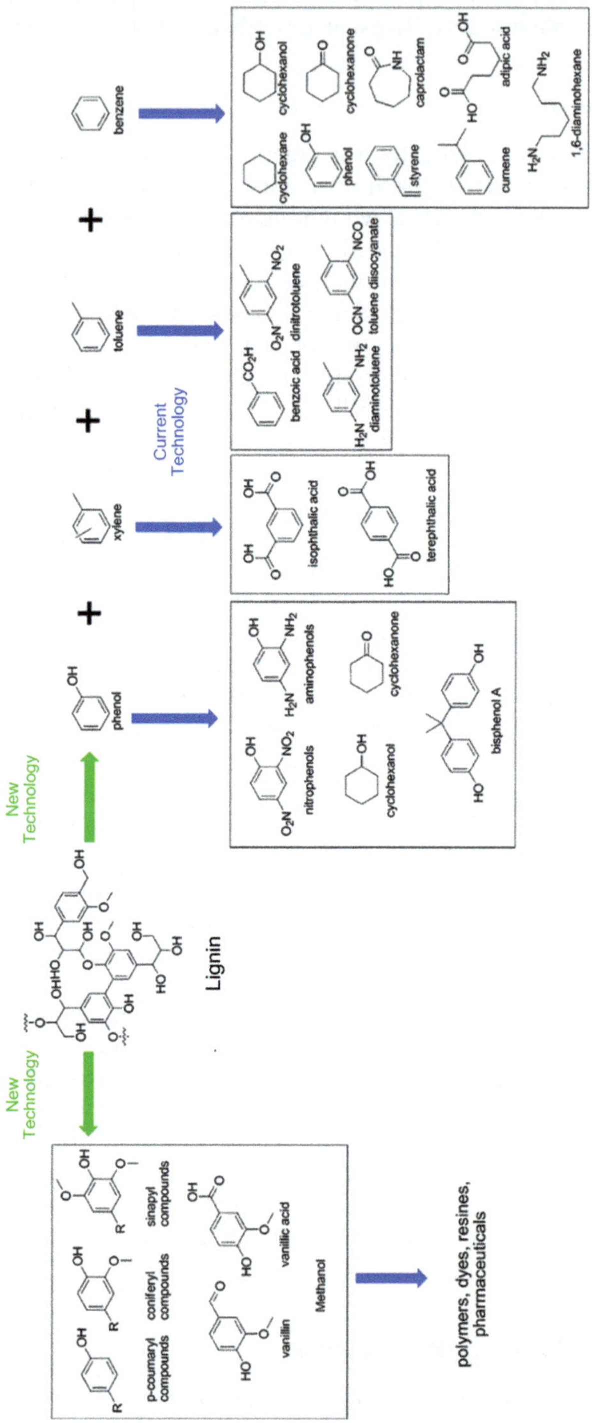

Figure 7.7 Two potential pathways for catalytic lignin degradation (adapted with permission from ref. 53. Copyright (2010) American Chemical Society.)

lignin into simple aromatic compounds that serve as platform chemicals for other chemical commodities and fuels. The other pathway directly produces valuable chemicals and fuels from lignin in a one step process; however, such a conversion of lignin would require a highly selective enzyme or catalyst. The main obstacle to direct lignin depolymerization into useful products is the complex structure of lignin, which is not entirely known because it varies with biomass species, growing conditions and extraction treatments. Enzymatic degradation of lignin is not practical currently, since desired selectivity and conversion have not been achieved, and the techniques used are impractical for large-scale processing and not economically feasible. Currently, catalytic processes require multiple steps including pretreatment steps, and have high energy demands due to the high temperatures required to operate the current catalytic systems.

7.3.1 Catalytic Depolymerization

Considering the unique structure and chemistry of lignin, there is a great potential for catalytically converting lignin into a wide range of chemicals *via* either catalytic hydrolysis, catalytic reduction or catalytic oxidation reactions.[53–56] Catalytic oxidation of lignin allows for the synthesis of chemicals with a broad range of functionalities that can directly be used as fine chemicals or can be processed further into secondary chemicals. For example, Wu *et al.*[56] have shown that steam-exploded hardwood lignin oxidized using oxygen at 170 °C formed a mixture consisting of syringaldehyde, hydroxybenzaldehydes, and vanillic aldehyde in the presence of copper and iron catalysts.

Fargues *et al.*[55] obtained about 11% vanillin using lignin derived from Pine (*Pinus* sp.). Similar results were reported by Villar *et al.*[57] wherein it was demonstrated that hardwood lignin was converted into syringaldehyde, vanillin, and their corresponding acids *via* copper and cobalt catalysts at 150 °C. Sales *et al.*[58,59] evaluated catalytic wet oxidation (100–140 °C and 2–10 atm) of sugar cane bagasse into aromatic aldehydes using alumina-impregnated palladium catalysts. The authors noted that palladium catalysts selectively oxidized lignin into aldehydes while minimizing the formation of undesirable end products. In a separate study by Crestini *et al.*,[60] methyltrioxorhenium was evaluated as a catalyst for oxidizing model lignin compounds and lignin obtained from sugar cane, red spruce, and hardwood using hydrogen peroxide as an oxidant. They found that 95% of the model compound (vanillyl alcohol) was oxidized into vanillin, 4-hydroxy-3 methoxy-benzoic acid, and other chemicals at room temperature. Bhargava *et al.*[61] screened various metals as catalysts for the catalytic wet oxidation of lignin (ferulic acid) and concluded that copper/manganese/alumina exhibited optimum activity and stability. Zhang *et al.*[62] also investigated catalytic oxidation of lignin derived from corn stalks, using Perovskite-type oxide catalysts. The authors reported significantly enhanced yields of aromatic aldehydes in the presence of catalysts at 120 °C.

Catalytic reduction typically uses a metal hydrogenation catalyst, such as platinum, palladium, ruthenium, rhodium, osmium, gallium, copper or nickel, to perform hydrogenolysis and hydrodeoxygenation reactions. Many of the studies using these supported hydrogenation catalysts make use of lignin model compounds, so to narrow the scope, the following examples are looking at more recent studies using lignin/processed lignin feedstocks. Parsell *et al.*[63] used a milled poplar washed with water and ethanol *via* soxhlet extraction for a feedstock with a Pd–Zn/C (Pd nanoparticles, Zn^{II} sites and carbon support) catalyst system. The reaction was performed in a methanol solvent with 500 psig H_2 gas at 225 °C. Based on the lignin content of the biomass, 40–54% yield of propylguaiacol and propylsyringol monomers was achieved. Song *et al.*[64] made use of a Ni/C (nickel on active carbon support) to convert extracted lignin from birch sawdust (soxhlet extraction using an ethanol and benzene mixture). Using a methanol solvent (where the alcohol solvent provides the active hydrogen species) at 200 °C, a 50% conversion of the lignin was achieved with 97% selectivity to monomeric phenols (mainly propylguaiacol and propylsyringol). Singh *et al.*[65] used a Ni/HZSM-5 (nickel on acid zeolite support) catalyst system to convert Kraft lignin purified from black liquor. The reaction was performed in a mixture of 1 : 1 water–methanol with NaOH at 220 °C and 96.8–98.5% lignin conversion was obtained (~16% water soluble portion, ~40% ethyl acetate soluble portion, ~40% gas products, <1% char) with a 15.4% yield of alkyl phenols (from the water soluble portion). The basic environment showed a synergistic effect with the catalyst system. Again the alcohol solvent was used to provide hydrogen for the metal hydrogenation catalyst instead of using gaseous hydrogen.

Zhao and Lercher[66] used Ni/HZSM-5 at 250 °C, with 5 MPa H_2 gas in a water-only solvent to convert a pyrolysis oil from pine (extracted with *n*-hexanes). Hydrodeoxygenation of the furans, alcohols, ketones, and phenols present in the pyrolysis oil mixture resulted in a mixture of gasoline range hydrocarbons free of oxygen. This reaction system was also used on lignin model compounds with success in complete deoxygenation of the models (monomers and dimers), but a mixture of saturated and aromatic rings. This reaction system was not applied to a native lignin.

Zhang *et al.*[67] and Konnerth *et al.*[68] both worked with organosolv lignin and a Ni_7/Au_3 bimetallic system, which is a colloidal catalyst system instead of the typical supported metal catalyst system, along with 10 bar H_2 gas in water solvent. Performing the reaction at 170 °C for 1 to 12 h resulted in 4% to 14% monomer yield, mainly guaiacyl and syringyl propanols, respectively.[67] A similar experiment performed for 4 h at 130–160 °C, but with the addition of NaOH to the system, resulted in 10.9% monomer yield, again mainly guaiacyl and syringyl propanols.[68] The addition of the base promoter increased the amount of aromatic monomer yield for both organosolv lignin and lignin model compound reactions, and reduced the amount of undesired hydrogenation/saturation.[68] Another important note is the use of water solvent (instead of organic or organic–water mixture solvent) with lignin feedstock. The use of only water is not only beneficial with regards to cost (no need to remove water/moisture from lignin feedstocks), but also from an environmental standpoint since water is an environmentally friendly solvent.

The main obstacle for the catalytically reductive pathway that needs to be overcome for effective lignin depolymerisation is the solubility/access of the lignin to the catalyst system. The examples given make use of different methods including solvent mixtures (methanol and water) and addition of base, which typically increases the solubility of lignin in water (in addition to the synergistic catalytic effects of the base promoter). In the case of the last examples, a colloidal catalyst system was used as an attempt to allow better access of the catalyst to the insoluble lignin present in the reaction system.

7.3.2 Biological Depolymerization

Lignin is highly resistant to degradation. Its phenolic and non-phenolic regions have variable resistance to degradation; the non-phenolic regions comprise the majority of the structure (90%) and exhibit greater resistance to degradation. White-rot fungi are the most efficient biological lignin degraders and the only ones capable of complete mineralization, utilizing combinations of lignin peroxidase (LiP), manganese peroxidase (MnP), and laccase.

Lignin peroxidase (LiP) catalyzes the H_2O_2-dependent oxidative depolymerization of lignin. It has a high redox potential, up to 1.4 V, in the presence of H_2O_2, so that it oxidizes phenolic and non-phenolic lignin and also a range of aromatic compounds. Figure 7.8 (adapted from Wong[69]) shows LiP-catalyzed oxidation of the non-phenolic β-O-4 lignin model compound.

Manganese peroxidase (MnP) catalyzes the Mn-dependent reaction: $2Mn(\text{II}) + 2H^+ + H_2O_2 = 2Mn(\text{III}) + 2H_2O$. It oxidizes Mn(II) to Mn(III), which in turn oxidizes phenolic lignin (Figure 7.9, taken from Tuor *et al.*[70]). To degrade non-phenolic lignin, Mn(III) needs a second mediator such as oxalate and malonate to form M(III) chelators at high redox potentials (0.8–0.9 V).

Laccases belong to a family of multicopper enzymes. Laccase has a redox potential of 0.78–0.80 V, which is lower than MnP and LiP. Nevertheless, it can still oxidize phenolic lignin. With mediators such as 2,2′-azinobis-(3-ethylbenzthiazoline-6-sulfonate) (ABTS) and 1-hydroxybenzotriazole (HBT), produced by the fungi, laccase can also degrade non-phenolic lignin.[71] Geng and co-workers also reported that *P. cinnabarinus*, which produces only laccase, degraded more than 40% of the non-phenolic lignin and 70% of the phenolic lignin.[72]

In addition to traditional enzyme systems identified in fungal systems, strategies involving oxidative delignification have been developed recently. Wiermans *et al.*[73] report the use of lipases in a peracid-mediated lignin depolymerization process, utilizing dimethyl carbonate (DMC) as the solvent and an acyl-donor reagent, hydrogen peroxide, and lipase. The lipase

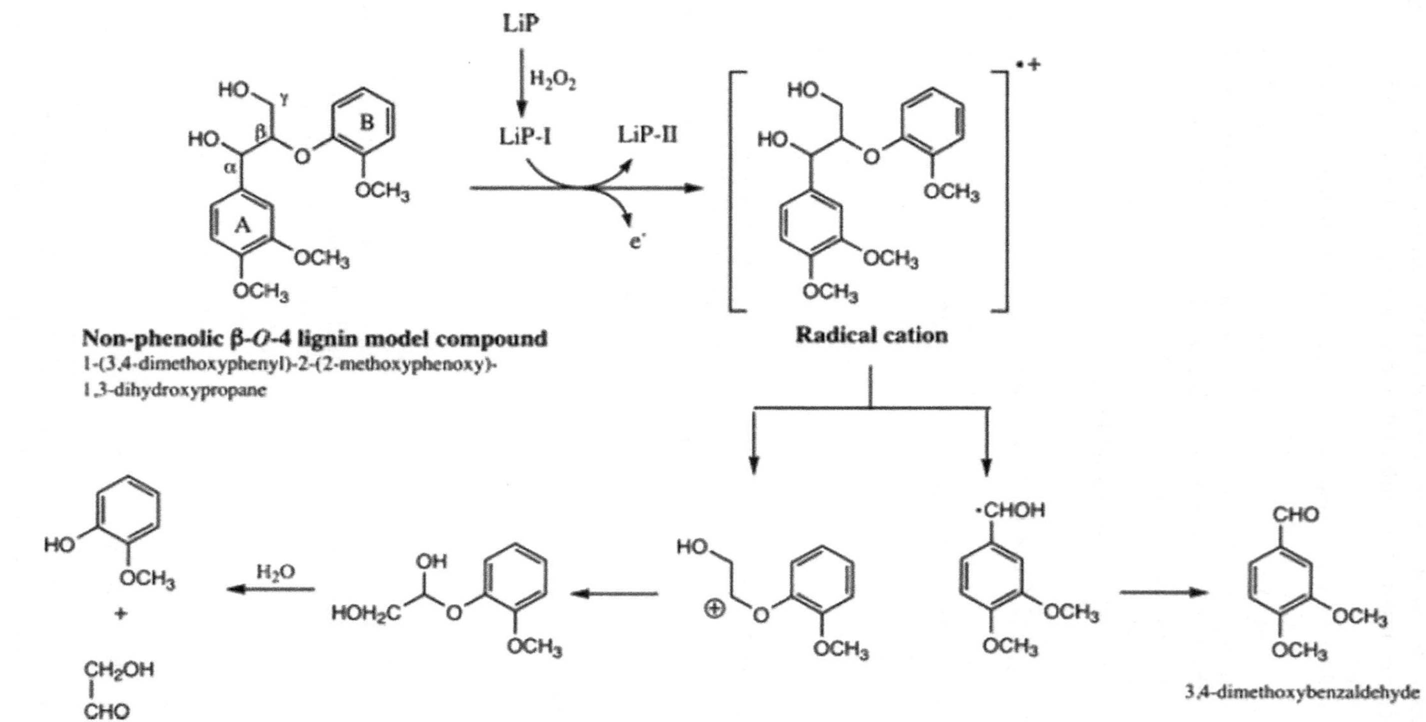

Figure 7.8 LiP-catalyzed oxidation of a non-phenolic β-O-4 lignin model compound (adapted from ref. 69, with kind permission from Springer Science and Business Media).

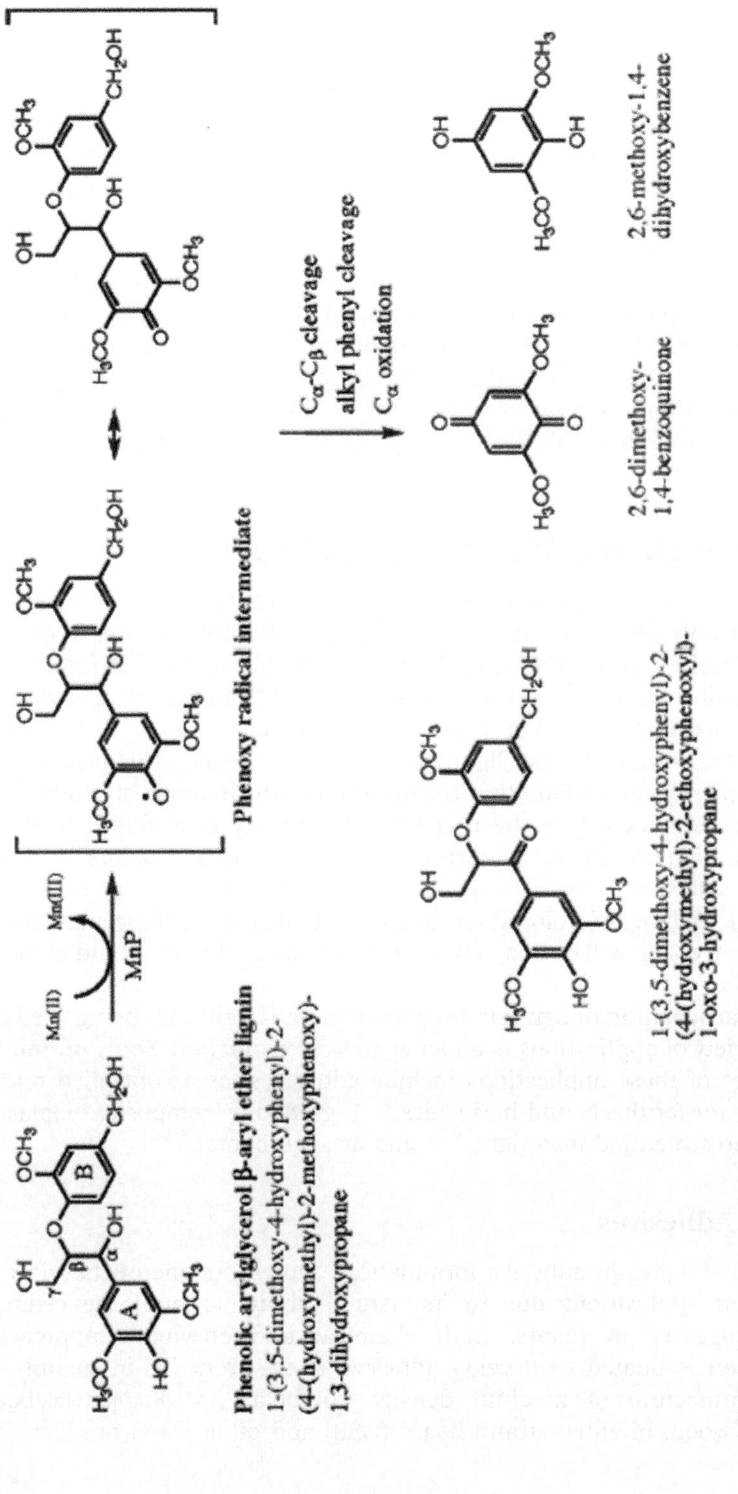

Figure 7.9 MnP-catalyzed oxidation of phenolic lignin compound (reprinted with permission from ref. 70. Copyright (1992) American Chemical Society).

catalyzed the formation of a peracid from DMC and H_2O_2. Beachwood was dissolved in the mixture and digestion was performed over 9 h at 80 °C. This treatment generated a non-aromatic "lignin oil" with yields from 50% to 60%.

Co-immobilization of laccase and horseradish peroxidase (HRP) has also been evaluated as a means to effect oxidative lignin modification.[74] Laccase and HRP were immobilized onto 2 mm alumina particles, silanized, and activated by glutaraldehyde treatment. They were then coated sequentially with polyallyl amine hydrochloride (PAA) and polystyrene sulfonate. Wheat milled lignin (WL) was treated with combinations of laccase and HRP immobilized separately or coimmobilized, in the presence and absence of 1-hydroxybenzotriazole (HBT; an oxidation mediator), in acetate buffer containing H_2O_2. Results indicate that polymerization and *exo*-depolymerization reactions were occurring for the co-immobilized multienzyme system, generating insoluble high Mw and Mn species, and oligomeric, low Mw and Mn soluble material.

7.4 Functional Materials from Lignin

It is estimated that approximately 50 million tons of lignin are generated per year, but only 2% has been commercialized for the formulation of dispersants, adhesives, and surfactants.[75] Another possible application for lignin is as a starting material for some bulk chemicals (vanillin, syringaldehyde, toluene benzene and other phenolic compounds).[76] Some of the major reasons that hinder a wider lignin application for added value products are the complexity and variability of the macromolecule in terms of composition and structure caused by the existence of a variety of different inter-unit linkages, high affinity for the formation of a condensed structure, and poor product selectivity. However, lignin transformation plays a key role in the economic viability of a biorefinery and the establishment of mechanisms for lignin conversion will open a new route for aromatic chemicals and biobased materials.

The participation of lignin in functional materials that are being used in a wide variety of applications is under aggressive evaluation. Some prominent examples of these applications include adhesives, as a controlled release polymer for fertilizers and herbicides,[77–79] grafted or composite bioplastics and nano-structured materials,[79–87] and as adsorbents.[88–90]

7.4.1 Adhesives

The use of lignin in adhesive formulations represents one of the most investigated applications due to its natural ability to bind the cellulose fibers together in plants. Both chemical and enzymatic approaches have been evaluated to develop adhesive resins from lignin, mainly for the manufacture of medium density fiberboard (MDF), particleboard (PB), plywood, oriented strand board (OSB) and other wood products.[91,92]

Phenol-formaldehyde resins (PF) are widely used for producing different types of boards for interior and exterior uses. However, this kind of phenol-based resin has a high impact on the production cost mainly due to the fluctuations of petroleum prices, while formaldehyde is considered as a major indoor pollutant and harmful to human health.

Due to its phenolic composition, lignin has been considered as a logical phenol replacement compound. However, its use in adhesive formulations is dependent on its structural characteristics and reactivity. The simplest way to use lignin is as a partial replacement of phenol in PF resins.[93] Lignins from different origins and in various percentages have been evaluated.[94–98] Early work by Tejado[95] using three lignins determined that the degree of activation (phenolic groups or other reactive substituents) of the aromatic rings in the processed lignin was the primary determinant of effectiveness of lignin-phenol-formaldehyde (LPF) resins.

The degree of variability of LPF resin properties was elucidated recently by Zhang *et al.*[94] They substituted one of four biorefinery residues (from ethanol, butanol, xylitol and lactic acid biorefineries; denoted as ER, BR, XR and LR, respectively) for a fraction of the phenol in residue-phenol-formaldehyde (RPF) resins. Table 7.1 indicates the major performance metrics for the resins and for plywood manufactured using these resins.

The performance of RPF resins differs significantly from that of the PF resin. Viscosity increased, likely due to the higher MW of the lignin relative to phenol, free formaldehyde levels were higher, as were formaldehyde emissions, and bonding strength decreased. The latter result is easily explained: on a mass basis, lignin has fewer reactive hydroxyl groups than phenol. Within the lignin RPF resins, it was expected that the bond strength would be strongly correlated with the relative hydroxyl group content. Instead, the correlation was strongest with the ash content of the residues, a confounding result.

A review of the strategies to increase the reactivity of lignin toward formaldehyde[99] identified two major activities: demethylation and phenolation. Sulfur compounds are used to remove methyl groups from the aromatic rings in lignin, increasing its reactivity and generating DMSO.

Alonso *et al.*[100] demonstrated that phenol groups can be attached to aromatic rings in lignin. This increases the *ortho* and *para* reactive sites on the grafted phenol groups. Utilizing this basic approach, substituted phenols have been attached to lignin, providing it with new properties.[101] Qiao *et al.*[102] recently combined and extended the work of Zhang[94] and Alonso.[100] They selectively extracted lignin using furfuryl alcohol from cornstalk that had been steam exploded and enzymatically digested.[103] The lignin was then phenolated in an alkaline environment at 90 °C. This lignin then replaced a fraction of the phenol in PF resins. This treatment regimen (enzymatic hydrolysis followed by phenolation) gave superior bonding strength in 30% and 50% substitution levels with phenol, with strength essentially indistinguishable from commercial PF resins.[102]

Yuan *et al.*[104] sought to improve the performance of lignin-based adhesives through the use of the chemical binding agents modified

Table 7.1 Lignin-substituted adhesive performance (adapted from ref. 94 with permission. Copyright © 2012 Elsevier B.V.).

	Mass ratio	Resin performance					Plywood performance	
Adhesive	Phenol – residue	pH	Viscosity (mPas)	Solids (%)	% Free formaldehyde	% Free phenol	Bonding strength (MPa)	Formaldehyde emissions (mg L^{-1})
PF	100 : 0	12.1	100	48.9	0.10	0.65	1.65 ± 0.21	0.13 ± 0.01
ERPF 50	50 : 50	11.6	160	49.5	0.47	0.26	1.31 ± 0.22	0.11 ± 0.04
BRPF 50	50 : 50	10.9	300	48.1	0.54	0.49	0.88 ± 0.12	0.23 ± 0.02
XRPF 50	50 : 50	11.1	1080	47.9	0.76	0.49	0.84 ± 0.09	0.32 ± 0.03
LRPF 50	50 : 50	11.8	>20 000	46.6	1.10	0.74	1.52 ± 0.20	0.14 ± 0.05

ammonium lignosulfonate (MAL) and polyethylenimine (PEI). The MAL/PEI composites exhibited higher storage modulus and tan δ than the unmodified ammonium lignosulfonate (UMAL)/PEI composites, indicating improved bonding strength characteristics.

7.4.2 Copolymers or Polymer Additives

Olefins constitute the vast majority of commodity plastics; examples include polyethylene (PE) and polypropylene (PP). Their low density limits their strength, stiffness and dimensional stability, disqualifying them for load-bearing applications.[105] Given that composites of olefins that exhibit elevated strength would be of value, lignin has been considered as a potential polymer additive.

Previous studies investigating the use of lignin as a composite reinforcement[106–113] have primarily used lignin from pulp and paper processing—softwood Kraft lignin and hardwood Kraft lignin or organosolv lignin. The key benefits of organosolv lignin over sulfonated and Kraft lignin include high purity, low molecular weight, and a greater ability to be derivatized (*i.e.* larger fraction of aromatics unsubstituted in the reactive *ortho*-position).[109]

The use of pulp and paper derived lignins in polymeric systems has been investigated for several decades. Challenges that remain for these materials include oil absorption, color retention, scratch resistance, and biodegradation. An early study[114] compared biocomposite panels with popular wood-based products—plywood, oriented strand board, particle board, standard hardboard, and medium density fiberboard. Wood-filled thermoplastic composites performed similarly to conventional wood products in the areas of tensile strength, hardness, and shear and are more moisture resistant. Biocomposite panel performance was inferior to wood products for bending modulus of elasticity, bending modulus of rupture, specific gravity, and coefficient of thermal expansion. Organosolv lignin and derivatives are ideal for coating because they have a small particle size, are hydrophobic, and have the ability to penetrate cellulose surfaces.[109]

Five lignin co-products (resulting from the Kraft, lignosulfate, soda-anthraquinone, organosolv, and steam explosion with simultaneous saccharification and fermentation pretreatment processes) were incorporated into phenolic resins. The variation in the weight percent of the phenolic and aliphatic hydroxyl groups was shown to play a critical role in compatibilization and adhesion.[115] It was also demonstrated that the total amount of lignin and the ratio between acid insoluble and acid soluble lignin was a function of the pretreatment process. Feldman[116] reviewed the use of a variety of lignins obtained from various processes (Kraft lignin, lignosulfonates, organosolv lignin, hydrolytic lignin, steam exploded lignin, and hydroxypropyl lignin) in several polymer matrices (phenol-formaldehyde, epoxy, poly(dimethyl siloxane), low and high density polyethylene, polypropylene, poly(ethylene-*co*-vinyl acetate), poly(vinyl chloride), and a variety of rubbers). Feldman concluded that the primary challenge was that the processes to isolate the lignin decreased its chemical reactivity.

Polyvinyl chloride (PVC) and vinyl-chloride-vinyl-acetic acid (VC-VAc) copolymers are often stiff and rigid. These physical characteristics can be altered by addition of plasticizers.[117] However, plasticizers are susceptible to microbial attacks; fungi produce enzymes that are capable of breaking down the ester linkages in most plasticizers.[117] Lignin is resistant to fungal degradation. Using lignin as a plasticizer in PVC, it was found that plasticization with the lignin was enhanced by the addition of VC-Vac.[117] Using prehydrolysis lignin in PP films, Kosikova[118] found from rheological data that lignin addition up to 10% w/w had a good compatibility with PP and concluded that these UV absorbing films could potentially be used in the food packing industry.

Recently, industrial lignins (Kraft, Lignoboost, Indulin AT, wheat straw Soda, and pyrolytic) were evaluated for their suitability in composite production.[119] The conclusions of this work were wide-ranging:

> *"The thermal properties of all studied lignins seem to be convenient...therefore, lignins could be used in biocomposites based on polyolefins.... Kraft lignin seems to be on the whole, the most convenient lignin for biocomposites application.... However, due to their polar nature..., lignins might need to be modified to improve their compatibility with polyolefins. One way to modify lignins is the esterification for which the higher hydroxyl content of lignins is favorable, which is the case of Kraft lignins.... the high content of p-hydroxyphenyl units in pyrolytic and particularly in Soda Protobind 2400 lignins would be advantageous for resin and adhesive applications."*[119]

7.4.3 Foams and Gels

Lignin is also a focus of research surrounding the production of biodegradable hydrogels. Hydrogels have the intrinsic ability to swell in water/aqueous solvents up to thousands of times their dry weight in water without being dissolved.[120] Biopolymeric hydrogels are well-suited to biological applications,[121,122] are relatively easily degraded, and exhibit tunable properties.[123] The polymer networks formed in hydrogels are generally cross-linked and are rich in hydrophilic domains.[124] The polymer composition of the hydrogel has a significant effect on the swelling performance and strength of the gel. Common polymers used in hydrogels include poly(2-hydroxyethyl methacrylate) (PHEMA), poly(ethyl methacrylate) (PEMA), polyacrylamide (PAAm), poly(methacrylic acid) (PMA), poly(acrylic acid) (PAA), poly(glucosylethyl methacrylate) (PGEMA) and poly(hydroxypropyl methacrylamide) (PHPMA).[125] The inclusion of hydrophobic regions in the polymer network can enhance the stability of the resulting hydrogel. The potential of lignin to replace these synthetic polymers provides a degree of biocompatibility that makes modified lignin a highly desirable component of hydrogels.[126,127]

Alkaline (AL) and Kraft lignins (KL) have been used in the preparation of hydrogels. The lignins were involved in graft polymerization reactions with acrylamide and polyvinyl alcohol (PVA).[127] The AL hydrogels exhibited higher swelling ratios and slower water uptake rates than KL hydrogels, indicating the influence of lignin composition on gel performance. Subsequent studies utilizing black liquor (mixture of lignin, cellulose, and hemicellulose) to produce hydrogels following graft copolymerization indicate that this is also an effective hydrogel precursor.[128a]

Temperature-sensitive gels have also been produced using lignin. Poly-(*N*-isopropylacrylamide) (PNIPAAm) hydrogels exhibit a sharp transition temperature that is close to body temperature.[128b] Copolymerization of hydrophobic or hydrophilic monomers with *N*-isopropylacrylamide (NIPAAm) can result in improved properties. Feng and coworkers[129] copolymerized acetic acid lignin (AAL; precipitated from acetic acid pulping waste liquor) and NIPAAm in the presence of *N*,*N'*-methylenebisacrylamide (MBAAm) as the crosslinker and H_2O_2 as the initiator. The resulting hydrogel exhibited a lower critical solution temperature of approximately 31 °C. The copolymer contained pore sizes between 20 and 100 μm that were tunable by altering the lignin concentration.

Porous polymeric materials (macroporous foams) are also used in a number of fields, including as insulation, in adsorption and separation processes,[130] and as layers in capacitors.[131] Aerogels and foams are the two major forms of macroporous polymeric material. Aerogels are generally formed through sol–gel condensation of thermosetting resins,[132] while foams are generally formed by carbonization under inert atmospheres at high temperature and pressure.[133] Lignin, with its large number of phenolic units, reacts readily with formaldehyde and can be used as a component in the production of thermosetting resins. One example of this was work by Seo and co-workers,[134] who synthesized foams using lignin-resorcinol-formaldehyde (LRF) networks with poly(methyl methacrylate) (PMMA) particles. Results indicated that pore structure can be effectively controlled by the amount of PMMA, that lignin addition improved the thermal stability, and that carbon yield was greater with lignin-substituted foam precursors relative to those formed using other biomaterials.[134]

7.4.4 Carbon Fibers

Carbon fibers (CF) exhibit a unique combination of properties (lightweight, have high strength, flexibility and fatigue resistance) that make them suitable for a wide range of applications. Polyacrylonitrile (PAN) and regenerated cellulose are the primary precursors of current carbon fiber production.[135] Carbon fiber manufacturing involves melt or wet-spinning, oxidative stabilization and carbonization under an inert atmosphere, followed by surface treatment and sizing.[136]

Because of its low cost, lignin has drawn significant interest as a raw material for the production of carbon fibers, with a potential application in passenger vehicles.[137–141] The primary stumbling block to such applications is the difficulty in generating high quality fibers from such a complex, heterogeneous material.[142,143]

Different lignins have been evaluated for the production of CF. Kadla *et al.*[138] used Kraft lignin without chemical modification, the first report of lignin-based carbon fiber production utilizing a commercially-available lignin. They removed carbohydrates and inorganic compounds from lignin and decreased the hydroxyl content by condensing the lignin.

Fiber spinning has been facilitated by the addition of poly(ethylene oxide) PEO (3–5%) to lignin.[138,144] Microfibers of Kraft lignin blended with poly(ethylene oxide) (PEO) produced by electrospinning a solution of lignin and high molecular weight poly(ethylene oxide) (PEO) in alkaline water and *N*,*N*-dimethylformamide (DMF) with different polymer concentrations were compared.[144] Fibers spun from aqueous solutions had lower PEO content and were found to have larger diameters, both characteristics that enhanced CF formation.

Additional studies of electrospun fibers produced from lignin indicate a promising performance. Jin *et al.*[145] produced carbon nanofibers from PAN and refined lignin (RL) extracted from hardwood lignosulfonate and used the material to fabricate electrodes. These electrodes exhibited good overall performance, including a high reversible capacity, high initial efficiency, good rate capability and excellent cycling stability. Schreiber *et al.*[146] utilized a reactive electrospinning process to create sodium carbonate lignin-chitosan (SCL-CS) polyelectrolyte fibers. PEO was introduced to enhance the spinnability of the solution, and subsequently removed *via* washing. Polyelectrolyte complex (PEC) fibers were successfully produced. A 4 : 3 SCL–CS ratio produced the best fiber morphology; PEO and sodium ions present in the fibers were removed by soaking the fibers in water, which had the added benefit of increasing the thermal stability of the fibers.

In a separate study, Schreiber *et al.*[147] evaluated cellulose acetate-lignin (CA-LL) fibers spun from solutions of CA, hardwood organosolv lignin, acetone and *N*,*N*-dimethylacetate. To maintain fiber morphology during carbonization, the fibers were treated with iodine vapor. Electropsun carbon fibers were produced from blends of lignin and CA up to a LL–CA acetate ratio of 4 : 1. The iodine treatments significantly enhanced the morphology retention of lignin and the blend fibers. Overall, the lignin-cellulose acetate fibers exhibited good thermal and morphological stability, showing significant promise toward improving the production of inexpensive and green carbon fibers.

Nordström *et al.*[142] successfully utilized softwood Kraft lignin (SKL) and hardwood Kraft lignin (HKL) in a melt spinning process after developing softening agents based on permeate from the ultra-filtration of black liquor from hardwood or softwood Kraft lignin (HKLP, SKLP). Spinning of SKL softened with HKLP, or of SKLP directly, resulted in solid and smooth carbon fibers, representing a breakthrough in the ability to use SKL for CF production.

7.5 Final Thoughts

Given its aromatic structure, as well as the high density of phenolic hydroxyl groups, lignin is an attractive starting material for the production of a wide range of fine chemicals and high-performance materials. Its intrinsic chemistry presents a tableau for a variety of chemical modifications, once the degree of condensation of processed lignins can be reliably reduced. It is also economically attractive as a waste byproduct of wood pulping whose current utility is as a boiler fuel. Research efforts are going beyond model compounds, and significant progress has been achieved. The recent upswing in research efforts involving heterogeneous catalysts, and mixed enzyme systems, offers great promise for the development of commercially viable lignin conversion processes.

References

1. J. Conti, P. Holtberg, S. Napolitano and A. M. Schaal (2014) International Energy Outlook 2014 (US Energy Information Administration), (DOE).
2. A. J. Ragauskas, G. T. Beckham, M. J. Biddy, R. Chandra, F. Chen, M. F. Davis, B. H. Davison, R. A. Dixon, P. Gilna, M. Keller, P. Langan, A. K. Naskar, J. N. Saddler, T. J. Tschaplinski, G. A. Tuskan and C. E. Wyman, Lignin Valorization: Improving Lignin Processing in the Biorefinery, *Science*, 2014, **344**, 6185.
3. P. Sannigrahi and A. J. Ragauskas, Characterization of Fermentation Residues from the Production of Bio-Ethanol from Lignocellulosic Feedstocks, *J. Biobased Mater. Bioenergy*, 2011, **5**(4), 514–519.
4. W. Boerjan, J. Ralph and M. Baucher, Lignin Biosynthesis, *Annu. Rev. Plant Biol.*, 2003, **54**, 519–546.
5. M. B. Kleinert, Towards a Lignincellulosic Biorefinery: Direct One-Step Conversion of Lignin to Hydrogen-Enriched Biofuel, *Energy Fuels*, 2008, **22**(2), 1371–1379.
6. A. J. Ragauskas, C. K. Williams, B. H. Davison, G. Britovsek, J. Cairney, C. A. Eckert, W. J. Frederick, J. P. Hallett, D. J. Leak and C. L. Liotta, The path forward for biofuels and biomaterials, *Science*, 2006, **311**, 484–489.
7. B. A. Simmons, D. Logue and J. Ralph, Advances in modifying lignin for enhanced biofuel production, *Curr. Opin. Plant Biol.*, 2010, **13**(3), 313–320.
8. G. Henriksson, Lignin, in *Wood Chemistry and Biotechnology, Pulp and Paper Chemistry and Technology*, ed. M. Ek, G. Gellerstedt and G. Henriksson, De Gruyter, Berlin, 2009, vol. 1.
9. F. Chen, Y. Tobimatsu, D. Havkin-Frenkel, R. A. Dixon and J. Ralph, A polymer of caffeyl alcohol in plant seeds, *Proc. Natl. Acad. Sci. U. S. A.*, 2012, **109**(5), 1772–1777.
10. F. Chen, Y. Tobimatsu, L. Jackson, J. Ralph and R. A. Dixon, Novel seed coat lignins in the Cactaceae: structure, distribution and implications for the evolution of lignin diversity, *Plant J.*, 2012, **73**(2), 201–211.

11. Y. Tobimatsu, F. Chen, J. Nakashima, L. L. Escamilla-Treviño, L. Jackson, R. A. Dixon and J. Ralph, Coexistence but independent biosynthesis of catechyl and guaiacyl/syringyl lignin polymers in seed coats, *Plant Cell*, 2013, **25**(7), 2587–2600.
12. C.-T. Do, B. Pollet, J. Thévenin, R. Sibout, D. Denoue, Y. Barrière, C. Lapierre and L. Jouanin, Both caffeoyl Coenzyme A 3-O-methyltransferase 1 and caffeic acid Omethyltransferase 1 are involved in redundant functions for lignin, flavonoids and sinapoyl malate biosynthesis in Arabidopsis, *Planta*, 2007, **226**(5), 1117–1129.
13. N. D. Bonawitz and C. Chapple, The Genetics of Lignin Biosynthesis: Connecting Genotype to Phenotype, *Ann. Rev. Genet.*, 2010, **44**, 337–363.
14. R. Vanholme, K. Morreel, C. Darrah, P. Oyarce, J. H. Grabber, J. Ralph and W. Boerjan, Metabolic engineering of novel lignin in biomass crops, *New Phytol.*, 2012, **196**(4), 978–1000.
15. E. Miedes, R. Vanholme, W. Boerjan and A. Molina, The role of the secondary cell wall in plant resistance to pathogens, *Front. Plant Sci.*, 2014, **5**, 358.
16. J. Rencoret, G. Marques, A. Gutierrez, D. Ibarra, J. Li, G. Gellerstedt, J. I. Santos, J. Jimenez-Barbero, A. T. Martinez and J. C. del Rio, Structural characterization of milled wood lignins from different eucalypt species, *Holzforschung*, 2008, **62**(5), 514–526.
17. A. Guerra, I. Filpponen, L. Lucia and D. Argyropoulos, Comparative evaluation of three lignin isolation protocols for various wood species, *J. Agric. Food Chem.*, 2006, **54**(26), 9696–9705.
18. E. Sjöström, *Wood Chemistry: Fundamentals and Applications*, Academic Press, 1993.
19. M. P. Pandey and C. S. Kim, Lignin Depolymerization and Conversion: A Review of Thermochemical Methods, *Chem. Eng. Technol.*, 2011, **34**(1), 29–41.
20. C. J. Biermann, *Essentials of Pulping and Papermaking*, Academic Press, 1993.
21. P. Sannigrahi, S. J. Miller and A. J. Ragauskas, Effects of organosolv pretreatment and enzymatic hydrolysis on cellulose structure and crystallinity in Loblolly pine, *Carbohydr. Res.*, 2010, **345**(7), 965–970.
22. S. M. Shevchenko, R. P. Beatson and J. N. Saddler, The nature of lignin from steam explosion/ enzymatic hydrolysis of softwood: structural features and possible uses, scientific note, *Appl. Biochem. Biotechnol.*, 1999, 77–**79**, 867–876.
23. P. Sannigrahi, A. J. Ragauskas and S. Miller, Effects of two-stage dilute acid pretreatment on the structure and composition of lignin and cellulose in loblolly pine, *BioEnergy Res.*, 2008, **1**(3), 205–214.
24. G. E. Fredheim, S. M. Braaten and B. E. Christensen, Molecular weight determination of lignosulfonates by size-exclusion chromatography and multi-angle laser light scattering, *J. Chromatogr. A*, 2002, **942**(1–2), 191–199.

25. R. F. Bucholz, J. A. Neal and J. L. McCarth, Some Properties if Pauci-disperse Gymnosperm Lignin Sulfonates of Different Molecular Weights, *J. Wood Chem. Technol.*, 1992, **12**(4), 447–469.
26. O. Ringena, B. Saake and R. Lehnen, Isolation and fractionation of lignosulfonates by amine extraction and ultrafiltration: A comparative study, *Holzforschung*, 2005, **59**(4), 405–412.
27. K. M. Askvik, S. A. Gundersen, J. Sjoblom, J. Merta and P. Stenius, Complexation between lignosulfonates and cationic surfactants and its influence on emulsion and foam stability, *Colloids Surf., A*, 1999, **159**(1), 89–101.
28. Y. Ge, D. Li and Z. Li, Effects of Lignosulfonate Structure on the Surface Activity and Wettability to a Hydrophobic Powder, *Bioresources*, 2014, **9**(4), 7119–7127.
29. W. Zhu, G. Westman and H. Theliander, Investigation and Characterization of Lignin Precipitation in the LignoBoost Process, *J. Wood Chem. Technol.*, 2014, **34**(2), 77–97.
30. M. Brebu and I. Spiridon, Co-pyrolysis of LignoBoost (R) lignin with synthetic polymers, *Polym. Degrad. Stab.*, 2012, **97**(11), 2104–2109.
31. M. Nagy, M. Kosa, H. Theliander and A. J. Ragauskas, Characterization of CO2 precipitated Kraft lignin to promote its utilization, *Green Chem.*, 2010, **12**(1), 31–34.
32. Q. Yang, J. Shi, L. Lin, L. Peng and J. Zhuang, Characterization of changes of lignin structure in the processes of cooking with solid alkali and different active oxygen, *Bioresour. Technol.*, 2012, **123**, 49–54.
33. Y. X. Pang, F. Gao, H. M. Lou and Y. H. Deng, Application Performance of Lignin-modified Amino-sulfonic Acid-based Dispersant on the Ceramic Slurry, in *Advances in Chemical Engineering III*, ed. L. Yu, J. Guo, G. Yi and Q. Yu, Advanced Materials Research, 2013, Pts 1–4, pp. 952–956.
34. H. Hergert, Developments in Organosolv Pulping–An Overview, in *Environmentally Friendly Technologies for the Pulp and Paper Industry*, ed. R. Young and M. Akhtar, Wiley, 1998.
35. L. Pazner, Catalyzed Alcohol Organosolv Pulping, in *Environmentally Friendly Technologies for the Pulp and Paper Industry*, ed. R. Young and M. Akhtar, Wiley, 1998.
36. X. J. Pan, C. Arato, N. Gilkes, D. Gregg, W. Mabee, K. Pye, Z. Z. Xiao, X. Zhang and J. Saddler, Biorefining of softwoods using ethanol organosolv pulping: Preliminary evaluation of process streams for manufacture of fuel-grade ethanol and co-products, *Biotechnol. Bioeng.*, 2005, **90**(4), 473–481.
37. H. J. Huang, S. Ramaswamy, U. W. Tschirner and B. V. Ramarao, A review of separation technologies in current and future biorefineries, *Sep. Purif. Technol.*, 2008, **62**(1), 1–21.
38. R. El Hage, N. Brosse, L. Chrusciel, C. Sanchez, P. Sannigrahi and A. Ragauskas, Characterization of milled wood lignin and ethanol

organosolv lignin from miscanthus, *Polym. Degrad. Stab.*, 2009, **9**(10), 1632–1638.
39. M. Vallejos, F. E. Felissia, A. S. Curvelo, M. D. Zambon, L. Ramos and M. C. Area, Chemical and physico-chemical characterization of lignins obtained from ethanol-water fractionation of bagasse, *BioResources*, 2011, **6**(2), 1158–1171.
40. T. J. McDonough, The Chemistry of Organosolv Delignification, *Tappi J.*, 1993, **76**(8), 186–193.
41. M. Leschinsky, H. Sixta and R. Patt, Detailed mass balances of the autohydrolysis of Eucalyptus globulus at 170 °C, *BioResources*, 2009, **4**(2), 687–703.
42. C. K. Nitsos, K. A. Matis and K. S. Triantafyllidis, Optimization of hydrothermal pretreatment of lignocellulosic biomass in the bioethanol production process, *ChemSusChem*, 2013, **6**(1), 110–122.
43. Y. Zhao, W. Lu, J. Chen, X. Zhang and H. Wang, Research progress on hydrothermal dissolution and hydrolysis of lignocellulose and lignocellulosic waste, *Front. Environ. Sci. Eng.*, 2014, **8**(2), 151–161.
44. R. M. Boehm, The Masonite Process, *Ind. Eng. Chem.*, 1930, **22**(5), 493–497.
45. I. Ballesteros, M. J. Negro, J. M. Oliva, A. Cabañas, P. Manzanares and M. Ballesteros, Ethanol production from steam-explosion pretreated wheat straw, *Appl. Biochem. Biotechnol.*, 2006, **130**(1–3), 496–508.
46. H. Kahr, A. Jäger and C. Lanzerstorfer, Bioethanol Production from Steam Explosion Pretreated Straw, in *Bioethanol*, ed. M. A. P. Lima and A. P. P. Natalense, InTech, 2012.
47. R. Martin-Sampedro, E. A. Capanema, I. Hoeger, J. C. Villar and O. J. Rojas, Lignin changes after steam explosion and laccase-mediator treatment of eucalyptus wood chips, *J. Agric. Food Chem.*, 2011, **59**(16), 8761–8769.
48. F. Araya, E. Troncoso, R. T. Mendonça and J. Freer, Condensed lignin structures and re-localization achieved at high severities in autohydrolysis of Eucalyptus globulus wood and their relationship with cellulose accessibility, *Biotechnol. Bioeng.*, 2015, **112**(9), 1783–1791.
49. M. Leschinsky, H. K. Weber, R. Patt and H. Sixta, Formation of insoluble components during autohydrolysis of Eucalyptus globulus, *Lenzinger Ber.*, 2009, **87**, 16–25.
50. M. Leschinsky, G. Zuckerstatter, H. K. Weber, R. Patt and H. Sixta, Effect of autohydrolysis of Eucalyptus globulus wood on lignin structure. Part 1: Comparison of different lignin fractions formed during water prehydrolysis, *Holzforschung*, 2008, **62**(6), 645–652.
51. M. Leschinsky, G. Zuckerstatter, H. K. Weber, R. Patt and H. Sixta, Effect of autohydrolysis of Eucalyptus globulus wood on lignin structure. Part 2: Influence of autohydrolysis intensity, *Holzforschung*, 2008, **62**(6), 653–658.

52. J. Li, G. Gellerstedt and K. Toven, Steam explosion lignins; their extraction, structure and potential as feedstock for biodiesel and chemicals, *Bioresour. Technol.*, 2009, **100**, 2556–2561.
53. J. B. Zakzeski, C. A. Pieter, A. L. Jongerius and B. M. Weckhuysen, The Catalytic Valorization of Lignin for the Production of Renewable Chemicals, *Chem. Rev.*, 2010, **110**(6), 3552–3599.
54. M. B. Hocking, Vanillin: Synthetic flavoring from spent sulfite liquor, *J. Chem. Educ.*, 1997, **74**(9), 1054–1059.
55. C. Fargues, A. Mathais and A. Rodrigues, Kinetics of Vanillin production from Kraft lignin oxidation, *Ind. Eng. Chem. Res.*, 1996, **35**, 28–36.
56. G. Wu, M. Heitx and E. Chornet, Improved Alkaline Oxidation Process for the Production of Aldehydes (Vanillin and Syringaldehyde) from Steam-Explosion Hardwood Lignin, *Ind. Eng. Chem. Res.*, 1994, **33**, 718–723.
57. J. C. Villar, A. Caperos and F. Garcia-Ochoa, Oxidation of hardwood kraft-lignin to phenolic derivatives with oxygen as an oxidant, *Wood Sci. Technol.*, 2001, **35**, 245–255.
58. F. G. Sales, L. C. A. Maranhao, N. M. L. Filho and C. A. M. Abreu, Kinetic evaluation and modeling of lignin catalytic wet oxidation to selective production of aromatic aldehyde, *Ind. Eng. Chem. Res.*, 2006, **45**, 6627–6631.
59. F. G. Sales, L. C. A. Maranhao, N. M. L. Filho and C. A. M. Abreu, Experimental evaluation and continuous catalytic process for fine aldehyde production from lignin, *Chem. Eng. Sci.*, 2007, **62**, 5386–5391.
60. C. Crestini, P. Pro, V. Neri and R. Saladino, Methyltrioxorhenium: a new catalyst for the activation of hydrogen peroxide to the oxidation of lignin and lignin model compounds, *Bioorg. Med. Chem.*, 2005, **13**(7), 2569–2578.
61. S. Bhargava, H. Jani, J. Tardio, D. Akolekar and M. Hoang, Catalytic wet oxidation of ferulic acid (a model lignin compound) using heterogeneous copper catalysts, *I&EC Res.*, 2007, **46**(25), 8652–8656.
62. J. Zhang, H. Deng and L. Lin, Wet aerobic oxidation of lignin into aromatic aldehydes catalyzed by a Perovskite-type oxide: LaFe1-xCuxO3 (x = 0, 0.1, 0.2), *Molecules*, 2009, **14**, 2747–2757.
63. T. Parsell, S. Yohe, J. Degenstein, T. Jarrell, I. Klein, E. Gencer, B. Hewetson, M. Hurt, J. I. Kim, H. Choudhari, B. Saha, R. Meilan, N. Mosier, F. Ribeiro, W. N. Delgass, C. Chapple, H. I. Kenttamaa, R. Agrawal and M. M. Abu-Omar, A synergistic biorefinery based on catalytic conversion of lignin prior to cellulose starting from lignocellulosic biomass, *Green Chem.*, 2015, **17**(3), 1492–1499.
64. Q. Song, F. Wang, J. Y. Cai, Y. H. Wang, J. J. Zhang, W. Q. Yu and J. Xu, Lignin depolymerization (LDP) in alcohol over nickel-based catalysts via a fragmentation-hydrogenolysis process, *Energy Environ. Sci.*, 2013, **6**(3), 994–1007.

65. S. K. Singh and J. D. Ekhe, Towards effective lignin conversion, HZSM-5 catalyzed one-pot solvolytic depolymerization/hydrodeoxygenation of lignin into value added compounds, *RSC Adv.*, 2014, **4**(53), 27971–27978.
66. C. Zhao and J. A. Lercher, Upgrading Pyrolysis Oil over Ni/HZSM-5 by Cascade Reactions, *Angew. Chem., Int. Ed.*, 2012, **51**(24), 5935–5940.
67. J. G. Zhang, H. Asakura, J. van Rijn, J. Yang, P. Duchesne, B. Zhang, X. Chen, P. Zhang, M. Saeys and N. Yan, Highly efficient, NiAu-catalyzed hydrogenolysis of lignin into phenolic chemicals, *Green Chem.*, 2014, **16**(5), 2432–2437.
68. H. Konnerth, J. Zhang, D. Ma, M. H. G. Prechtl and N. Yan, Base promoted hydrogenolysis of lignin model compounds and organosolv lignin over metal catalysts in water, *Chem. Eng. Sci.*, 2015, **123**, 155–163.
69. D. W. S. Wong, Structure and action mechanism of lignolytic enzymes, *Appl. Biochem. Biotechnol.*, 2009, **157**, 174–209.
70. U. Tuor, H. Wariishi, H. E. Schoemaker and M. H. Gold, Oxidation of phenolic arylglycerol β-aryl ether lignin model compounds by manganese peroxidase from Phanerochaete chrysosporium: Oxidative cleavage of an α-carbonyl model-compound, *Biochemistry*, 1992, **31**, 4986–4995.
71. X. Geng, K. Li and F. Xu, Investigation of hydroxamic acids as laccase-mediators for pulp bleaching, *Applied Microbiol. Biotechnol.*, 2004, **64**(4), 493–496.
72. X. Geng and K. Li, Degradation of non-phenolic lignin by the white-rot fungus *Pycnoporus cinnabarinus*, *Appl. Microbiol. Biotechnol.*, 2002, **60**(3), 342–346.
73. L. Wiermans, M. Perez-Sanchez and P. D. de Maria, Lipase-Mediated Oxidative Delignification in Non-Aqueous Media: Formation of De-Aromatized Lignin-Oil and Cellulase-Accessible Polysaccharides, *ChemSusChem*, 2013, **6**, 251–255.
74. C. Crestini, F. Melone and R. Saladino, Novel multienzyme oxidative biocatalyst for lignin bioprocessing, *Bioorg. Med. Chem.*, 2011, **19**, 5071–5078.
75. S. Laurichesse and L. Averous, Chemical modification of lignins: Towards biobased polymers, *Prog. Polym. Sci.*, 2014, **39**(7), 1266–1290.
76. J. Zakzeski, P. C. A. Bruijnincx, A. L. Jongerius and B. M. Weckhuysen, The catalytic valorization of lignin for the production of renewable chemicals, *Chem. Rev.*, 2010, **110**(6), 3552–3599.
77. W. J. Mulder, R. J. A. Gosselink, M. H. Vingerhoeds, P. F. H. Harmsen and D. Eastham, Lignin based controlled release coatings, *Ind. Crops Prod.*, 2011, **34**(1), 915–920.
78. D. Xueyu, J. Li and M. E. Lindström, Modification of industrial softwood kraft lignin using Mannich reaction with and without phenolation pretreatment, *Ind. Crops Prod.*, 2014, **52**, 729–735.
79. X. Wang and J. Zhao, Encapsulation of the Herbicide Picloram by Using Polyelectrolyte Biopolymers as Layer-by-Layer Materials, *J. Agric. Food Chem.*, 2013, **61**(16), 3789–3796.

80. Y.-L. Chung, J. V. Olsson, R. J. Li, C. W. Frank, R. M. Waymouth, S. L. Billington and E. S. Sattely, A Renewable Lignin–Lactide Copolymer and Application in Biobased Composites, *ACS Sustainable Chem. Eng.*, 2013, **1**(10), 1231–1238.
81. B. Del Saz-Orozco, M. Oliet, M. V. Alonso, E. Rojo and F. Rodriguez, Formulation optimization of unreinforced and lignin nanoparticle-reinforced phenolic foams using an analysis of variance approach, *Compos. Sci. Technol.*, 2012, **72**(6), 667–674.
82. C. Frangville, M. Rutkevičius, A. P. Richter, O. D. Velev, S. D. Stoyanov and V. N. Paunov, Fabrication of Environmentally Biodegradable Lignin Nanoparticles, *ChemPhysChem*, 2012, **13**(18), 4235–4243.
83. L. Klapiszewskia, M. Nowackaa, G. Milczarekb and T. Jesionowskia, Physicochemical and electrokinetic properties of silica/lignin biocomposites, *Carbohydr. Polym.*, 2013, **94**(1), 345–355.
84. D. Stewart, Lignin as a base material for materials applications: Chemistry, application and economics, *Ind. Crops. Prod.*, 2008, **27**(2), 202–207.
85. M. Tortora, F. Cavalieri, P. Mosesso, F. Ciaffardini, F. Melone and C. Crestini, Ultrasound Driven Assembly of Lignin into Microcapsules for Storage and Delivery of Hydrophobic Molecules, *BioMacromolecules*, 2014, **15**(5), 1634–1643.
86. B.-L. Xue, J.-L. Wen and R.-C. Sun, Lignin-Based Rigid Polyurethane Foam Reinforced with Pulp Fiber: Synthesis and Characterization, *ACS Sustainable Chem. Eng.*, 2014, **2**(6), 1474–1480.
87. C. Yang and P. Liu, Water-Dispersed Conductive Polypyrroles Doped with Lignosulfonate and the Weak Temperature Dependence of Electrical Conductivity, *I&EC Res.*, 2009, **48**(21), 9498–9503.
88. Z.-W. He, Q.-F. Lu and J.-Y. Zhang, Facile Preparation of Hierarchical Polyaniline-Lignin Composite with a Reactive Silver-Ion Adsorbability, *ACS Appl. Mater. Interfaces*, 2012, **4**(1), 369–374.
89. Z.-W. He, L.-H. He, J. Yang and Q.-F. Lü, Removal and Recovery of Au(III) from Aqueous Solution Using a Low-Cost Lignin-Based Biosorbent, *I&EC Res.*, 2013, **52**(11), 4103–4108.
90. C. M. Fierroa, J. Górkaa, J. A. Zazob, J. J. Rodriguez, J. Ludwinowicza and M. Jaronieca, Colloidal templating synthesis and adsorption characteristics of microporous–mesoporous carbons from Kraft lignin, *Carbon*, 2013, **62**, 233–239.
91. N. E. El Mansouri, A. Pizzi and J. Salvado, Lignin-based polycondensation resins for wood adhesives, *J. Appl. Poly. Sci.*, 2006, **103**, 1690–1699.
92. N. E. El Mansouri and J. Salvadó, Structural characterization of technical lignins for the production of adhesives: Application to lignosulfonate, kraft, soda-anthraquinone, organosolv and ethanol process lignins, *Ind. Crops Prod.*, 2006, **24**, 8–16.
93. A. Effendi, H. Gerhauser and A. V. Bridgwater, Production of renewable phenolic resins by thermochemical conversion of biomass: A review, *Renewable Sustainable Energy Rev.*, 2008, **12**(8), 2092–2116.

94. W. Zhang, Y. F. Ma, C. P. Wang, S. H. Li, M. M. Zhang and F. X. Chu, Preparation and properties of lignin-phenol-formaldehyde resins based on different biorefinery residues of agricultural biomass, *Ind. Crops Prod.*, 2013, **43**, 326–333.
95. A. Tejado, C. Pena, J. Labidi, J. M. Echeverria and I. Mondragon, Physico-chemical characterization of lignins from different sources for use in phenol–formaldehyde resin synthesis, *Bioresour. Technol.*, 2007, **98**, 1655–1663.
96. B. Danielson and R. Simonson, Kraft lignin in phenol formaldehyde resin. Part 1. Partial replacement of phenol by kraft lignin in phenol formaldehyde adhesives for plywood, *J. Adhes. Sci. Technol.*, 1998, **12**(9), 923–939.
97. S. H. Ghaffar and M. Fan, Lignin in straw and its applications as an adhesive, *Int. J. Adhes. Adhes.*, 2014, **48**, 92–101.
98. P. Widsten and A. Kandelbauer, Adhesion improvement of lignocellulosic products by enzymatic pre-treatment, *Biotechnol. Adv.*, 2008, **26**, 379–386.
99. L. Hu, H. Pan, Y. Zhou and M. Zhang, Methods to improve lignin's reactivity as a phenol substitute and as replacement for other phenolic compounds: A brief review, *BioResources*, 2011, **6**(3), 3515–3525.
100. M. Alonso, M. Oliet, F. Rodriguez, J. Garcia, M. A. Gilarranz and J. J. Rodriguez, Modification of ammonium lignosulfonate by phenolation for use in phenolic resins, *Bioresour. Technol.*, 2005, **96**(9), 1013–1018.
101. M. Funaoka, Sequential transformation and utilization of natural network polymer "LIGNIN", *React. Funct. Polym.*, 2013, **73**(2), 396–404.
102. W. Qiao, S. Li, S. Guo, S. Ren and Y. Ma, Synthesis and characterization of phenol-formaldehyde resin using enzymatic hydrolysis lignin, *J. Ind. Eng. Chem.*, 2015, **21**, 1417–1422.
103. S. Li, G. W. Guo, X. Nan, Y. L. Ma, S. X. Ren and S. Y. Han, Selective Liquefaction of Lignin from Bio-ethanol Production Residue Using Furfuryl Alcohol, *Bioresources*, 2013, **8**(3), 4563–4573.
104. Y. Yuan, M. H. Guo and F. Y. Liu, Preparation and Evaluation of Green Composites Using Modified Ammonium Lignosulfonate and Polyethylenimine as a Binder, *Bioresources*, 2014, **9**(1), 836–848.
105. A. K. Mohanty, M. Misra and L. T. Drzal, Sustainable bio-composites from renewable resources: Opportunities and challenges in the green materials world, *J. Polym. Environ.*, 2002, **10**(1-2), 19–26.
106. P. Alexy, B. Kosikova and G. Podstranska, The effect of blending lignin with polyethylene and polypropylene on physical properties, *Polymer*, 2000, **41**(13), 4901–4908.
107. P. Alexy, B. Kosikova, G. Crkonova, A. Gregorova and P. Martis, Modification of lignin-polyethylene blends with high lignin content using ethylene-vinylacetate copolymer as modifier, *J. Appl. Poly. Sci.*, 2004, **94**(5), 1855–1860.

108. M. Canetti and F. Bertini, Supermolecular structure and thermal properties of poly(ethylene terephthalate)/lignin composites, *Compos. Sci. Technol.*, 2007, **67**(15–16), 3151–3157.
109. W. Doherty, P. Halley, L. Edye, D. Rogers, F. Cardona, Y. Park and T. Woo, Studies on polymers and composites from lignin and fiber derived from sugar cane, *Polym. Adv. Technol.*, 2007, **18**(8), 673–678.
110. H. Nitz, H. Semke and R. Mulhaupt, Influence of lignin type on the mechanical properties of lignin based compounds, *Macromol. Mater. Eng.*, 2001, **286**(12), 737–743.
111. R. Pucciariello, V. Villani, C. Bonini, M. D'Auria and T. Vetere, Physical properties of straw lignin-based polymer blends, *Polymer*, 2004, **45**(12), 4159–4169.
112. W. Thielemans, E. Can, S. S. Morye and R. P. Wool, Novel applications of lignin in composite materials, *J. Appl. Poly. Sci.*, 2002, **83**(2), 323–331.
113. G. Toriz, F. Denes and R. A. Young, Lignin-polypropylene composites. Part 1: Composites from unmodified lignin and polypropylene, *Polym. Compos.*, 2002, **23**(5), 806–813.
114. R. H. Falk, D. J. Vos, S. M. Cramer and B. W. English, Performance of fasteners in wood flour-thermoplastic composite panels, *For. Prod. J.*, 2001, **51**(1), 55–61.
115. N. E. El Mansouri, X. Farriol and J. Salvado, Structural modification and characterization of lignosulfonate by a reaction in an alkaline medium for its incorporation into phenolic resins, *J. Appl. Poly. Sci.*, 2006, **102**(4), 3286–3292.
116. D. Feldman, Lignin and its polyblends – A review, in *Chemical Modification, Properties, and Usage of Lignin*, ed. T. Hu, Kluwer Academic, New York, NY, 2002, pp. 81–99.
117. D. Feldman, D. Banu and A. El-Aghoury, Plasticization effect of lignin in some highly filled vinyl formulations, *J. Vinyl Addit. Technol.*, 2007, **13**(1), 14–21.
118. B. Kosikova, V. Demianova and M. Kacurakova, Sulfur-free Lignins as Composites of Polypropylene Films, *J. Appl. Poly. Sci.*, 1993, **47**(6), 1065–1073.
119. D. Schorr, P. N. Diouf and T. Stevanovic, Evaluation of industrial lignins for biocomposites production, *Ind. Crops Prod.*, 2014, **52**, 65–73.
120. A. S. Hoffman, Hydroigels for biomedical applications, *Adv. Drug Delivery Rev.*, 2012, **64**(Supplement), 18–23.
121. A. M. Karasian, M. A. Tshabalala and G. Buschle-Diller, Wood hemicellulose/chitosan-based semi: Interpenetrating network hydrogels: Mechanical, swelling and controlled drug release properties, *Bioresources*, 2010, **5**(2), 1036–1054.
122. X. Liang, Z. Huang, Y. Zhang, H. Hu and Z. Liu, Synthesis and properties of novel superabsorbent hydrogels with mechanically activated sugarcane bagasse and acrylic acid, *Polym. Bull.*, 2013, **70**(6), 1781–1794.

123. V. J. H. Sewalt, W. de Oliviera, W. G. Glasser and J. P. Fontenot, Lignin Impact on Fibre Degradation: 2—A Model Study Using Cellulosic Hydrogels, *J. Sci. Food Agric.*, 1999, **71**(2), 204–208.
124. N. Bhattarai, J. Gunn and M. Zhang, Chitosan-based hydrogels for controlled, localized drug delivery, *Adv. Drug Delivery Rev.*, 2010, **62**(1), 83–99.
125. V. K. Thakur and M. K. Thakur, Recent Advances in green hydrogels from lignin: a review, *Int. J. Biol. Macromol.*, 2015, **72**, 834–847.
126. M. Kumari and G. S. Chauhan, Adsorption capacity, kinetics, and mechanism of copper(II) uptake on gelatin-based hydrogels, *Appl. Polym. Sci.*, 2010, **119**(1), 363–370.
127. W. K. El-Zawawy, Preparation of hydrogel from green polymer, *Polym. Adv. Technol.*, 2004, **16**(1), 48–54.
128. (a) W. K. El-Zawaway and M. M. Ibrahim, Preparation and characterization of novel polymer hydrogel from industrial waste and copolymerization of poly(vinyl alcohol) and polyacrylamide, *J. Appl. Poly. Sci.*, 2012, **124**(5), 4362–4370; (b) M. Temtem, T. Casimiro, J. F. Mano and A. Aguiar-Ricardo, Green Synthesis of a temperature sensitive hydrogel, *Green Chemistry*, 2007, **9**(1), 75–79.
129. Q. Feng, F. Chen and H. Wu, Preparation and Characterization of a Temperature-Sensitive Lignin-Based Hydrogel, *Bioresources*, 2011, **6**(4), 4942–4952.
130. S. Han, K. Sohn and T. Hyson, Fabrication of New Nanoporous Carbons through Silica Templates and Their Application to the Adsorption of Bulky Dyes, *Chem. Mater.*, 2000, **12**(11), 3337–3341.
131. R. W. Pekala, J. C. Farmer, C. T. Alviso, T. D. Tran, S. T. Mayer, J. M. Miller and B. Dunn, Carbon aerogels for electrochemical applications, *J. Non-Cryst. Solids*, 1998, **225**, 74–80.
132. E. J. Zanto, S. A. Al-Muhtaseb and J. A. Ritter, Sol – Gel-Derived Carbon Aerogels and Xerogels: Design of Experiments Approach to Materials Synthesis, *I&EC Res.*, 2002, **41**(13), 3151–3162.
133. C. Chen, E. B. Kennel, A. H. Stiller, P. G. Stansberry and J. W. Zondio, Carbon foam derived from various precursors, *Carbon*, 2006, **44**(8), 1535–1543.
134. J. Seo, H. Park, K. Shin, S. H. Baeck, Y. Rhym and S. E. Shim, Lignin-derived macroporous carbon foams prepared by using poly(methyl methacrylate) particles as the template, *Carbon*, 2014, **76**, 357–367.
135. S. Maradui, C. H. Kim, S. Y. Kim, B.-H. Kim, W. C. Kim and K. S. Yang, Preparation of carbon fibers from a lignin copolymer with polyacrylonitrile, *Synth. Met.*, 2012, **162**(5-6), 453–459.
136. I. Norberg, Y. Nordstrom, R. Drougge, G. Gellerstedt and E. Sjoholm, A new method for stabilizing softwood kraft lignin fibers for carbon fiber production, *J. Appl. Poly. Sci.*, 2012, **128**(6), 3824–3830.
137. D. Baker and T. G. Rials, Recent advances in low-cost carbon fiber manufacture from lignin, *J. Appl. Poly. Sci.*, 2013, **130**(2), 713–728.

138. J. F. K. Kadla, S. Venditti, R. A. Gilbert, R. D. Compere and A. L. Griffith, Lignin-based carbon fibers for composite fiber applications, *Carbon*, 2002, **40**(15), 2913–2920.
139. M. Langholtz, M. Downing, R. Graham, F. Baker, A. Compere, W. Griffith, R. Boeman and M. Keller, Lignin-derived carbon fiber as a co-product of refining cellulosic biomass, *SAE Tech. Pap.*, 2014, 7(1), 115–121.
140. A. Wheatley, D. Warren and S. Das, Low-Cost Carbon Fibre: Applications, Performance and Cost Models, in *Advanced Composite Materials for Automotive Applications: Structural Integrity and Crashworthiness*, ed. A. Elmarakbi, Wiley, 2013, pp. 405–434.
141. A. Wheatley, D. Warren and S. Das, Development of Low-Cost Carbon Fibre for Automotive Applications, in *Advanced Composite Materials for Automotive Applications: Structural Integrity and Crashworthiness*, ed. A. Elmarakbi, Wiley, 2013, pp. 51–73.
142. Y. Nordström, I. Norberg, E. Sjoholm and R. Drougge, A new softening agent for melt spinning of softwood kraft lignin, *J. Appl. Poly. Sci.*, 2013, **129**(3), 1274–1279.
143. D. Baker, N. C. Gallego and F. S. Baker, On the characterization and spinning of an organic-purified lignin toward the manufacture of low-cost carbon fiber, *J. Appl. Poly. Sci.*, 2012, **124**(1), 227–234.
144. V. Poursorkhabi, A. K. Mohanty and M. Misra, Electrospinning of Aqueous Lignin/Poly(ethylene oxide) Complexes, *J. Appl. Poly. Sci.*, 2015, **132**, 41260–41269.
145. J. Jin, B.-J. Yu, Z.-Q. Shi, C.-Y. Wang and C.-B. Chong, Lignin-based electrospun carbon nanofibrous webs as free-standing and binder-free electrodes for sodium ion batteries, *J. Power Sources*, 2014, **272**, 800–807.
146. M. Schreiber, S. Vivekanandhan, P. Cooke, A. K. Mohanty and M. Misra, Electrospun green fibres from lignin and chitosan: a novel polycomplexation process for the production of lignin-based fibres, *J. Mater. Sci.*, 2014, **49**, 7949–7958.
147. M. Schreiber, S. Vivekanandhan, P. Cooke, A. K. Mohanty and M. Misra, Iodine Treatment of Lignin – Cellulose Acetate Electrospun Fibers: Enhancement of Green Fiber Carbonization, *ACS Sustainable Chem. Eng.*, 2015, **3**, 33–41.
148. S. Dutta, K. C.-W. Wu and B. Saha, Emerging strategies for breaking the 3D amorphous network of lignin, *Catal.: Sci. Technol.*, 2014, **4**(11), 3785–3799.

CHAPTER 8

Update on Research and Development of Microbial Oils

YANNA LIANG

1230 Lincoln Dr., Department of Civil and Environmental Engineering, Southern Illinois University Carbondale, Carbondale, IL 62901, USA
Email: liangy@siu.edu

8.1 Introduction

Microbial oils, also known as single cell oils (SCOs), are those that are accumulated intracellularly by microbial strains. The oils typically in the form of oil droplets or lipid bodies (LBs) consist mainly (80–90%) of triglycerides. LBs comprise a lipid core encased in a phospholipid monolayer, within which many proteins with diverse biochemical properties are embedded.[1] Microorganisms containing more than 20% of cell dry weight as oils are defined as oleaginous.[2] Since microbial oils have many advantages over plant oils, such as a shorter life cycle, less labor required, they are affected less by venue, season and climate, and are easier to scale up.[3] Producing microbial oils to replace oils from oilseeds has been a dream of countless researchers, which has been pursued since the first half of the twentieth century.

Between the periods of the two World Wars, researchers, especially those in Germany, started to explore oils produced by oleaginous microbes. Unfortunately, these early efforts did not end up with commercial scale production of microbial oils due to the absence of industrial scale fermentation technology and genetic tools to modify target microbial strains.[4] In the 1980s and 1990s, a significant amount of research was once again devoted to oleaginous microorganisms, yeasts and microalgae, with the US Department of Energy's (DOE)

RSC Green Chemistry No. 43
Commercializing Biobased Products: Opportunities, Challenges, Benefits, and Risks
Edited by Seth W. Snyder

Published by the Royal Society of Chemistry, www.rsc.org

Aquatic Species Program (ASP) being a well-known example. However, except those aiming for the production of high value arachidonic (ARA, 20 : 4, *n*-6) and docosahexanoic acid (DHA, 22 : 6, *n*-3) that have been commercialized since the 1980s, all other work seeking to produce low value biofuels and biochemicals stopped suddenly due to cost-related issues.

During the past 10 years, owing to concerns with energy shortages, energy security and the adverse environmental impact of non-renewable oil resources, microbial oils have drawn great attention from the scientific community and policy decision makers. These oils have been targeted as alternative biofuel feedstocks for producing renewable hydrocarbon biofuels, such as gasoline, diesel, and jet fuel through hydrotreatment, decarboxylation and other upgrading approaches. If produced at large scales, these oils can potentially replace conventional oil: vegetable oil, yellow grease and animal fats and be used anywhere the plant and animal-based oil sources have been used for producing various materials and chemicals.

Microorganisms that have been investigated for these purposes include yeasts, microalgae, fungi, and bacteria. However, except certain high value specialty oils, such as DHA and Eicosapentaenoic acid (EPA) that have been commercially produced from microalgal strains, the pathway from microbial oils to biofuels and bioproducts faces a tremendous economic hurdle at industrial scales.[5] The cost for producing 1 ton of microbial oils is reported to be \$3000. In the case of soybean oil, it is \$1247–1263.[6] To overcome the cost barrier and to make bioproduct production from microbial oils economically viable, four approaches need to be explored: (1) identify microbial species possessing high lipid-accumulating capability; (2) find low- or zero-value carbon sources to support heterotrophic microbial growth; (3) improve biomass and lipid productivities; and (4) optimize the process of converting microbial oils to bio-based products. In view of the facts that: (1) numerous research articles and reviews have already focused on different type of oleaginous microbial species and (2) a broad range of products can be derived from microbial oils, this chapter only reviews recent progress made on the second and third approach.

8.2 Identifying Low- or Zero-value Carbon Sources

The cost of feedstocks or carbon sources required for the production of microbial oils accounts for 60–75% of the total expense.[7] Thus, inexpensive carbon sources with either low or negative values have been eagerly sought. Low value carbon sources typically have low selling prices and need to be treated to a certain extent before they can be sold at a higher price. Examples of these include crude glycerol, molasses and cheese whey. Negative value carbon sources are those that are commonly referred to as wastes, such as wastewater from either municipalities or industries. Besides these carbon sources that are generated from various industries, in recent years, fermentable sugars derived from lignocellulosic feedstocks have been investigated for microbial oil production.

8.2.1 Carbon Sources from Industry Waste Streams

As shown in Table 8.1, crude glycerol has been studied extensively for various yeast strains for the purpose of producing microbial oils.[8,9–12] Crude glycerol has been a by-product of fat saponification and alcoholic beverage fabrication units.[11] During recent years, with the renewed interest and rapid expansion of biodiesel production, the market is being flooded with crude glycerol.[13] In general, ten gallons of crude glycerol is generated for every 100 gallons of biodiesel produced.[14] The glycerol glut and its decreasing price led Dow and Procter & Gamble to shut down their glycerol production plants in 2006.[15] Due to the presence of color, odor contaminants as well as other minor compounds, crude glycerol is a low value commodity with a high cost for refining to pure glycerol.[16] Presently, in various European countries, due to the obligatory requirement for biodiesel production, crude glycerol which is in huge excess is simply treated as a new kind of industrial wastewater.[17]

The composition of crude glycerol varies from source to source and depends on the oil feedstocks used, condition for transesterification and downstream processes after the esterification reaction. For example, if methanol is not recovered after the reaction, the crude glycerol may contain glycerol, soap, methanol, and water as 48.7%, 3.0%, 22.7%, and 25.6%, respectively.[18] If methanol is recovered, the composition of the crude glycerol may be 65% of glycerol, 4–5% of potassium and sodium salts, 1% methanol, 1% heavy metals and lignin, 0.5% of non-glycerol organic materials and 28% of water.[11]

Crude glycerol has substantial commercial value if purified to United States Pharmacopoeia (USP) grade. However, the cost to refine this product is approximately $0.20 per lb.[14] For small and medium sized operations, this process is expensive.[19] In addition to refining, several avenues have been explored regarding the utilization of crude glycerol, for example: (1) thermochemical conversion of glycerol into propylene glycol[20–22] or acetol[23] and (2) fermentation processes to produce a myriad of products such as 1,3-propanediol by *Clostridium acetobutylicum*[24] and *C. butyricum*,[15] citric acid and erythritol by *Yarrowia lipolytica* Wratislavia K1,[25] pigments by microalgae *Spirulina platensis*,[26] and a mixture of succinic acid, butanol, ethanol, and hydrogen by *E. coli*.[27] Thus, producing microbial oils from crude glycerol will need to compete with these alternatives for the feedstock material.

Use of crude glycerol for producing microbial oils necessitates at least pH adjustment. For some microorganisms, for example, a DHA-producing microalgae, *Schizochytrium limacinum* SR21, methanol and soap present in crude glycerol had negative effects on cell growth and DHA production. The biomass productivity dropped from 1.92 to 0.94 g L^{-1} per day when the concentration of methanol added to pure glycerol was from 0 to 20 g L^{-1}. Within the same range of methanol, DHA productivity decreased from 0.28 to 0.12 g L^{-1} per day. For the same algal strain, besides an inhibitory effect on cell growth and DHA production, soap containing crude glycerol was

Table 8.1 Microbial oil production from industrial wastes through fermentation of various oleaginous yeast strains.

Yeast strain	Carbon source	Nitrogen source	Lipid content (%, w/w)	Total dry weight ($g L^{-1}$)	Lipid productivity ($g L^{-1}$ per day)	Lipid yield ($g g^{-1}$)	Culture mode	Ref.
Cryptococcus albidus	Volatile fatty acids	Ammonium chloride	27.80	1.20	0.08	0.17	Flask	43
Cryptococcus curvatus ATCC 20509	Crude glycerol	Ammonium chloride	52.90	32.90	1.45	0.21	Two-stage fed-batch	8
Cryptococcus curvatus 20509	Food waste hydrolysates diluted by municipal wastewater	Nitrogen in wastewater	28.60	7.50	0.36	NA	Flask	37
Cryptococcus curvatus ATCC 20508	Crude glycerol	Corn steep liquor, baker's yeast autolysate; malt extract	43.00	58.90	4.54	0.54[a]	Fed-batch 6 L fermentor	9
Cryptococcus curvatus ATCC 20508	Crude glycerol	Baker's yeast autolysate	43.00	52.30	4.03	0.54	Fed-batch 6 L fermentor	9
Cryptococcus curvatus ATCC 20508	Crude glycerol	Deoiled Cryptococcus lysate	48.00	69.20	5.95	0.53	Fed-batch 6 L fermentor	9
Cryptococcus curvatus ATCC 20508	Crude glycerol	Corn steep liquor, *C. curvatus* lysate	45.00	50.40	4.06	0.64	Fed batch 26 L fermentor	9
Cryptococcus curvatus KCTC 17162	Pretreated waste activated sludge with glucose (40 $g L^{-1}$)	Ammonium chloride	23.00	9.84	2.26	NA	Flask	38
Cryptococcus curvatus ATCC 20509	Whey permeate	Ammonium chloride and yeast extract	33.00	91.40	23.88	NA	2 L Fermentor with partial recycling	40
Cryptococcus curvatus strain D	Whey permeate	Ammonium hydroxide	57.00	26.80	5.09	0.27	14 L Fermentor	41

Table 8.1 (*Continued*)

Yeast strain	Carbon source	Nitrogen source	Lipid content (%, w/w)	Total dry weight (g L^{-1})	Lipid productivity (g L^{-1} per day)	Lipid yield (g g^{-1})	Culture mode	Ref.
Cryptococcus curvatus ATCC 20509	Hydrogen production effluent	None	13.53	9.50	1.10	NA		39
Rhodosporidium toruloides AS2.1389	Crude glycerol from alkaline-catalysed biodiesel production process	Yeast extract, ammonium sulfate	47.70	19.20	1.37	NA	Flask	10
Rhodosporidium toruloides AS2.1389	Crude glycerol from an enzyme-catalysed biodiesel production process	Yeast extract, ammonium sulfate	42.90	20.10	1.29	NA	Flask	10
Rhodosporidium toruloides AS2.1389	Crude glycerol from alkaline-catalysed biodiesel production process	Yeast extract, ammonium sulfate	69.50	26.70	2.27	NA	5 L Fermentor	10
Rhodosporidium toruloides AS2.1389	Crude glycerol from an enzyme-catalysed biodiesel production process	Yeast extract, ammonium sulfate	74.10	18.00	1.63	NA	5 L Fermentor	10
Yarrowia lipolytica	Palm oil mill effluent	Nitrogen in mill effluent	48.00	3.54	0.57	NA	Flask	42
Yarrowia lipolytica	Waste oil from frying vegetable with glucose	Yeast extract, ammonium sulfate	57.89	7.56	0.73	0.61[b]	Flask	44
Yarrowia lipolytica LGAM S(7)1	Crude glycerol	Yeast extract, ammonium sulfate	43.00	8.10	2.88	NA	Continuous culture	11
Yarrowia lipolytica LGAM S(7)1	Glucose, crude glycerol, stearin	Yeast extract, ammonium sulfate	30.00	11.40	0.63	NA	Flask	12

[a]Oil yield was based on glycerol consumption. Carbon sources in yeast extract and others were not considered.
[b]Oil yield was calculated based on glucose consumption. Oil consumed was not taking into consideration.

shown to affect fatty acid composition. In cells grown on a soap-free medium, the proportion of C18 fatty acids was negligible ($\sim$1% of total fatty acids (TFAs)). But for cells cultivated on a soap-containing medium, significant proportions of C18 fatty acids (25 – 35% of TFAs) were observed.[28] Similar results on the same algal strain were reported by another study where soap- and methanol-containing crude glycerol led to lower biomass and lipid productivities than those grown on treated crude glycerol.[18]

However, for an oleaginous yeast strain, *Rhodosporiduim toruloides* AS2.1389, the results were different. As shown in Table 8.1, this yeast strain was able to grow on untreated crude glycerol derived from either alkali- or enzyme-catalyzed transesterification reactions. Interestingly, when sodium oleate (0–2 g L^{-1}, representing soap), sodium chloride (0–16 g L^{-1}, representing salts), methyl oleate (0–2 g L^{-1}, representing fatty acid methyl ester (FAME)), or glycerol monooleate (0–2 g L^{-1}, representing glyceride) was added to refined glycerol medium, increased biomass and lipid concentrations were detected with the increase of concentrations of these added compounds. Methanol, however, was found to be inhibitory. Cell density, lipid concentration and lipid yield decreased by about 0–5%, 6–24% and 6–23%, respectively, when methanol was added at concentrations from 4 to 16 g L^{-1}.[10] Thus, the use of crude glycerol for microbial fermentation needs to be evaluated specifically for different strains.

Like any commodity, the market price of crude glycerol has fluctuated from time to time. Although crude glycerol is commonly recognized as an industrial byproduct with a large volume, a careful calculation reveals that oil production from crude glycerol is really limited. Using the US as an example, the total production volume of biodiesel was 1359 million gallons (MG) in 2013.[29] Generally, for every 10 gallons of biodiesel produced, 1 gallon of crude glycerol will be released. Thus, the total volume of crude glycerol generated from US biodiesel producers was $\sim$136 MG in 2013. Theoretically, the maximum lipid yield from glucose is 0.33 g g^{-1}. Considering the synthesis of other non-lipid cellular components, the highest practical conversion is 0.20–0.22 g lipid per g glucose.[6] If assuming lipid yield from glycerol is 0.2 g lipid per g glycerol, then the total oil yield is 27.2 MG. This amount of oil can be reacted to generate the same amount of biodiesel, which is 27.2 MG. Compared to the total biodiesel volume of 1359 MG, biodiesel produced from microbial oils accumulated on crude glycerol is only 2% of the total. Therefore, the contribution and impact to the biodiesel industry is minimal.

In contrast to crude glycerol, wastewater has almost unlimited volume. Considering the fact that wastewater contains a variety of organic matters, nitrogen, phosphorus and other nutrients that can support microbial growth, significant efforts have been dedicated to growing cells of microalgae[30–33] and yeasts[34–36] for the dual purpose of producing microbial oils and treating wastewater. Besides conventional wastewater, streams or hydrolysates generated from other industry processes, such as food waste,[37] activated sludge,[38] hydrogen production,[39] whey permeate,[40,41] palm oil mill,[42] oil wastes containing fatty acids[43] and waste oils from frying vegetables,[44] have been evaluated in terms of generating microbial oils. However, although this approach looks attractive, at least two challenges that prevent this strategy from being scaled up exist: (1) as shown in Table 8.1 and reviewed by Leiva-Candia,[45] the yields of microbial cells and oils are rather low compared to those from other sources and (2) due to slow microbial growth, glucose and/or nitrogen sources need to be provided to ensure reasonable cell growth.

Even if a robust microbial strain or a defined microbial consortium that yields high microbial oil productivity is identified, this approach still cannot compete with the conventional activated sludge process for treating wastewater. This latter process adopting a suite of mixed microbial species has been developed well enough to treat domestic and some industry wastewater in the outdoor environment. The resulting sludge which is rich in organic matters can be liquefied to crude bio-oil through a hydrothermal liquefaction process initially prototyped by the US Environmental Protection Agency (EPA)[46] and further scaled up in Japan to a demonstration level.[47] In contrast, using selected microbial strains for treating wastewater while generating microbial oils needs to address these questions: (1) as almost all studies on this aspect have been carried out in a well-controlled laboratory environment, can the chosen microbes survive in the open environment with constantly changing temperature and exposure to potential contaminants and predators? (2) Can the chosen elite microbial strains compete with those originally in wastewater? If not, a step of sterilization is needed. At industrial scales, however, this sterilization will cause serious issues and bring along significant expense. Therefore, although the idea of producing microbial oil from wastewater is interesting, whether it can be successfully commercialized and remain economically viable is still a tough question to answer.

8.2.2 Carbon Sources from Agricultural Waste Streams

Similar to wastewater, lignocellulosic biomass is also abundantly available. According to the US DOE's billion ton study,[48] under the baseline scenario, the US had a total of 473 million tons (MT) of feedstocks including forest and agricultural biomass and waste resources in 2012. Under the high yield scenario, the total feedstocks available was 555 MT in the same year. In 2030, there will be 1094 MT and 1374–1633 MT of feedstocks available for use under the baseline and high yield scenario, respectively. However, though the amount of biomass feedstocks is not limiting, utilizing these materials has faced great obstacles. The ultimate reason for these challenges arises from the nature of lignocellulosic materials, which are made to be sturdy, recalcitrant and resistant to any processes that may disintegrate the tight cell wall structures.

The cell wall of lignocellulosic biomass is composed of cellulose fibrils embedded in less well-organized hemicellulose matrices which, in turn, are surrounded by an outer lignin seal. Cellulose is a linear condensation homopolymer of anhydrous glucose molecules connected through β(l–4)-glycosidic linkages.[49] Every other glucose molecule is rotated by approximately 180° along the axis of the cellulose chain compared with its two neighboring monomeric units. This rotation causes cellulose to be highly symmetrical, since each side of the chain has an equal number of hydroxyl groups. Coupling of adjacent cellulose molecules by hydrogen bonds and van der Waal's forces results in a parallel alignment and a crystalline structure that contribute to biomass recalcitrance.[50,51] Hemicellulose is a heteropolymer composed of sugar monomers, including xylose, mannose, glucose, galactose, and others. Given the presence of side chains and the lack of repeating β-1,4-glycosidic bonds, hemicellulose does not have a crystalline structure and is much more susceptible to degradation and hydrolysis compared to cellulose. Lignin is a complex compound with a high molecular weight and an even more random structure than hemicellulose. The three most prevalent monomers are: *p*-coumaryl alcohol, coniferyl alcohol, and sinapyl alcohol.

Typically, plant cell walls contain 28–45% of cellulose, 18–36% of hemicellulose, and 19–25% of lignin. Converting low-cost lignocellulosic feedstocks to fermentable sugars offers outstanding benefits to the nation through: (1) improved strategic security, (2) decreased trade deficits, (3) healthier rural economies, (4) improved environmental quality, (5) technology exports, and (6) a sustainable energy resource supply.[52–65] To unlock sugars out of cellulose and hemicellulose through the biochemical conversion platform, at least two processes—pretreatment and enzymatic hydrolysis—are needed. Over the years, explosive research and development have been devoted to this effort. Interested readers are recommended to consult recent review articles.[51,66–70]

Using bagasse of sweet sorghum as an example, different pretreatment strategies have been reported such as sulfuric acid,[71,72] hydrochloric acid,[73] phosphoric acid,[74] steam,[75] dilute ammonia hydroxide,[76] ammonia fiber explosion,[77] liquid hot water,[78] microwave,[79] and lime.[80] Upon pretreatment followed by enzymatic hydrolysis, the hydrolysates developed from sorghum bagasse can be readily used for fermentation without conditioning, washing, or adding extra nutrients. It is not only ethanol that can be directly produced from the hydrolysates,[72,76] microbial oils are accumulated in oleaginous yeast strains, too[80–82] (Table 8.2).

For other hydrolysates derived from other feedstock materials, however, direct fermentation may not be possible.[83–91] During pretreatment, as a result of high temperature and/or pressure, degradation of released sugars may occur. Glucose and xylose can be degraded to furfural and 5-hydroxymethylfurfural (5-HMF), respectively. When furfural and HMF are degraded, formic acid is formed. Levulinic acid is formed by degradation of HMF and phenolic compounds are generated from the partial breakdown of lignin.[92] Besides these listed here, hundreds of other compounds have been identified in cellulosic hydrolysates.[92,93] To some

Table 8.2 Comparison of lipid productivity among oleaginous yeast strains grown on hydrolysates developed from different biomass feedstocks.

Strain	Feedstock	Biomass pretreatment	Detoxification	Cell density ($g\ L^{-1}$)	Oil content (%)	Lipid productivity ($g\ L^{-1}$ per day)	Lipid coefficient ($g\ g^{-1}$ sugar)	Ref.
Cryptococcus curvatus	Sorghum bagasse	Lime	No	10.80	73.30	1.08	0.22	80
Cryptococcus curvatus	Sorghum bagasse	Microwave	No	15.50	64.00	2.48	NA	81
Cryptococcus curvatus	Corn fiber	Lime	No	6.60	49.80	0.83	0.27	83
Cryptococcus curvatus	Corn stover	Ionic liquid	Yes[a]	NA	NA	3.00	0.11[b]	84
				16.50	43.40	3.60	0.14[c]	84
Cryptococcus curvatus	Wheat straw	Dilute sulfuric acid	No	17.20	33.50	0.96	NA	91
			Yes	15.60	27.10	0.60	NA	91
Lipomyces starkeyi	Corn cob	Dilute sulfuric acid	Yes[d]	17.20	47.00	1.01	0.21	85
Lipomyces starkeyi	Wheat straw	Dilute sulfuric acid	Yes	12.70	29.10	0.53	NA	91
			No	14.70	31.20	0.66	NA	91
Rhodotorula graminis	Corn stover	Dilute acid	No	48.00	34.00	5.0[e]	NA	86
Rhodotorula glutinis	Wheat straw	Dilute sulfuric acid	Yes	11.80	20.70	0.35	NA	91
			No	13.80	25.00	0.49	NA	91
Trichosporon fermentans	Rice straw	Dilute acid	No	11.40	14.60	0.21	NA	87
			Yes[f]	28.60	40.10	1.44	NA	87
Trichosporon dermatis	Corn cob	Degradation by *T. reesei*[g]	No	24.40	40.10	1.40	0.17	88
Trichosporon cutaneum	Corn stover	Dilute sulfuric acid	Yes[h]	19.40	39.20	1.87	0.15	89
Trichosporon cutaneum	Corn cob	Dilute sulfuric acid	Yes[i]	22.10	36.00	0.00	NA	90

[a]Wash by methanol and water.
[b]SSELP: simultaneous saccharification and enhanced lipid production.
[c]SHELP: separated hydrolysis and enhanced lipid production.
[d]Overliming and adsorption with activated carbon.
[e]Stirred-tank bioreactor with addition of yeast extract, corn steep solids and salts.
[f]Overliming, concentration, adsorption by Amberlite™ XAD-4.
[g]*Trichoderma reesei*.
[h]Concentration, overliming, filtration.
[i]Overliming and adsorption with activated charcoal.

fermentation processes and microbial strains, the presence of these chemicals causes significant inhibition and necessitates expensive steps for removal.

Thus, although lignocellulosic feedstocks are rich in sugars and abundantly available, use of these materials is not straightforward and cheap. Besides the general treatment train—pretreatment, enzymatic hydrolysis and fermentation—other unexpected and specific steps may be needed to enhance fermentation performance. All of these add cost to the final desired products. Hence, if microbial oils are the only products from fermenting cellulosic hydrolysates, then the economic aspect is going to be limited. To improve the economic competitiveness, besides increasing the oil productivity, at least two approaches need to be considered: increasing the value of the final oil rather than using it for biofuel production and using oleaginous strains that can produce other products in addition to intracellular microbial oils. Without these considerations, producing only intracellular microbial oils from lignocellulosic biomass will face unsurmountable challenges to be economically viable.

8.3 Maximizing Oil Productivity through Biochemical Approaches

The process economics of microbial oil fermentations are mainly determined by two costs: substrate and fermentor. The former depends on the price of substrates and lipid yield (g lipid per g substrate). The latter is strongly dependent on lipid production rate (g L^{-1} per time).[40] To minimize the cost of producing microbial oils, besides finding low-cost substrates as discussed earlier, extensive work has been conducted on improving oil yield and oil production rate. This work is centered on identifying the optimal concentration of carbon (C) and nitrogen (N) source, the optimal C–N ratio, and the best pH and temperature for batch operations.

Specific to oleaginous microbes under nutrient limitation, three growth phases are commonly observed: (1) cell proliferation or the exponential growth phase; (2) lipid accumulation phase where growth slows down as a result of nutrient limitation and reaching maximal lipid synthesis; and (3) late accumulation phase where lipids continue to accumulate, but β-oxidation, the catabolic pathway is kicked off in an effort to remobilize the carbon stored.[4] Based on this growth characteristic, two other cultivation modes—continuous culture and fed-batch—have been investigated. While the C–N ratio is initially fixed and modification is not allowed during batch cultivation, continuous culture allows a constant C–N ratio throughout the fermentation, and regulation of substrate concentration leads to the fine tuning of the growth rate. The fed-batch mode allows precise control of nutrient and substrate flow rates during fermentation and is suggested as the most accurate and reliable approach to use for control of lipid accumulation.[4]

Since numerous studies have aimed at enhancing the oil production rate for different oleaginous species, this section will just focus on one representative yeast strain, *Cryptococcus curvatus.* A study on *C. curvatus* was started in the 1980s. The original purpose was to produce enough yeast oils to replace cocoa butter since the fatty acid compositions of the two are very similar.[94] Following years of biochemical and genetic work, a fermentation process was developed and tested at both pilot scale (0.5 m^3) and industrial scale (8.2 m^3). The manufacturing cost of refined yeast oil was estimated as $800–1000 per ton.[95]

Along this line of research, *C. curvatus* was found to convert various sugars—glucose, galactose, cellobiose, sucrose, lactose and xylose—into fatty acids.[96,97] Oat hulls, bean blanch water, corncobs, wood cellulose, ripe bananas and molasses have also been tested as substrates.[98] To enhance oil production, nitrogen-limited and carbon-limited media with different pure sugars or ethanol under batch or continuous culture modes have been studied.[97] In another study, a lack of ferrous ions was found to have no effect on cell growth, oil production and fatty acid composition in batch cultures. The highest biomass concentration of 70 g L^{-1} was achieved in 172 h with an oil content of 53% in a fed-batch culture. A C–N ratio between 20 and 50 was identified to be optimal for oil accumulation and a ratio of >50 became crucial for both growth and oil production in fed-batch cultures. An oil yield of 0.14 was reported from this study.[99] A higher cell density of 118 g L^{-1} was observed when *C. curvatus* was cultivated on pure glycerol in a 50-h two-phase fed-batch fermentation operated at a constant pH of 5.5 at 28 °C. The oil production rate was 14.2 g L^{-1} per day. The cellular lipid content was 25%.[94]

On substrates rather than pure compounds, the highest cell yield of 69.2 g L^{-1} was reported by Thiru.[9] In this study, *C. curvatus* was cultivated in a super-rich medium containing: glycerol, 10 g L^{-1}; corn steep liquor (CSL), 20 g L^{-1}; baker's yeast autolysate, 5 g L^{-1} (local bakery); and malt extract (HiMedia Labs, India), 2 g L^{-1}.[9] With the presence of all three nitrogen sources, the cell density reached 58.9 g L^{-1} in 134 h. The lipid content was 43% with a lipid yield of 0.54. When malt extract was eliminated from the medium, the cell concentration was 52.3 g L^{-1}. Lipid content and yield remained the same. When deoiled, *C. curvatus* cells were used to replace yeast autolysate; within the same experimental period, cell dry weight was 69.1 g L^{-1} with a lipid content of 48%. The lipid yield was 0.53. Finally, the whole process was scaled up in a 26-liter fermentor. The cell density, lipid content and lipid yield were reported to be 50.4 g L^{-1}, 45% and 0.64, respectively. Obviously, all lipid yields (g lipid per g glycerol) were above the theoretical value of 0.22.[6] A closer look at CSL revealed that CSL contains high concentrations of lactic acid (26 g L^{-1}), phytic acid (7.8 g L^{-1}), and reducing sugars (2.5 g L^{-1}) besides 47% of crude protein.[100] Thus, glycerol was not the only carbon source for cell growth. An oil yield of g oil per g glycerol would exaggerate the actual one on substrates if acids and sugars present in the CSL were not taken into consideration in doing the calculation.

During recent years, cultivating *C. curvatus* on cellulosic hydrolysates has been studied by several groups of researchers. Unlike *Rhodosporidium toruloides,* which does not tolerate the inhibitors in the hydrolysates of wheat straw[91] and corn stover[101] and *Yarrowia lipolytica* whose preferred substrates are hydrophobic, such as fatty acid and alkane,[1,102,103] *C. curvatus* grows well on unconditioned hydrolysates of sweet sorghum bagasse[80,81] and tolerates inhibitory compounds well.[91] In particular, among several oleaginous yeast species, *Rhodotorula glutinis, R. toruloides, Lipomyces starkeyi,* and *Y. lipolytica, C. curvatus* had the highest lipid yield on both the detoxified and non-detoxified hydrolysates of wheat straw after dilute sulfuric acid pretreatment.[91] In addition, this yeast has good lipid productivity on effluent from hydrogen production from food waste,[39] municipal wastewater,[34] crude glycerol,[8,9] and wheat straw pretreated by ozonation and alkaline peroxide[104] (Tables 8.1 and 8.2).

As shown in Table 8.2, the highest lipid production rate of 5.04 g L^{-1} per day was obtained through cultivating *R. graminis* on undetoxified corn stover hydrolysates derived from dilute acid pretreatment followed by enzymatic hydrolysis.[86] This fermentation was conducted in a 20 L stirred-tank bioreactor with dissolved oxygen maintained at more than 30% of saturation level. Unlike other investigations where yeast growth was purely on biomass hydrolysates, this study added corn steep solids (5 g L^{-1}), yeast extract (2 g L^{-1}) and four different salts to the fermentor. The presence of these nutrients may be the reason for the higher than normal lipid production rate and the highest cell density of 48 g L^{-1} reported so far.

Very recently, in order to decrease the cost of producing microbial oils from cellulosic hydrolysates, simultaneous saccharification and fermentation (SSF) has been adopted to replace the conventional route of separate hydrolysis and fermentation (SHF). Since both SSF and SHF have been tightly related to ethanol production from lignocellulosic materials, two different terms—simultaneous saccharification and enhanced lipid production (SSELP) and separate hydrolysis and enhanced lipid production (SHELP)—have been coined for producing microbial oils from plant-based feedstocks. Theoretically, SSELP should not work well owing to the different temperature preference of oleaginous yeast strains (30 °C) and hydrolytic cellulase enzymes (50 °C). However, as reported by Gong *et al.*,[84] *C. curvatus* grew well at 37 °C for the SSELP process even though it does not survive on agar plates at this temperature. When cultivated on corn stover pretreated by ionic liquid and washed by methanol and water, the oil yield was 6.0 g L^{-1} after 48 h with an oil coefficient of 0.112 g g^{-1} corn stover. For this SSELP process, the loading of cellulase, cellobiase, and xylanase was 4.0 filter paper unit (FPU) per g corn stover; 8.0 cellobiase unit (CBU) per g corn stover; and 5.0 mg g^{-1} corn stover, respectively. It needs to be noted that no nutrients were supplemented for this oil production and the observed oil yield was higher than 3.2 g L^{-1} when *Trichosporon cutaneum* was fermented simultaneously with saccharification of corn stover pretreated by dilute acid and detoxified.[105] Under the SHELP mode, higher oil concentration and oil coefficient—7.2 g L^{-1} and 0.138 g g^{-1} corn stover, respectively—were observed during the same experimental duration. However, for this strategy, twice as much enzyme and extra time and space are required to prepare the corn stover hydrolysates.

Even with SSELP, the low oil yield will still make the overall process prohibitively costly to be commercialized. Unlike pure carbon substrates, producing microbial oils from lignocellulosic sugars is just at the beginning stage. Considering the complicated nature of cellulosic hydrolysates, controlling the C–N ratio will be difficult if not impossible. Thus, maximizing microbial oil productivity from cellulosic sugars still requires significant improvement.

8.4 Maximizing Oil Productivity through Molecular Biology Techniques

From the perspective of genetic/metabolic engineering of microbial species for producing microbial oils, there are two approaches. The first one is to engineer well-studied microbial strains, such as *E. coli*, to improve microbial oil yield. The second strategy is to maximize oil production from oleaginous microorganisms beyond the biochemical approaches discussed earlier.

8.4.1 Increasing Production of Fatty Acids through a Systems Biology Approach

This strategy is based on our understanding of the fatty acid synthesis pathway. The type II fatty acid synthesis system in bacteria begins by carboxylation of acetyl-CoA which results in the formation of a three-carbon intermediate, malonyl-CoA. Elongation of fatty acid molecules proceeds through the iterative addition of two carbon units derived from malonyl-CoA with the elimination of CO_2. Generally, this process is complete when the chain length reaches 16 carbons (*i.e.*, palmitic acid). However, other longer or shorter fatty acids may also be formed, with varying degrees of saturation. Due to the energy-intensive nature of the fatty acid biosynthesis, this pathway is highly regulated to prevent accumulation of excessive lipids. Any excess lipids are rapidly degraded by oxidative cleavage to acetyl-CoA. Thus, if over-production of free fatty acids is desired, these regulatory mechanisms need to be overcome.[106]

The motivation for increasing the cellular content of free fatty acids (FFAs) lies in these facts: (1) FFAs are energy-rich molecules and are a part of intracellular lipid bodies (*e.g.*, oils, fats) as a form of storage;[107] (2) microbially-produced FFAs are attractive chemical feedstocks that could replace plant oils, especially in cases where specific chain lengths or modifications (*e.g.*, branches, desaturations) are desired. In the oleochemical industry, FFAs are precursors to compounds used as agrochemicals, biocidal agents, textile processing agents, soaps, surfactants, and polymer additives.[108,109]

In the biofuel arena, FFAs can be converted to renewable diesel blendstock through hydrotreating.[106]

Over the years, a significant amount of effort has been devoted to this field of research. Through quantitative analysis of fatty acid synthesis in a cell-free system, it was determined that fatty acid synthesis is highly dependent on the availability of malonyl-CoA. A 10-fold increase of malonyl-CoA concentration is required to reach the theoretical yield of FFAs of 0.3–0.4 g g^{-1} glucose.[110] To increase production of FFAs, various approaches, such as knocking out fatty acid degradation, expression of plant thioesterase, increasing malonyl-CoA production, and decreasing feedback inhibition by over-expressing an endogenous thioesterase, have been investigated.[111–113] The highest titer of extracellular FFAs was reported to be 7.0 g L^{-1} with a yield of 0.28 g FFAs per g glucose.[114] This was achieved through the *fadD* knockout in strain RB03 to prevent re-utilization of the synthesized fatty acids. The choice of thioesterase allowed control over both length and functionality of the fatty-acid side chain. For example, C16 and C18 saturated fatty acids were the only products when *FadM* was overexpressed, while *YciA* and *TesA* overexpression supported the synthesis of 3-hydroxy (C14 : 3OH) and unsaturated (C18 : 1) fatty acids, respectively. Concentration of both extracellular and intracellular FFAs in strain RB03 [*fadBA-fadM+*] and the corresponding controls showed that the engineered reversal of the β-oxidation cycle contributed to the synthesis of 90–95% of the total FFAs. The second highest titer of FFAs was 4.8 g L^{-1}, achieved through both strain improvements and cultivation process optimization.[115] But even with this FFAs concentration, the yield was only 0.044 g g^{-1} glucose or 12.1% of the theoretical yield.

To eliminate the need for maintaining a specific concentration of dissolved oxygen during cultivation and maximize the proportion of reducing equivalents directed to FFA synthesis rather than ATP production, producing FFAs through engineered strains under anaerobic conditions has been tested. However, under this growth condition, a redox imbalance, such as a relatively high ratio of NADH–NAD^+, could have detrimental effects on cell physiology, both at a regulatory and at the enzyme function levels. Several researchers have attempted to address this challenge by identifying homologs of *E. coli FabG*, an essential reductase involved in fatty acid synthesis. *E. coli* strains with overexpressed *FabG* homologs did produce more FFAs than the control ones. However, the titer under anaerobic conditions was much lower than those in an aerobic environment.[116]

Over the years, although significant progress has been made with regard to modifying and engineering the fatty acid synthesis pathway, the barriers that prevent FFA production to reach near-theoretical yields remain unknown. Lower titers are likely to be the result of new challenges such as balancing additional enzymes, energy (ATP), and reducing power (NADH) needed on top of those targeted for FFA production. Alternatively, the presence of futile cycles (*e.g.*, thioesterase activity on acyl-CoAs), product toxicity, and/or product accumulation could reduce final yields.[117]

8.4.2 Increasing Production of Microbial Oils through Engineering Oleaginous Species

Besides the *de novo* synthesis of fatty acids, the Kennedy pathway for synthesizing triacylglycerol (TAG) is also well-known for all microorganisms. But the genetic factors that control the extent to which lipid accumulates in a particular organism remain largely unclear. For example, under the same nitrogen-limiting conditions, *C. curvatus* and *Rhodotorula* sp. can accumulate 40–70% of cell biomass as lipid while *Saccharomyces cerevisiae* and the food yeast, *Candida utilis* are unable to produce lipids larger than 5–10% of cell mass.[118] In particular, the process of lipid accumulation in oleaginous microorganisms where the lipid content may reach 70% of cellular biomass, and possibly higher, is still poorly understood.[119] This statement is true for both microalgal[120–122] and yeast species.[2,119,123,124] Specific to oleaginous microorganisms, increased lipid accumulation is often tightly related to nutrient depletion as described earlier. The common observation is that when nutrient is limiting (nitrogen being a general example), the cells begin to channel the carbon source directly to lipid synthesis. Non-oleaginous species, however, do not accumulate lipids even when placed in the same nitrogen-deficient medium. Throughout the years, in light of the precursors (acetyl-CoA, malonyl-CoA, and glycerol) and energy (ATP, NADPH) required for lipid synthesis, three enzymes have been proposed to be crucial for lipid accumulation besides AMP deaminase, which is activated by the exhaustion of nitrogen in the medium.[124] These three enzymes are: ATP:citrate lyase (ACL), acetyl-CoA carboxylase (ACC), and malic enzyme (ME).

ACL provides large amount of acetyl-CoA as the initial biosynthetic unit for fatty acid synthesis by cleaving citrate to give oxaloacetate and acetyl-CoA. However, there is no correlation between ACL activity and lipid accumulation. Over-expressing of ACL in oleaginous *Y. lipolytica* did not increase lipid content at all.[1] In addition, low lipid content in the majority of non-oleaginous species, such as *S. cerevisiae* and *C. utilis,* does not seem to be a result of low ACL activity.[2]

ACC catalyzes the conversion of acetyl-CoA to malonyl-CoA, the elongation unit for fatty acid synthesis. Attempts to over-express ACC for the purpose of improving lipid production have given both encouraging and disappointing results. While increased lipid content was observed for *Y. lipolytica* as detailed later, no enhanced lipid accumulation was detected for diatom *Cyclotella cryptica* despite a two- or three-fold increase in ACCase activity.[125] Additionally, no data are available to verify whether the enzyme is indeed a limiting step for oleaginous yeast species.

Besides an abundant supply of acetyl-CoA, oleaginous microorganisms must be able to produce a sufficient supply of NADPH as the essential reductant for fatty acid synthesis considering the highly reducing nature of fatty acids. For example, to synthesize 1 mol of C18 fatty acid, 16 mols of NADPH are needed.[124] Thus, ME was proposed to be the major supplier of

NADPH by carrying out the irreversible decarboxylation of malate to pyruvate with the formation of NADPH from $NADP^+$.

Recently, the hypothesis that the rate-limiting step in fatty acid synthesis in oleaginous species is the supply of NADPH, which is generated specifically and solely by ME, has been strengthened by experimental results.[119] In the reported study, the whole-length ME gene from dimorphic fungus *Mucor circinelloides* was cloned into a leucine auxotroph of the same strain and placed under the control of the constitutive glyceraldehyde-3-phosphate dehydrogenase gene promoter. After confirming correct expression of the ME genes, two recombinant and wild type strains were grown in a nitrogen limiting medium with a high C–N ratio. The activities of ME were increased by two- to three-fold in the recombinant strains. At the end of the 96 h incubation, lipid contents in the transformants were 2.5-fold higher than that in the wild type. It appears that increased lipid content was due to ME activity. However, a close examination of this study led to several questions: (1) the wild type strain used in this study is a leucine auxotroph. At the end of the four-day experiment, the lipid content in this strain was only 12%, which disqualifies it from being an oleaginous species. But it still had ME activity; (2) as clearly demonstrated by the figures presented in the article, during the experimental period, ME activity for both the transformants and the wild type strain decreased with time, but lipid content increased within the same time frame; (3) on the fourth day, ME activity completely ceased for all three strains while the lipid content was the highest among daily samples. Therefore, it is not convincing that there is a positive correlation between ME activity and lipid accumulation. The authors obviously realized this point as they stated that: "It is, however, clear that there are still other factors that are limiting the continued activity of ME even though it is now being expressed quasi-constitutively" and "one possible explanation for this is that there could be an enzyme that specifically degrades ME (an ME converting enzyme?) once the cells are nitrogen limited and have commenced lipid storage". In other words, there must be other enzymes that can furnish the NADPH for increased lipid synthesis to compensate decreased supply of NADPH caused by diminished ME activity.

Besides the fact that the role of ME in lipid accumulation is still not clear, this enzyme does not appear to be ubiquitous among all lipid-accumulating species and may be absent in some oleaginous yeasts, such as *Lipomyces* sp. and some *Candida* sp.[124] A global proteomic study done by 2-D LC/MS/MS for *Lipomyces starkeyi* did not reveal the existence of ME.[126] Instead, 6-phosphogluconate dehydrogenase was up-regulated drastically. This enzyme catalyzes an NADPH-regenerating reaction in the pentose phosphate pathway and was also found in oleaginous yeast *R. toruloides.*[127] Additionally, ME overproduction in *Y. lipolytica* indicated that NADPH concentration was not limiting for lipid accumulation in this yeast strain.[103] Furthermore, transcription of ME in *R. toruloides* was down-regulated.[128] Thus, to claim that ME is the rate-limiting step in fatty acid synthesis or that ME is the controlling factor for increased lipid accumulation is speculative. ME could have a major function in lipid synthesis, but this does not exclude other factors/enzymes that may have more critical roles in the complicated lipid accumulation process.

The complexity of lipid accumulation in oleaginous cells placed in a nitrogen-limiting environment is well demonstrated by several very recent studies. First, a proteomic analysis of *L. starkeyi*[126] indicated that nitrogen deficiency induced a global response and led to changes of many proteins. A great deal of proteins involved in glycolysis were down-regulated while the majority of up-regulated proteins were related to proteolysis, peptidolysis, carbohydrate and lipid metabolism. Second, a multi-omic study of *R. toruloides* revealed that lipid accumulation under nitrogen-limited conditions correlated with the induction of lipogenesis, nitrogenous compound recycling, macromolecule metabolism, and autophagy.[128] In contrast to results from *L. starkeyi*, the glycolysis pathway was up-regulated at lipid-production stages. In addition, several key genes related to lipogenesis, such as *ACL1*, *ACC1*, *ME1*, were not transcriptionally regulated. Third, lipid accumulation in *Y. lipolytica* appeared to be very different from other oleaginous yeasts. While ME is not critical for lipid accumulation for this yeast, deletion of a gene coding for glycerol-3-phosphate dehydrogenase (*GUT2*) resulted in a strong boost in the final lipid content of the yeast.[102] Fourth, a comparative genomic analysis between oleaginous and non-oleaginous strains showed that there may be other routes leading to acetyl-CoA synthesis and leucine metabolism and lysine degradation may be heavily involved in the process.[129]

Thus, the mechanism for extensive lipid accumulation in oleaginous species remains an unsolved puzzle. Oleaginous strains may share some common strategies for accumulating lipids, but the major machinery may very much depend on individual cells. The following section describes briefly *Y. lipolytica*, the oleaginous yeast strain that has been intensively and successfully studied and engineered.

The yeast *Y. lipolytica*, an obligate aerobe, is often isolated from environments rich in hydrophobic substrates (HS), such as alkanes and lipids. This yeast species can use HS as the sole carbon source[130] and store oils as high as 90% as TAGs. Its superior capability in accumulating oils when grown on HSs is probably related to protrusions formed on cell surfaces, which may facilitate uptake of HSs from the medium.[131] Besides HSs, *Y. lipolytica* can also utilize sugars, alcohols, starch hydrolysates and crude glycerol for growth.

Lipid accumulation in microbes that can use HS as the carbon source takes place through two pathways. First, similar to other microorganisms that do not possess the capability of utilizing HSs, intracellular oils are accumulated through a *de novo* synthesis pathway. Second, unique to these HS-consuming microbial strains, microbial oils are accumulated through the *ex novo* accumulation pathway. This pathway enables the cells to uptake fatty acids, oils and TAGs from the extracellular medium through processes, such as hydrolysis of HSs, transport of released fatty acids into the cells and re-assembly of these molecules to form TAGs.[1]

The genome of *Y. lipolytica* (ATCC 20460) was sequenced ten years ago. Since then, extensive studies have been performed in terms of enhancing lipid productivity through genetic and metabolic engineering approaches. Five different pathways have been explored. The first pathway starts from citric acid whose elevated level under excessive carbon but limiting nitrogen leads to production of acetyl-CoA due to activities of ACL. Part of this pool is converted to malonyl-CoA by the ACC. The fatty acid synthase (FAS) then adds successively acetyl-CoA to elongate the acyl chain by two carbons at each cycle. Through overexpressing ACC[132] and increasing citric acid production by inactivating the 2-methylcitrate dehydrogenase,[7] increased lipid production was observed. The second pathway involves synthesis of glycerol-3-phosphate (G3P), the TAG backbone. *Y. lypolytica* was found to have a unique glycerol metabolism, dedicated to the synthesis of G3P, but not glycerol. Deletion of glycerol-3-phosphate dehydrogenase gene (*GUT2*) prevented conversion of G3P to dihydroxyacetone phosphate (DHAP) and resulted in a three-fold increase in lipid accumulation.[102] The additional overexpression of *GPD1* gene catalyzing the conversion of DHAP to G3P[133] led to a four-fold increase in lipid production.

The third pathway focuses on diacylglycerol (DAG), the direct precursor to TAG. This molecule can be either produced by the glycerol-3-phosphate pathway or by the monoacylglycerol (MAG) pathway. In the latter, DAG is synthesized by direct acylation of MAG by an acyl-CoA:diacylglycerol acyltransferase (DGAT). Overexpression of the *DGA2* gene under the control of acyl-CoA oxidase (*POX2*) constitutive promoter in *Y. lipolytica* cultivated in oleic acid containing medium resulted in an overall increase of 63% in lipid accumulation.[134] Combining overproduction of acyl-CoA by overexpression of ACC and overexpression of *DGA2* have led to increased TAG synthesis.[132] The fourth pathway aims to improve TAG accumulation by abolishing degradation of FFAs and acyl-CoA through inactivation of the β oxidation. Deletion of six *POX* genes encoding for acyl-CoA oxidase together with deletion of the *GUT2* gene and overexpression of the *GPD1* gene led to obese *Y. lipolytica* cells with an intracellular lipid content of 80%.[134] The fifth pathway is to modify the fatty acid profiles of the TAG accumulated. In *Y. lipolytica*, there are two fatty acid desaturases: Δ9 and Δ12. Through overexpression of homologous and heterologous desaturases and elongases from different microorganisms, engineered *Y. lipolytica* strains were developed for the production of oil enriched in EPA and ARA.[7] For details regarding lipid metabolism in oleaginous yeasts, recent reviews by Nicaud's group in France are highly recommended.[1,4,7,102,103,134]

8.5 Concluding Remarks

Production of microbial oils has faced ups and downs throughout its short history. Technically, this approach is attractive and feasible. However, it is still too expensive to be applied broadly. In the short term, producing high value microbial oils beyond omega-3 fatty acids will have a bright future.

In the long term, microbial oils can be produced at industrial scale only if these challenges are addressed:

(1) *Lower feedstock prices:* since sugars derived from lignocellulosic feedstocks possess the feature of being abundant and renewable, more research and development will be devoted to cultivating wild or engineered oleaginous strains or other microbial species on cellulosic sugars for oil production. This area of research will benefit from the ongoing work intensively dedicated to maximizing sugar release and decreasing the process cost incurred during pretreatment and enzymatic hydrolysis. Thus, cellulosic sugars may be a cost-effective choice for feeding oil-producing microorganisms.
(2) *Robust microbial strains:* lignocellulosic hydrolysates generally consist of non-sugar compounds that may be detrimental to microbial growth. While some oleaginous yeast strains can tolerate these inhibitory chemicals, their lipid production titer may still be affected negatively. In addition, for engineered non-oleaginous strains, such as *E. coli*, the toxicity issue must be addressed. Furthermore, if wastewater is used as the carbon source, then robust microbes, either a single strain or a community, must be resilient enough to face challenges encountered in an open environment.
(3) *Effective downstream processes:* oils accumulated in oleaginous strains are intracellular. Instead of using the traditional solvent extraction approach for oil extraction, innovative ways must be developed to maximally disrupt cell structures and release the oils. Currently, though various methods that are not reviewed here have been investigated, they all suffer the issues of high energy consumption and low efficiency.

References

1. A. Beopoulos, J. Cescut, R. Haddouche, J. L. Uribelarrea, C. Molina-Jouve and J. M. Nicaud, *Prog. Lipid Res.*, 2009, **48**, 375.
2. C. Ratledge and J. P. Wynn, *Adv. Appl. Microbiol.*, 2002, **51**, 1.
3. Q. Li, W. Du and D. Liu, *Appl. Microbiol. Biotechnol.*, 2008, **80**, 749.
4. A. Beopoulos and J.-M. Nicaud, *OCL*, 2012, **19**, 22.
5. Y. Chisti, *Trends Biotechnol.*, 2008, **26**, 126.
6. C. Ratledge and Z. Cohen, *Lipid Technol.*, 2008, **20**, 155.
7. F. Thevenieau and J.-M. Nicaud, *OCL*, 2013, **20**, D603.
8. Y. Liang, Y. Cui, J. Trushenski and J. W. Blackburn, *Bioresource Technol.*, 2010, **101**, 7581.
9. M. Thiru, S. Sankh and V. Rangaswamy, *Bioresour. Technol.*, 2010, **102**, 10436.
10. J. Xu, X. Zhao, W. Wang, W. Du and D. Liu, *Biochem. Eng. J.*, 2012, **65**, 30.
11. S. Papanikolaou and G. Aggelis, *Bioresour. Technol.*, 2002, **82**, 43.

12. S. Papanikolaou, L. Muniglia, I. Chevalot, G. Aggelis and I. Marc, *Curr. Microbiol.*, 2003, **46**, 124.
13. J. C. Thompson and B. B. He, *Appl. Eng. Agric.*, 2006, **22**, 261.
14. Z. P. Chi, D. Pyle, Z. Wen, C. Frear and S. Chen, *Process Biochem.*, 2007, **42**, 1537.
15. Asad-ur-Rehman, R. G. Saman Wijesekara, S. S. Nakao Nomura and M. Matsumura, *J. Chem. Technol. Biotechnol.*, 2008, **83**, 1072–1080.
16. K. S. Tyson, J. Bozell, R. Wallace, E. Petersen and L. Moens, *National Renewable Energy Laboratory, NREL/TP-510-34796*, 2004.
17. Y. Mu, Z.-L. Xiu and D.-J. Zhang, *Biochem. Eng. J.*, 2008, **40**, 537.
18. Y. Liang, N. Sarkany, Y. Cui and J. W. Blackburn, *Bioresour. Technol.*, 2010, **101**, 6745.
19. M. J. Haas, A. J. McAloon, W. C. Yee and T. A. Foglia, *Bioresour. Technol.*, 2005, **97**, 671.
20. C. W. D. Chiu, M. A. Dasari, W. R. Sutterlin and G. J. Suppes, *Ind. Eng. Chem. Res.*, 2006, **45**, 791.
21. A. Alhanash, E. F. Kozhevnikova and I. V. Kozhevnikov, *Catal. Lett.*, 2008, **120**, 307.
22. M. A. Dasari, P. Kiatsimkul, W. R. Sutterlin and G. J. Suppes, *Appl. Catal., A*, 2005, **281**, 225.
23. C. W. D. Chiu, M. A. Dasari, G. J. Suppes and W. R. Sutterlin, *AIChE J.*, 2006, **52**, 3543.
24. M. Gonzalez-Pajuelo, I. Meyinal-Salles, F. Mendes, J. C. Andrade, I. Vasconcelos and P. Soucaille, *Metab. Eng.*, 2005, 7, 329.
25. W. Rymowicz, A. Rywinska and W. Gladkowski, *Chem. Pap.*, 2008, **62**, 239–246.
26. M. S. Narayan, G. P. Manoj, K. Vatchravelu, N. Bhagyalakshmi and M. Mahadevaswamy, *Int. J. Food Sci. Nutr.*, 2005, **56**, 521.
27. Y. Dharmadi, A. Murarka and R. Gonzalez, *Biotechnol. Bioeng.*, 2006, **94**, 821.
28. D. J. Pyle, R. A. Garcia and Z. Wen, *J. Agric. Food Chem.*, 2008, **56**, 3933.
29. EIA. http://www.eia.gov/biofuels/biodiesel/production/table1.pdf.
30. W. Zhou, Y. Li, M. Min, B. Hu, P. Chen and R. Ruan, *Bioresour. Technol.*, 2011, **102**, 6909.
31. J. R. Benemann and W. J. Oswald, *NASA STI/Recon Technical Report N 95*, 1994, 19554.
32. M. Prathima Devi, G. Venkata Subhash and S. V. Mohan, *Renewable Energy*,**43**, 276.
33. S. Chinnasamy, A. Bhatnagar, R. W. Hunt and K. C. Das, *Bioresour. Technol.*, 2010, **101**, 3097.
34. D. Honda, T. Yokochi, T. Nakahara, M. Erata and T. Higashiharah, *Mycol. Res.*, 1998, **102**, 439.
35. C. Saenge, B. Cheirsilp, T. T. Suksaroge and T. Bourtoom, *Biotechnol. Bioprocess Eng.*, 2011, **16**, 23.
36. L. Huang, B. Zhang, B. Gao and G. Sun, *Environ. Technol.*, 2011, **32**, 1975–1981.

37. Z. Chi, Y. Zheng, A. Jiang and S. Chen, *Appl. Biochem. Biotechnol.*, 2011, **165**, 442–453.
38. Y. H. Seo, I. G. Lee and J. I. Han, *Bioresour. Technol.*, 2013, **135**, 304–308.
39. Z. Chi, Y. Zheng, J. Ma and S. Chen, *Int. J. Hydrogen Energy*, 2011, **36**, 9542–9550.
40. A. Ykema, E. C. Verbree, M. M. Kater and H. Smit, *Appl. Microbiol. Biotechnol.*, 1988, **29**, 211.
41. N. J. Moon, E. Hammond and B. A. Glatz, *J. Dairy Sci.*, 1978, **61**, 1537.
42. B. Cheirsilp and Y. Louhasakul, *Bioresour. Technol.*, 2013, **142**, 329.
43. Q. Fei, H. N. Chang, L. Shang, N. Kim and J. Kang, *Bioresour. Technol.*, 2011, **102**, 2695.
44. H. El Bialy, O. M. Gomaa and K. S. Azab, *World J. Microbiol. Biotechnol.*, 2011, **27**, 2791.
45. D. E. Leiva-Candia, S. Pinzi, M. D. Redel-Macías, A. Koutinas, C. Webb and M. P. Dorado, *Fuel*, 2014, **123**, 33.
46. P. Molton, A. Fassbender and M. D. Brown, US Environmental Protection Agency, Water Engineering Research Laboratory, 1986.
47. S. Itoh, A. Suzuki, T. Nakamura and S.-Y. Yokoyama, *Desalination*, 1994, **98**, 127.
48. M. Downing, L. M. Eaton, R. L. Graham, M. H. Langholtz, R. D. Perlack, A. F. Turhollow Jr, B. Stokes and C. C. Brandt, Oak Ridge National Laboratory (ORNL), 2011.
49. M. B. Sticklen, *Nat. Rev. Genet.*, 2008, **9**, 433.
50. A. H. Conner, ed. C.-S. Wu, Marcel Dekker, Inc., New York, Handbook of Size Exclusion Chromatography and Related Techniques, 1995, vol. 69, ch. 13, p. 331.
51. Y.-H. P. Zhang and L. R. Lynd, *Biotechnol. Bioeng.*, 2004, **88**, 797.
52. B. Kamm and M. Kamm, *Appl. Microbiol. Biotechnol.*, 2004, **64**, 137.
53. L. R. Lynd, *Annu. Rev. Energy Environ.*, 1996, **21**, 403.
54. Y.-H. P. Zhang, M. E. Himmel and J. R. Mielenz, *Biotechnol. Adv.*, 2006, **24**, 452.
55. K. Karim, L. T. Angenent, M. H. Al-Dahhan, B. A. Wrenn and R. Domiguez-Espinosa, *Trends Biotechnol.*, 2004, **22**, 477.
56. L. R. Lynd, P. J. Weimer, W. H. van Zyl and I. S. Pretorius, *Microbiol. Mol. Biol. Rev.*, 2002, **66**, 506.
57. L. R. Lynd, J. Cushman, R. J. Nichols and C. E. Wyman, *Science*, 1991, **251**, 1318.
58. L. R. Lynd, C. E. Wyman and T. U. Gerngross, *Biotechnol. Prog.*, 1999, **15**, 777.
59. C. E. Wyman, *Annu. Rev. Energy Environ.*, 1999, **24**, 189.
60. A. L. Demain, M. Newcomb and J. H. D. Wu, *Microbiol. Mol. Biol. Rev.*, 2005, **69**, 124.
61. A. Jain, K. Caldeira and M. I. Hoffert, *Science*, 2003, **299**, 2052.
62. K Caldeira, M. I. Hoffert, G. Benford, D. R. Criswell, C. Green, H. Herzog, A. K. Jain, H. S. Kheshgi, K. S. Lackner, J. Lewis *et al.*, *Science*, 2002, **298**, 981.

63. K Caldeira, M. I. Hoffert, A. K. Jain, E. F. Haites, L. D. D. Harvey, S. D. Potter, M. E. Schlesinger, S. H. Schneider, T. Wigley, R. G. Watts *et al.*, *Nature*, 1998, **395**, 881.
64. C. Gray, T. E. Wirth and J. D. Podesta, *Foreign Affairs*, 2003, **82**, 125.
65. C. Williams, A. J. Ragauskas, B. H. Davison, G. Britovsek, J. Cairney, C. A. Eckert, W. J. Frederick Jr., D. Leak, J. P. Hallett, C. L. Liotta *et al.*, *Science*, 2006, **311**, 484.
66. P. Alvira, E. Tomãs-Pejo, M. Ballesteros and M. J. Negro, *Bioresour. Technol.*, 2010, **101**, 4851.
67. V. S. Chang and M. T. Holtzapple, *Appl. Biochem. Biotechnol.*, 2000, **84–86**, 5.
68. J. Zhu and X. Pan, *Bioresour. Technol.*, 2010, **101**, 4992.
69. V. B. Agbor, N. Cicek, R. Sparling, A. Berlin and D. B. Levin, *Biotechnol. Adv.*, 2011, **29**, 675.
70. H. Tadesse and R. Luque, *Energy Environ. Sci.*, 2011, **4**, 3913.
71. J. T.-L. Simón, A. R. José and V. Manuel, *J. Sci. Food Agric.*, 2002, **82**, 505.
72. S. Mehmood, M. Gulfraz, N. F. Rana, A. Ahmad, B. K. Ahring, N. Minhas and M. F. Malik, *Afr. J. Biotechnol.*, 2009, **8**, 2857.
73. A. Herrera, S. J. Téllez-Luis, J. J. González-Cabriales, J. A. Ramírez and M. Vázquez, *J. Food Eng.*, 2004, **63**, 103.
74. M. Vázquez, M. Oliva, S. J. Téllez-Luis and J. A. Ramírez, *Bioresour. Technol.*, 2007, **98**, 3053.
75. B. Sipos, J. Reczey, Z. Somorai, Z. Kadar, D. Dienes and K. Reczey, *Appl. Biochem. Biotechnol.*, 2009, **153**, 151.
76. D. Salvi, G. Aita, D. Robert and V. Bazan, *J. Ind. Microbiol. Biotechnol.*, 2010, **37**, 27.
77. B.-Z. Li, V. Balan, Y.-J. Yuan and B. E. Dale, *Bioresour. Technol.*, 2010, **101**, 1285.
78. I. Dogaris, S. Karapati, D. Mamma, E. Kalogeris and D. Kekos, *Bioresour. Technol.*, 2009, **100**, 6543.
79. R. Choudhary, A. L. Umagiliyage, Y. Liang, T. Siddaramu, J. Haddock and G. Markevicius, *Biomass Bioenergy*, 2012, **39**, 218.
80. Y. Liang, T. Tang, T. Siddaramu, R. Choudhary and A. L. Umagiliyage, *Renewable Energy*, 2012, **40**, 130.
81. Y. Liang, T. Tang, A. L. Umagiliyage, T. Siddaramu, M. McCarroll and R. Choudhary, *Appl. Energy*, 2012, **91**, 451.
82. Y. Liang, K. Jarosz, A. T. Wardlow, J. Zhang and Y. Cui, *Appl. Bbiochem. Biotechnol.*, 2014, **173**, 2086.
83. Y. Liang, I. Perez, K. Goetzelmann and S. Trupia, *Biotechnol. Prog.*, 2014, **30**, 945.
84. Z. Gong, H. Shen, Q. Wang, X. Yang, H. Xie and Z. K. Zhao, *Biotechnol. Biofuels*, 2013, **6**, 1.
85. C. Huang, X.-F. Chen, X.-Y. Yang, L. Xiong, X.-Q. Lin, J. Yang, B. Wang and X.-D. Chen, *Appl. Biochem. Biotechnol.*, 2014, **172**, 2197.
86. S. Galafassi, D. Cucchetti, F. Pizza, G. Franzosi, D. Bianchi and C. Compagno, *Bioresour. Technol.*, 2012, **111**, 398.

87. C. Huang, M. Zong, H. Wu and Q. Liu, *Bioresour. Technol.*, 2009, **100**, 4535.
88. C. Huang, X. Chen, L. Xiong and L. Ma, *Bioresour. Technol.*, 2012, **110**, 711.
89. C. Hu, S. Wu, Q. Wang, G. Jin, H. Shen and Z. K. Zhao, *Biotechnol. Biofuels*, 2011, **4**, 25.
90. X.-f. Chen, C. Huang, L. Xiong and L.-l. Ma, *Biotechnol. Lett.*, 2012, **34**, 1025.
91. X. Yu, Y. Zheng, K. M. Dorgan and S. Chen, *Bioresour. Technol.*, 2011, **102**, 6134.
92. E. Palmqvist and B. Hahn-Hägerdal, *Bioresour. Technol.*, 2000, **74**, 25.
93. E. Palmqvist and B. r. Hahn-Hãgerdal, *Bioresour. Technol.*, 2000, **74**, 17.
94. P. Meesters, G. N. M. Huijberts and G. Eggink, *Appl. Microbiol. Biotechnol.*, 1996, **45**, 575.
95. D. J. Kyle and C. Ratledge, 1992. http://agris.fao.org/agris-search/search.do?recordID = US201300725603.
96. B. A. Glatz, E. G. Hammond, K. H. Hsu, L. Baehman, N. Bati, W. Bednarski, D. Brown and M. Floetenmeyer, *Biotechnol. Oils Fats Ind.*, 1984, 163–176.
97. C. T. Evans and C. Ratledge, *Lipids*, 1983, **18**, 623.
98. M. Hassan, P. Blanc, A. Pareilleux and G. Goma, *World J. Microbiol. Biotechnol.*, 1994, **10**, 534.
99. M. Hassan, P. J. Blanc, L. M. Granger, A. Pareilleux and G. Goma, *Process Biochem.*, 1996, **31**, 355.
100. Corn Refiners Association, volume chemical challenge program prepared for November, 2006.
101. X. Chen, Z. Li, X. Zhang, F. Hu, D. D. Y. Ryu and J. Bao, *Appl. Biochem. Biotechnol.*, 2009, **159**, 591.
102. A. Beopoulos, Z. Mrozova, F. Thevenieau, M. T. Le Dall, I. Hapala, S. Papanikolaou, T. Chardot and J. M. Nicaud, *Appl. Environ. Microbiol.*, 2008, **74**, 7779.
103. A. Beopoulos, T. Chardot and J. M. Nicaud, *Biochimie*, 2009, **91**, 692.
104. C. J. Greenwalt, J. B. Hunter, S. Lin, S. McKenzie and A. Denvir, *Int. J. Earth Space*, 2000, 7, 243.
105. W. Liu, Y. Wang, Z. Yu and J. Bao, *Bioresour. Technol.*, 2012, **118**, 13.
106. R. Davis, L. Tao, E. Tan, M. Biddy, G. Beckham, C. Scarlata, J. Jacobson, K. Cafferty, J. Ross and J. Lukas, National Renewable Energy Laboratory (NREL), Golden, CO., 2013.
107. T. C. Walther and R. V. Farese Jr, *Ann. Rev. Biochem.*, 2012, **81**, 687.
108. A. P. Desbois and V. J. Smith, *Appl. Microbiol. Biotechnol.*, 2010, **85**, 1629.
109. A. M. P. Jones, J. A. Klun, C. L. Cantrell, D. Ragone, K. R. Chauhan, P. N. Brown and S. J. Murch, *J. Agric. Food Chem.*, 2012, **60**, 3867.
110. T. Liu, H. Vora and C. Khosla, *Metabol. Eng.*, 2010, **12**, 378.
111. M. S. Davis, J. Solbiati and J. E. Cronan, *J. Biol. Chem.*, 2000, **275**, 28593.
112. H. Cho and J. E. Cronan, *J. Biol. Chem.*, 1995, **270**, 4216.

113. P. Jiang and J. Cronan, *J. Bacteriol.*, 1994, **176**, 2814.
114. C. Dellomonaco, J. M. Clomburg, E. N. Miller and R. Gonzalez, *Nature*, 2011, **476**, 355.
115. H. Liu, C. Yu, D. Feng, T. Cheng, X. Meng, W. Liu, H. Zou and M. Xian, *Microb. Cell Fact.*, 2012, **11**, 41.
116. P. Javidpour, J. H. Pereira, E.-B. Goh, R. P. McAndrew, S. M. Ma, G. D. Friedland, J. D. Keasling, S. R. Chhabra, P. D. Adams and H. R. Beller, *Appl. Environ. Microbiol.*, 2014, **80**, 497.
117. R. M. Lennen and B. F. Pfleger, *Trends Biotechnol.*, 2012, **30**, 659.
118. X. Meng, J. Yang, X. Xu, L. Zhang, Q. Nie and M. Xian, *Renewable Energy*, 2009, **34**, 1.
119. Y. Zhang, I. P. Adams and C. Ratledge, *Microbiology*, 2007, **153**, 2013.
120. J. L. Moseley, D. Gonzalez-Ballester, W. Pootakham, S. Bailey and A. R. Grossman, *Genetics*, 2009, **181**, 889.
121. K. Sugimoto, N. Sato and M. Tsuzuki, *FEBS Lett.*, 2007, **581**, 4519.
122. K. Sugimoto, T. Midorikawa, M. Tsuzuki and N. Sato, *Biochem. Biophys. Res. Commun.*, 2008, **369**, 660.
123. C. Ratledge, *Biochem. Soc. Trans.*, 2002, **30**, 1047.
124. C. Ratledge, *Biochimie*, 2004, **86**, 806.
125. T. G. Dunahay, E. E. Jarvis and P. G. Roessler, *J. Phycol.*, 1995, **31**, 1004.
126. H. Liu, X. Zhao, F. Wang, X. Jiang, S. Zhang, M. Ye, Z. K. Zhao and H. Zou, *FEMS Yeast Res.*, 2011, **11**, 42.
127. H. Liu, X. Zhao, F. Wang, Y. Li, X. Jiang, M. Ye, Z. K. Zhao and H. Zou, *Yeast*, 2009, **26**, 553.
128. Z. Zhu, S. Zhang, H. Liu, H. Shen, X. Lin, F. Yang, Y. J. Zhou, G. Jin, M. Ye and H. Zou, *Nat. Commun.*, 2012, **3**, 1112.
129. T. Vorapreeda, C. Thammarongtham, S. Cheevadhanarak and K. Laoteng, *Microbiology*, 2012, **158**, 217–228.
130. P. Fickers, P. H. Benetti, Y. Wache, A. Marty, S. Mauersberger, M. Smit and J. M. Nicaud, *FEMS Yeast Res.*, 2005, **5**, 527.
131. K. Mlíčková, E. Roux, K. Athenstaedt, S. d'Andrea, G. Daum, T. Chardot and J.-M. Nicaud, *Appl. Environ. Microbiol.*, 2004, **70**, 3918.
132. M. Tai and G. Stephanopoulos, *Metab. Eng.*, 2013, **15**, 1.
133. T. Dulermo and J.-M. Nicaud, *Metab. Eng.*, 2011, **13**, 482.
134. A. Beopoulos, J.-M. Nicaud and C. Gaillardin, *Appl. Microbiol. Biotechnol.*, 2011, **90**, 1193.

CHAPTER 9

Bioprocessing of Cost-competitive Biobased Organic Acids

YUPO J. LIN,*[a] JAMIE A. HESTEKIN,[b] MICHAEL P. HENRY[a] AND NORMAN SATHER[a]

[a] Argonne National Laboratory, 9700 S. Cass Avenue, Argonne, IL 50439, USA; [b] University of Arkansas, 3202 Bell Engineering Center, Fayetteville, AR 72701, USA
*Email: yplin@anl.gov

9.1 Introduction

This chapter discusses a new technology, called a separative bioreactor, which is being developed for the recovery of low-titer organic acids from fermentation broths. A separative bioreactor (SB) is a fermenter or other type of bioreactor with combined bioconversion and separations capability. The design of the separative bioreactor incorporates and integrates separations features such that product or products can be removed continuously from the bioreactor. This feature enables control of pH, product inhibition effects, and other operational factors that can limit the performance of the bioreactor. The potential of SB technology for economic bioproduction of organic acids is demonstrated for three different process designs in which the integral separations technology is electrodeionization. Electrodeionization (EDI) is an electrically-driven membrane separations technology for separating charged species, including organic acids, from liquids. The different SB-based

RSC Green Chemistry No. 43
Commercializing Biobased Products: Opportunities, Challenges, Benefits, and Risks
Edited by Seth W. Snyder

Published by the Royal Society of Chemistry, www.rsc.org

bioprocesses that will be discussed include the enzymatic conversion of glucose to produce dual products of gluconic acid and sorbitol, the anaerobic fermentation of glucose to produce succinic acid, and the aerobic production of gluconic acid from glucose. In addition to evaluating application of SB to a variety of bioconversion processes, a range of bioreactor operating conditions was also investigated to more fully demonstrate the performance of the SB concept. For each of the three SB-based processes the productivities and operational stabilities were measured with bench-scale equipment and subsequently validated at the pilot scale. Extended pilot test runs were conducted to investigate any long-term performance problems. Based on these results, a projection of the technical and economic feasibility of the SB concept for producing cost-competitive biobased organic acids is discussed.

9.2 Current Technology for Organic Acid Production with Bioconversion Processes

Biocatalysis (*e.g.* enzymatic conversion and fermentation) offers an attractive alternative to chemical processing because biocatalysts utilize renewable feedstocks under benign reaction conditions. One class of chemical products that could be produced in large volumes by biocatalysis is organic acids.[1–3] For example, citric acid is one of the most well developed fermentation pathways for industrial production with an annual global market value of over 2 billion dollars.[4] A number of articles have been published documenting the feedstock preparation, bioconversion, and purification of organic acids using various cultures including yeast,[5,6] *Aspergillus niger*,[7] and *Rhizopus orizae*.[8,9] The replacement of petrochemicals with biobased chemicals requires efficient bioprocesses, biocatalysts and product recovery. However, biocatalytic processes for producing organic acids typically result in dilute concentrations of the acid product that result from product inhibition and acidification that drives the bioreactor pH outside of the optimal range for the biocatalyst. Buffering or neutralization results in formation of the acid salt rather than the acid, which requires additional processing to recover the free acid product. It becomes evident that the most energy and cost intensive steps of the overall process are for product recovery and purification.

Recovery of organic products from fermentation broths in a continuous process is generally a two-step process. The first step separates the biocatalytic microbial cells from the broth using either multi-stage filtration or chromatography[10,11] and the second step strips the desired product or products from the cell-free broth. Once products are recovered from fermentation, purification is usually required. This step can take many forms and can be simple or complex depending on the nature of the organic product to be recovered. For organic acids, product recovery is typically accomplished by electrodialysis, liquid extraction, salting out, distillation, or crystallization. Electrodialysis (ED) is a charge-based separation process that selectively transports ions in solution from one compartment of an electrochemical cell to another, allowing the separation and concentration of ionic species, *e.g.*, organic acids, in solution. The advantages of ED are the selective ion transport and low energy requirements. Disadvantages include selectivity issues with membranes and limitations on final product concentrations due to water and co-ion transport, often resulting in a requirement for secondary purification steps. Liquid extraction is the most common recovery unit operation due to its simplistic approach and high recoveries and short retention times.[12,13] The main drawback is that harsh chemicals are often required, which produces large amounts of hazardous waste. Salting out is typically done by the addition of a stripping salt, which generates an organic rich aqueous phase. This is effective for quick recovery of organics but requires large quantities of salt and extraction efficiency is limited to 90%.[14,15] Crystallization is very common and beneficial because crystals can easily be recovered *via* filtration and dried to a final product. Limitations result when large amounts of waste salt are produced when crystals require dissolution to an acid state. Finally, distillation is a widely known process that is capable of extremely pure products (>99.9 wt%). Unfortunately, it is the one of the most cost and energy intensive of the available separation techniques.

Other product purification technologies that have been investigated for organic acid separations include reverse osmosis,[16] nanofiltration,[1,17] ion exchange,[18] ion chromatography,[19] Donnan dialysis,[20] and supercritical fluid extraction.[18] However, the generally *dilute concentration* of bioproduct organic acids renders these separations approaches economically impractical.[1]

Considerable reductions in product recovery costs and energy consumption, and the increase of bioconversion productivity, are required before bioproduced organic acids become cost-competitive with current petrochemical production routes. Typical electrodialytic separations require 0.2–3 kWh kg^{-1} product and cost about \$0.1–0.4 per kg acid recovered[21,22] depending on the salt or acid forms. For example, succinic acid production through biological pathways has been produced at \$0.85–2.20 per kg using electrodialysis and liquid–liquid extraction while the cost of lactic acid separation is approximately \$0.55 per kg.[2,23–25] Reliable cost information for organic acid purification by crystallization and liquid–liquid extraction is not available, but is likely to be high. Crystallization is a cumbersome operation involving refrigeration, filtration, centrifugation, and drying of crystals, and perhaps solvent extraction to remove impurities. The equipment, solvent, and power costs for pumping and mixing for product purification by liquid–liquid extraction are expected to be prohibitive for application to biochemicals.

9.3 Separative Bioreactors—Designs for Integrated Bioprocessing

The sensitivity to reduced pH and product inhibition are the two self-limiting constraints to most biological reactions. To overcome limitations

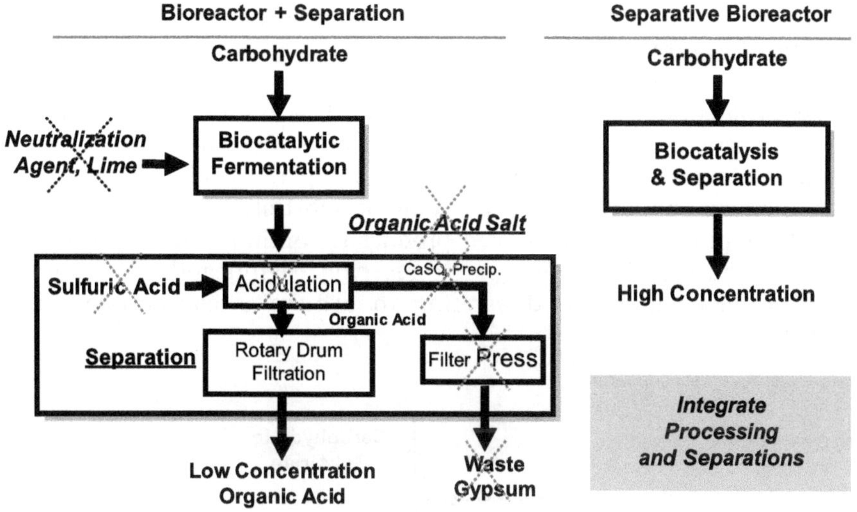

Figure 9.1 Schematic of process blocks of a batch fermentation with product separation and a separative bioreactor.

from the sensitivity to pH changes, fermentative bioprocessing generally requires the addition of chemicals (usually lime) to neutralize the organic acids so as to maintain optimal pH in the bioreactor, and this leads to the formation of organic salts, instead of organic acids. Limited by product inhibition, the concentrations of the organic salts are low. The diagram shown in the left of Figure 9.1 is a schematic of process blocks to produce and recover organic acids from conventional fermentation processes. One mole lime is required and one mole of waste gypsum is produced for each mole of organic acid formed in the fermenter. Equal moles of waste gypsum were produced for every mole production of organic acids in low product titers. For a biocatalytic reaction to be sustained without the need for acid neutralization, the organic acids need to be removed to fairly low levels from the bioreactor once they are produced to prevent product inhibition and to maintain optimum pH. The concept of *in situ* product separation during bioconversion is the principle of a separative bioreactor (SB) and the combined processes of bioconversion and separation are illustrated on the right of Figure 9.1.

9.3.1 The Separative Bioreactor

An integrated process that combines upstream bioconversion and downstream product recovery and purification offers a pathway for producing cost-competitive biobased organic acids. This is necessary to address the challenges of high-energy consumption and low acid titers associated with conventional bioproduction of organic acids. Reducing the total number of unit operations and recovering the dilute organic acids from the bioreactor at low cost are also key economic hurdles. As illustrated in the process flow diagram of Figure 9.1, several unit operations can be eliminated if the neutralization step can be avoided. A technology to support such a process design as illustrated in Figure 9.2 can provide a robust environment for biological activities without the limitations of product inhibition and pH toleration. In principle, the operation of a separative bioreactor can be continuous, as long as there is a supply of substrate. In addition, a concentrated ("purified") organic acid product is continuously stripped from the broth and captured in a separate reservoir, significantly reducing the need for energy-intensive dewatering to achieve a concentrated final product.

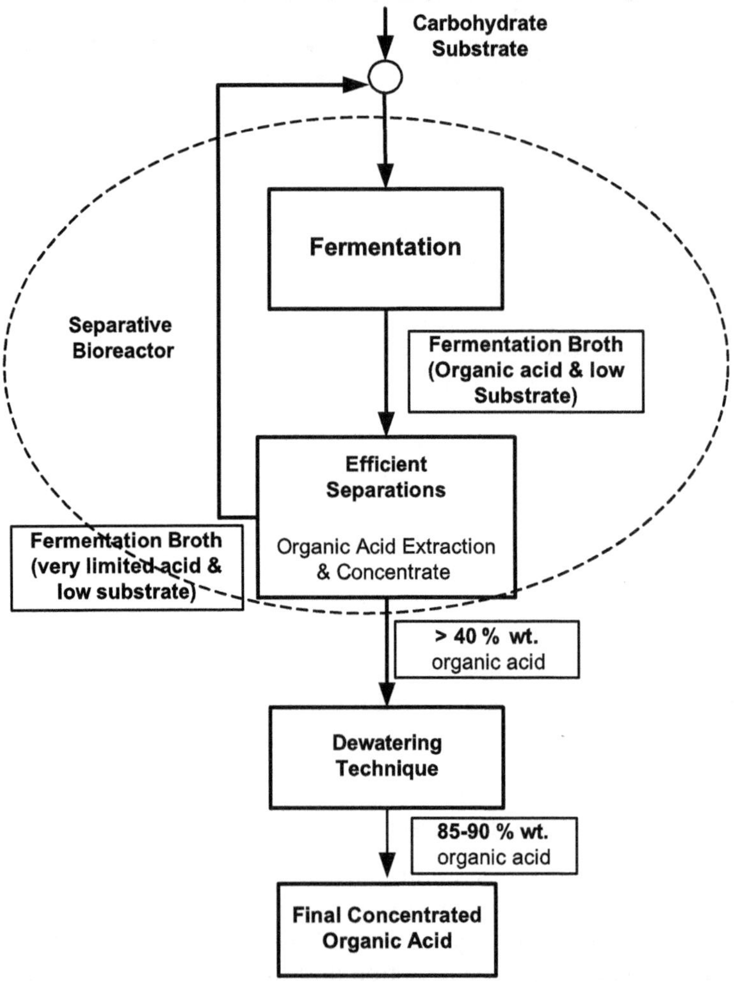

Figure 9.2 Schematic of the process blocks of organic acid production including the section of separative bioreactor.

9.3.2 Innovative Electrodeionization Technology to Capture Organic Acids

All these changes cannot be made without an efficient separations technology to effectively separate/recover the organic acids. Electrically-driven membrane separations, such as electrodialysis (ED) and electrodeionization (EDI), are considered to be energy and cost-effective processes. Due to the weakly acidic nature of most organic acids, EDI can be expected to have better energy efficiency for organic acid separations than ED. In EDI (see Figure 9.3), the organic acids are simultaneously ionized by the ion-exchange resin bed and selectively driven from the fermentation broth across the ion exchange membranes by electric field forces. EDI technology is used commercially to produce ultrapure water for the semiconductor and pharmaceutical industries. Recently, it has been applied in a seawater desalination pilot plant to produce a potable water supply.[26] The successful development of suitable resin wafers,[27] which are porous matrices containing immobilized ion-exchange resin beads, extends the potential applicability of EDI-based separation processes to relatively low-value products such as organic acids. Figure 9.4 shows the two forms of the ion exchange resin (resin wafer and loose resin beads) as seen filling the electrolyte space between the ion-exchange membranes of an EDI cell. Resin wafer electrodeionization (RW-EDI) technology is distinct from ED technology and is designed to reduce the electric energy consumption while maintaining high rates of charged species separation in low conductivity aqueous solutions.[27–29] Two key process parameters, the separation productivity (*i.e.*, the acid capture

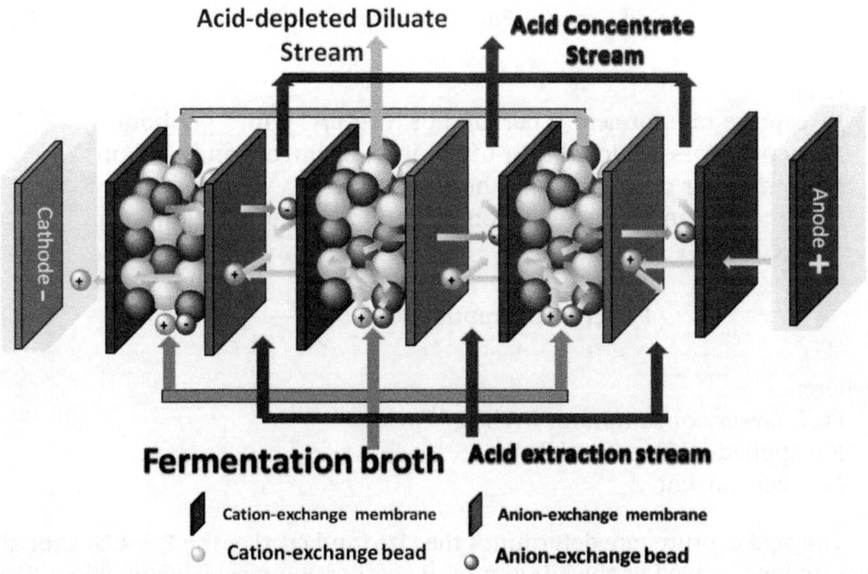

Figure 9.3 Illustration of the electrodeionization mechanism.

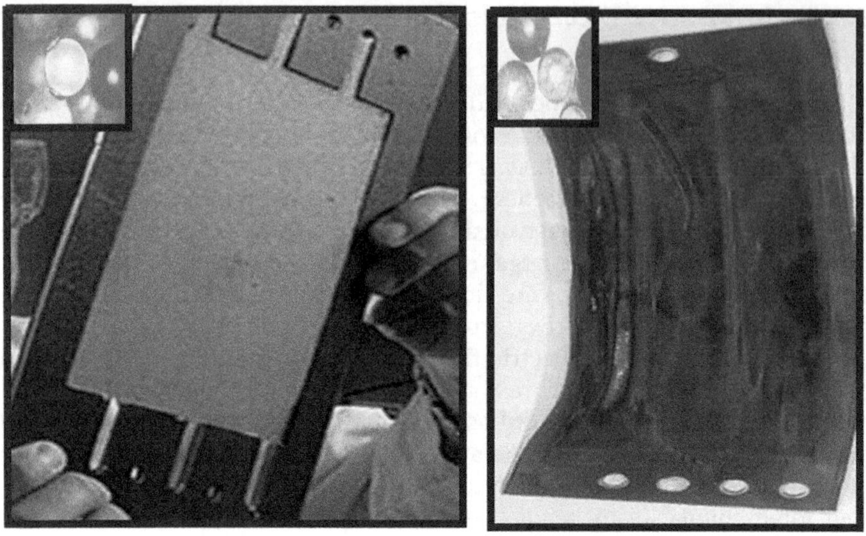

Figure 9.4 Pictures of the immobilized ion-exchange resin in porous wafer and the loose ion-exchange resin beads sealed in a pair of membranes.

rate in the case of organic acids production) and power consumption, are the main determinants of process performance and economics. These two quantities are calculated using the following formulas.

$$\text{Acid Capture Rate } (J) = 0.06 \cdot \frac{F \cdot (C_\text{i} - C_0)}{N_\text{CP} \cdot A} \tag{9.1}$$

where:

J = capture rate for acid separation or recovery, g m^{-2} per hour
A = active cross-sectional area of the ion-exchange membrane, m^2
N_CP = number of cell pairs in the RW-EDI stack
C_i, C_o = inlet and outlet acid concentrations in the solution, g kg^{-1}
F = feed rate, g min^{-1}

$$\text{Power Consumption (PC)} = \frac{V \cdot I}{J \cdot N_{CP} \cdot A} \tag{9.2}$$

where:

PC = power consumption, kWh kg^{-1} acid
V = applied stack potential, V
I = stack current, A

The acid capture rate determines the size (and cost) of the RW-EDI capital equipment, as well as the amount (and cost) of chemical consumables. The power consumption determines the electrical energy cost, which is the major component of energy usage in an EDI operation. These are the major cost items for EDI-based bioproduction of organic acids besides the feedstock cost. The economics of using SB for biobased organic acids recovery will be discussed in Section 9.4.3.

Operations and process performance of different SB configurations for producing organic acids are discussed next for three different example applications to enzymatic and fermentation bioconversion processes. In an enzymatic process the enzymes are immobilized inside the porous resin wafer of a RW-EDI stack by electrostatic forces. In the two fermentation applications of SB, a filtered slipstream of fermentation broth is pumped through the RW-EDI stack, where the organic acids are removed, and then returned to the fermenter. During the returning cycle, the bacteria are also recaptured from the retentate of the filtration step.

9.4 Demonstration of Separative Bioreactor Performance for Organic Acid Production

SB-based bioprocessing provides opportunities for improved process operation and control compared to those of conventional bioconversion processes for producing organic acids. First, it turns a batch process into a continuous process with the consequent advantages for process control and increased throughput. Second, the productivity of the bioreactor is enhanced because the continuous removal of organics greatly reduces product inhibition from elevated acid or salt concentrations. Instead, the acid capture rate in the SB implicitly and positively influences the bioconversion rate. Therefore, the design and scale-up of SB-based bioprocesses is greatly simplified compared to that of conventional bioprocesses for organic acid production. A scaled demonstration of the bench to pilot scales is discussed in the examples described next.

9.4.1 Enzyme Separative Bioreactor for the Production of Gluconic Acid and Sorbitol

In this example, gluconic acid and sorbitol are co-products produced by enzymatic conversion of glucose and fructose sugar substrates using the glucose–fructose oxidoreductase (GFOR) enzyme as a biocatalyst. The electrons flow from the oxidation of glucose to gluconic acid and the reduction of fructose to sorbitol and are shuttled and stored inside the GFOR enzyme.

The GFOR enzyme was immobilized within wafers in the RW-EDI stack by electrostatic forces associated with the slightly negative charge of the enzyme. Concentrated mixed glucose and fructose substrate was continuously fed to the electrodeionization separative bioreactor (EDI-SB). Figure 9.5 shows the process flow diagram.

Sugar substrate was charged with the recycled feed to the enzyme bioreactor. Gluconic acid produced in the bioreactor was transported by the

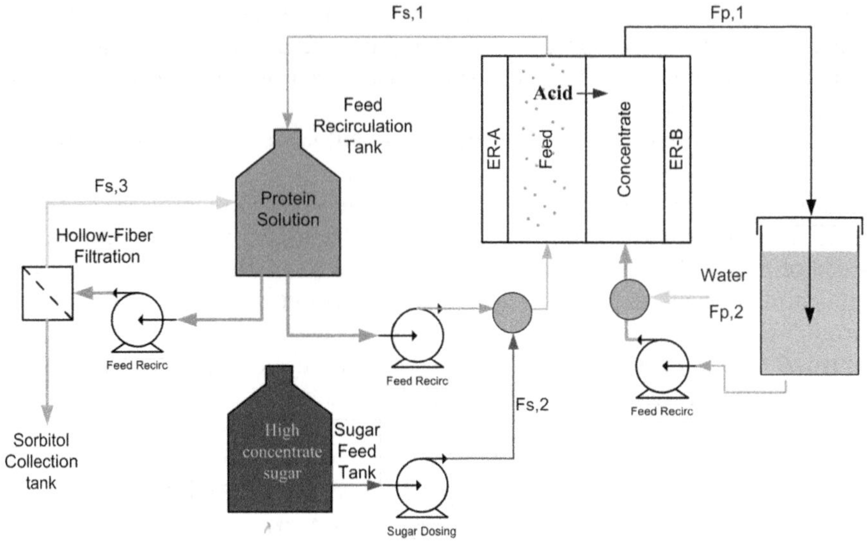

Figure 9.5 The process flow diagram for an enzyme EDI separative bioreactor.

electric field across the ion-exchange membranes and collected in the acid production tank. Another product, sorbitol, was collected in a separate tank. No pH buffering chemicals were used during the operation. The immediate removal of the gluconic acid as it was produced in the SB profoundly changed the outcomes of the biological reaction compared to a conventional bioreactor. The biggest surprise was that the effect of bulk pH in the bioreactor on biological activities was diminished. Normally, GFOR will be inactive when the pH drops below 3.5. This was not observed in the EDI-SB system. Figure 9.6 shows the enzyme activities and the solution pH for a laboratory-scale EDI-SB.

Enzyme activities were reduced by less than 20% during the entire 400 hours of continuous bioconversion despite a solution pH of about 3.0. About 6–7 g per hour of both gluconic acid and sorbitol were produced by the EDI-SB even at this low pH (see Figure 9.7).

400 g L^{-1} of purified gluconic acid was produced and extracted from the EDI-SB, corresponding to a productivity of approximately 50 g m^{-2} per hour (Figure 9.8). More than 90% of the glucose/fructose substrate was converted to the gluconic acid and sorbitol products. Greater than 99.5% of the gluconic acid produced was captured from the bioreactor in the gluconic acid recovery tank (see Figure 9.9). The final 389 g L^{-1} sorbitol product contained less than 0.5 g L^{-1} of gluconic acid.

To validate the performance of the enzymatic EDI-SB, the entire process, including enzyme production, was scaled up by a factor of 20 in a pilot-scale EDI-SB. During the pilot-scale tests the specific enzyme activity was seen to increase by nearly 10-fold. Figure 9.10 shows the comparison of enzyme activity measured in bench- and pilot-scale EDI-SB. During the pilot-scale

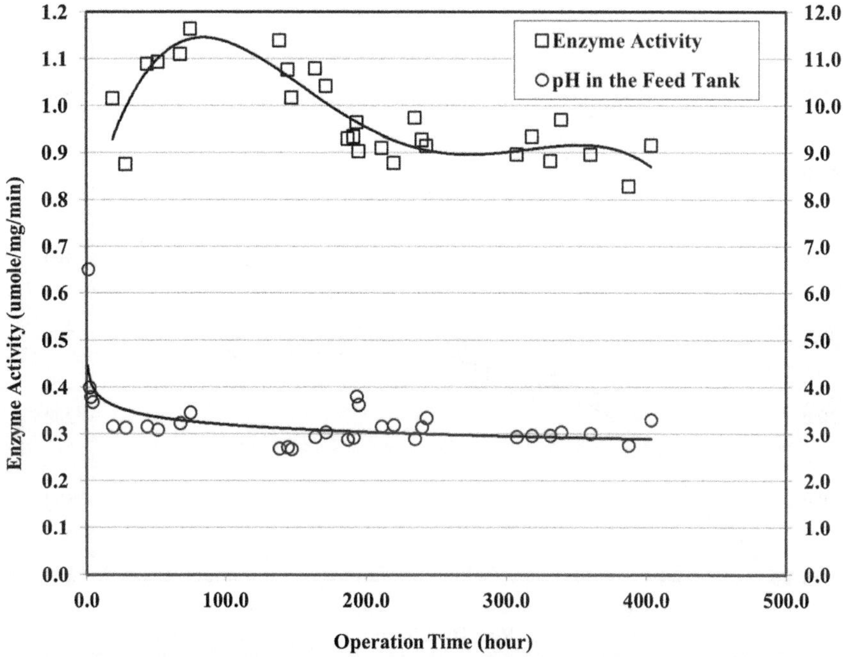

Figure 9.6 Enzyme activities and pH in the liquid phase of the SB.

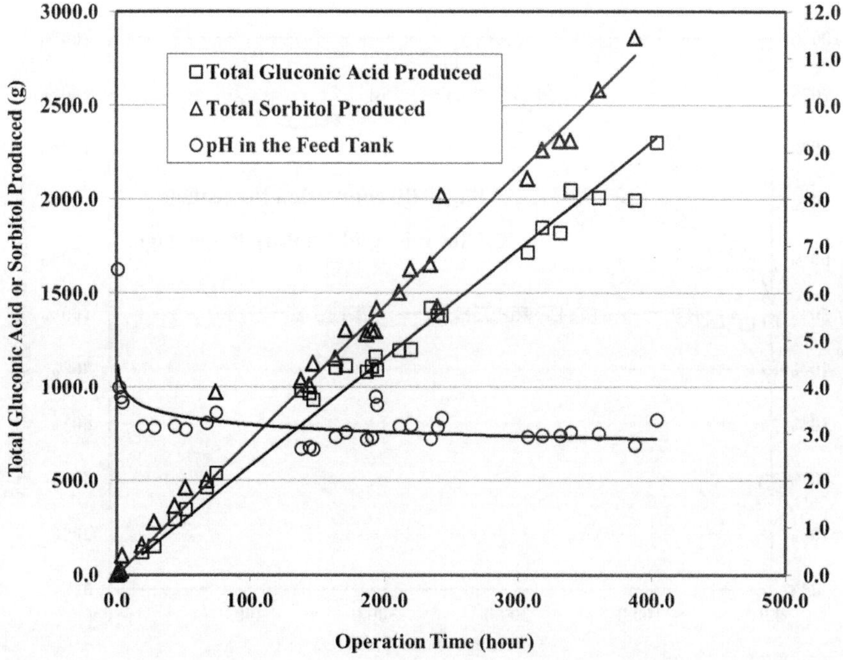

Figure 9.7 Production of gluconic acid and sorbitol and the solution pH in the EDI-SB.

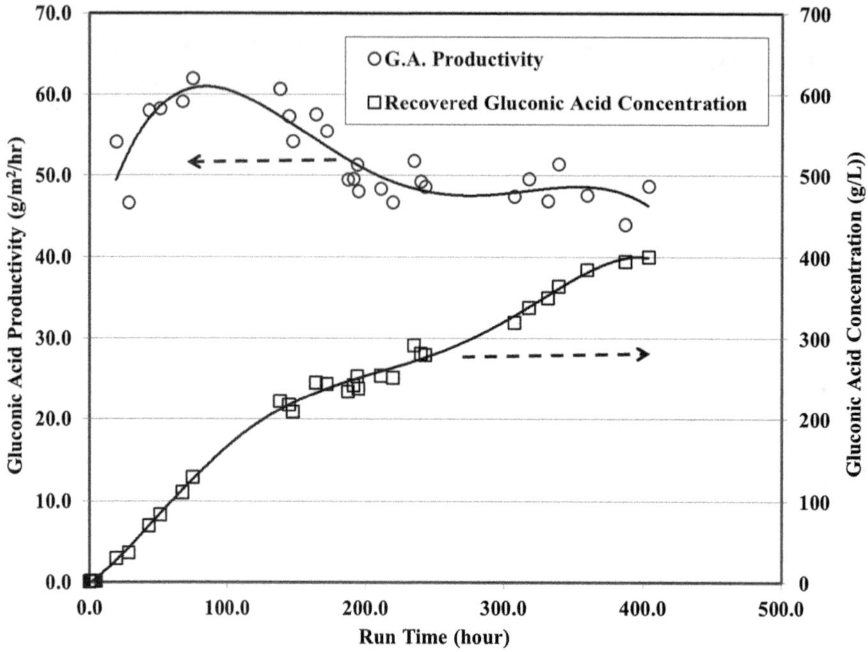

Figure 9.8 Gluconic acid productivity and concentration in enzymatic EDI-SB.

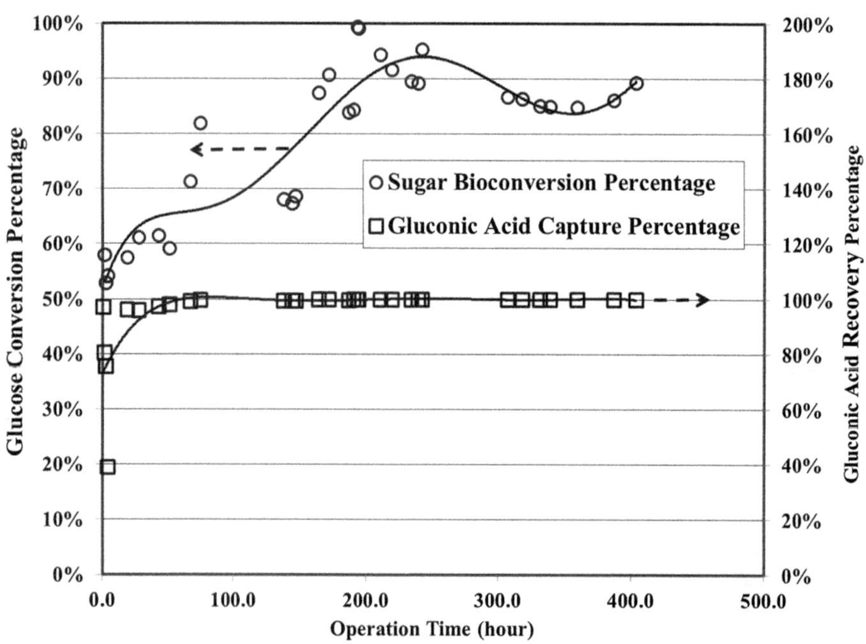

Figure 9.9 Sugar conversion percentage and gluconic acid capture ratio.

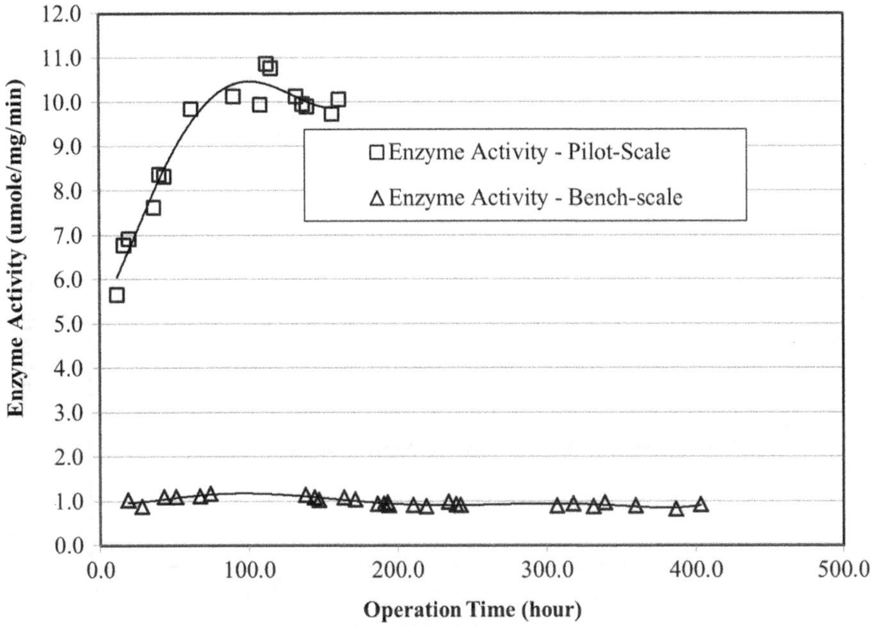

Figure 9.10 Enzyme reactivity in pilot and bench scale EDI-SB.

demonstration, there was no degradation of enzyme activity during 160 hours of continuous operation at pH 2.5 (see Figure 9.11). Both the productivity and final concentration of recovered gluconic acid were improved in pilot-scale EDI-SB compared to the bench-scale performance (see Figure 9.12). Key process performance parameters of the bench-scale and pilot-scale systems are listed in Table 9.1. The pilot-scale EDI-SB produced 1.75 lbs purified gluconic acid and 1.5 lbs sorbitol for every kWh of electricity used. The consistent process performance in both the bench- and pilot-scale systems demonstrates a linear scale-up factor for the process and robustness of EDI-SB. The stable enzyme activities and constant acid production rate shows that the pilot-scale EDI-SB can be expected to have a higher productivity if the enzyme loading can be increased.

9.4.2 Integrated Fermentation Separative Bioreactor

The coupling of fermentation and RW-EDI processes into an integrated system also simplifies the control strategy and operation compared to conventional fermentation. In addition, the control of pH is easier and more economical using an integrated fermentation and electrodeionization separative bioreactor (IF-EDI-SB) that recycles the buffering chemicals in the system. These factors and their impacts on process performance and economics are discussed next using, as examples, the anaerobic and aerobic fermentation of glucose for the production of succinic acid and gluconic acid, respectively.

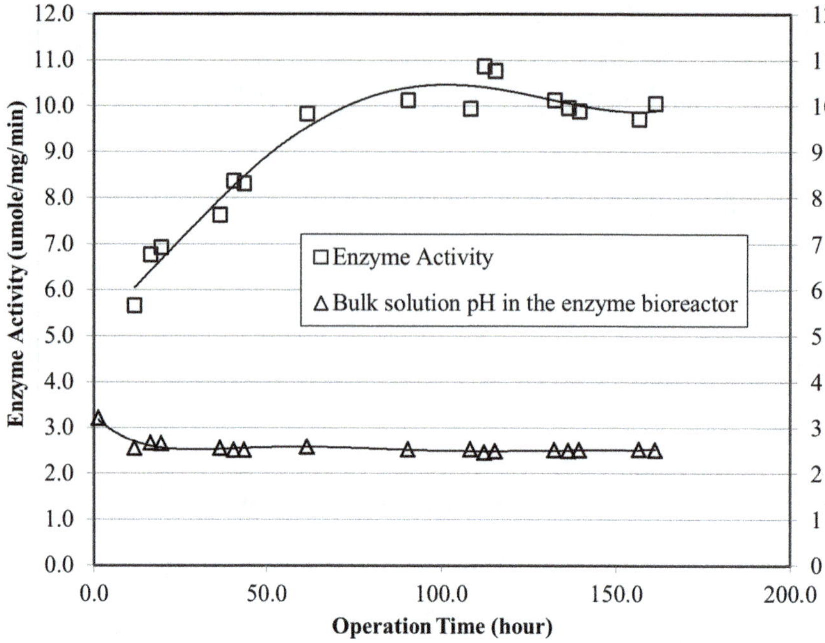

Figure 9.11 Enzyme reactivity and pH in pilot-scale EDI-SB.

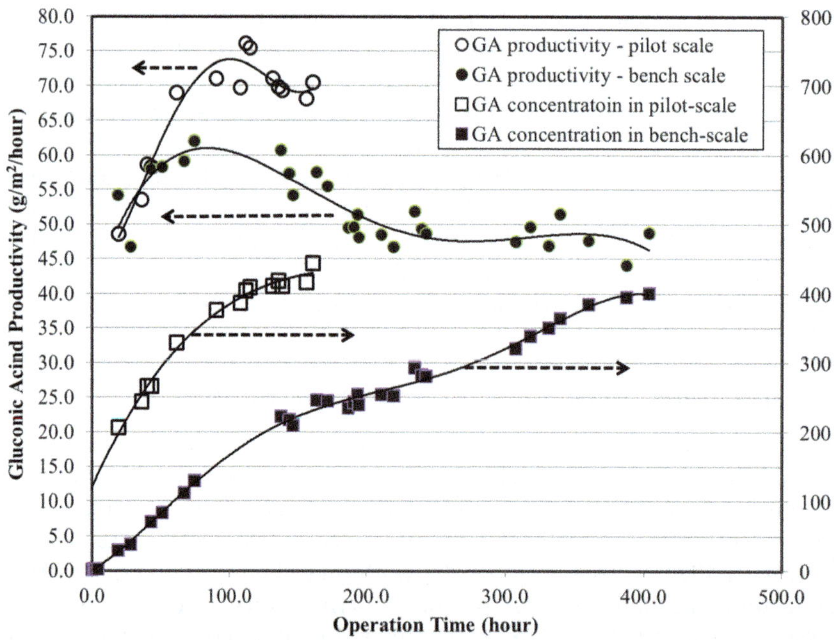

Figure 9.12 Process performance of gluconic acid production using enzyme EDI-SB.

Table 9.1 Summary of process performance of bench and pilot scale EDI-SB for gluconic acid and sorbitol production.

		Sugar conversion		Productivity		Power consumption		Product concentration		Production rate	
Run time (hour)	EDI SB scale	Gluconic acid (%)	Sorbitol (%)	Gluconic acid ($g\ m^{-2}\ h^{-1}$)	Sorbitol ($g\ m^{-2}\ h^{-1}$)	Gluconic acid ($kWh\ lb^{-1}$)	Sorbitol ($kWh\ lb^{-1}$)	Gluconic acid ($g\ L^{-1}$)	Sorbitol ($g\ L^{-1}$)	Gluconic acid ($g\ h^{-1}$)	Sorbitol ($g\ h^{-1}$)
404	Bench	89%	87%	49	61	0.81	0.66	401	389	6	7
161	Pilot	92%	85%	71	59	0.57	0.68	443	281	180	150

9.4.2.1 Anaerobic Fermentation to Produce Succinic Acid

Succinic acid is an important building-block chemical feedstock for polymers with many large market uses, including acyl halides, anhydrides, esters, amides, and nitriles for the application of drug, agriculture, and food products, and other industrial uses. It is a renewable source of maleic anhydride or butanediol. Current fermentation processes for making succinic acid produce succinate in the form of a salt, which requires many post-fermentation unit operations for conversion to succinic acid, as illustrated in Figure 9.1. The major uses of succinate salt are as a deicer and other specialty products. Application of the separative bioreactor to succinate fermentation can provide an economic way for direct production of purified succinic acid that can be sold as a commodity chemical in the marketplace.

Lab-scale tests were conducted with a mutant strain of *Escherichia coli*, AFP184 (pfl-, ldh-, and ptsG-), for the fermentation of glucose to succinic acid.[30] This strain contains mutations that enable the conversion of glucose to primarily succinic acid with only minor conversion to acetic acid and ethanol side products. The fermentation was set up as a two-stage fermentation with the facultative *E. coli* cells batch-grown aerobically first in Luria broth (LB) to produce cell mass, which were then centrifuged and re-suspended in deionized water for inoculation into an anaerobic fermenter to produce succinic acid. The anaerobic fermentation was conducted in a temperature-controlled, stirred bioreactor. The fermentation combines one mole CO_2 with one mole of glucose to produce one mole of succinic acid and other side products. CO_2 is provided from a CO_2 capture reactor in the form of carbonate, which is formed by bubbling CO_2 gas into a solution at basic pH. This neutralizes the succinic acid in the broth while also supplying the CO_2 needed for the fermentation pathway. When integrated with RW-EDI, the fermentation broth is sent to the RW-EDI device, which separates succinic acid from the solution into a purified stream and recycles the resultant basic solution to the CO_2 capture reactor. Figure 9.13 shows the process diagram for the IF-EDI-SB. In conventional fermentation, the cells are usually grown in LB; in contrast, the cells produced in the IF-EDI-SB process were able to produce succinic acid directly from a glucose solution containing no LB. This minimizes the introduction of salts and other by-products that are components of the LB that could interfere with the recovery of a purified of succinic acid product.

Figure 9.14 shows the two fermentation systems with and without LB growth medium for succinic acid production. The robust AFP-184 strain could produce succinic acid in deionized water. Although its production rate was lower than in a LB medium, this operating strategy was able to produce succinic acid with less by-product contamination, such as acetic acid. To extract the succinic salt in the fermentation broth using RW-EDI, a "bipolar" EDI membrane configuration was used to separate and convert succinates and other organic salts to their organic acid form. In the IF-EDI-SB process,

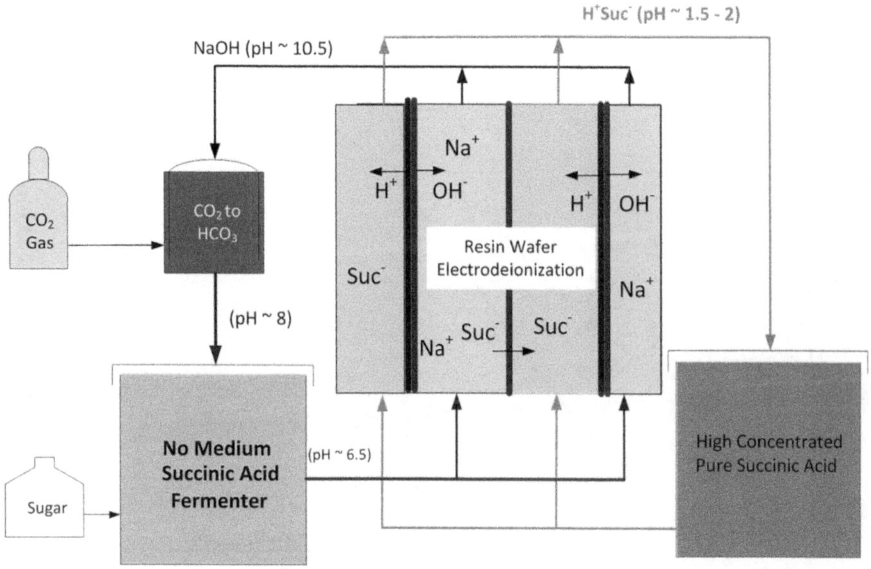

Figure 9.13 Process flow diagram of IF EDI-SB for succinic acid production.

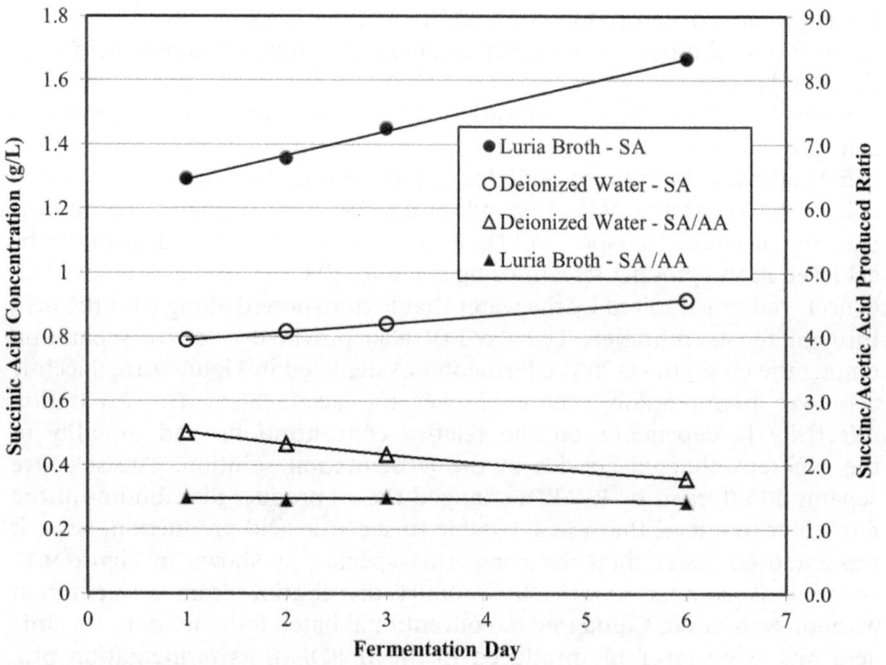

Figure 9.14 Succinic acid fermentation with/without nutrients.

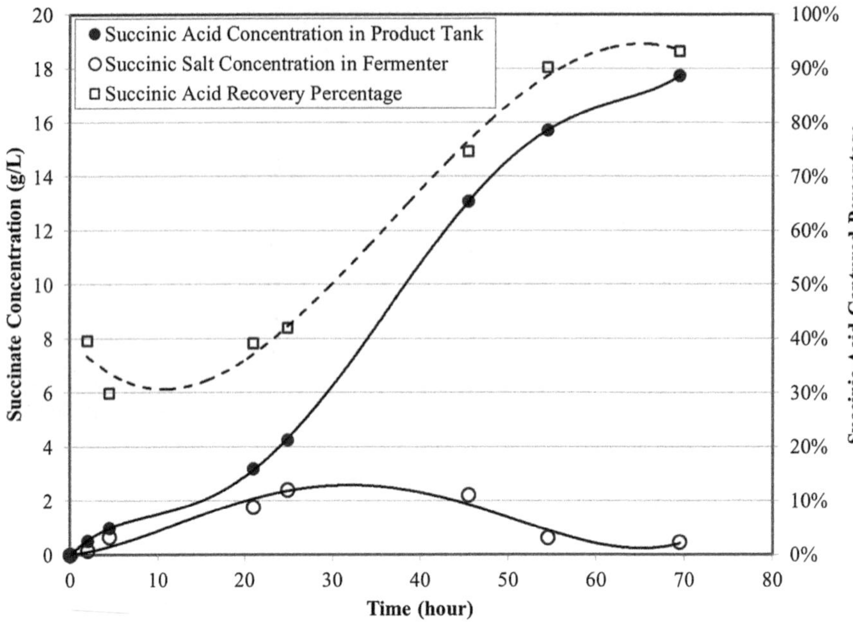

Figure 9.15 Succinic acid production and separation in IF EDI-SB.

the succinate concentration could be maintained at a very low level in the fermenter, and most of the succinate salts were captured as pure acid from the EDI device.

Figure 9.15 shows the concentrations of succinic acid in the fermenter and acid product tank. Final succinic acid concentrations in the product tank can be adjusted by appropriate selection of the initial liquid volume used to collect the separated acid. Depending on the anion-exchange membrane used for succinate transport in EDI, as much as 48 wt% succinic acid can be achieved in the product stream using standard ED membranes. The product concentration is limited by the water that is transported along with the acid through the membranes. The RW-EDI also provided selective separation among the co-products in the fermenter as depicted in Figure 9.16. Succinic acid was preferentially recovered over the acetic acid. The separation selectivity is dependent on the relative concentrations and mobility of the different charged species in the fermentation solution. The selective separation afforded by RW-EDI changed the co-product distribution in the fermenter to a level that was favorable to succinic acid production, since it was removed faster than the competing species, as shown in Figure 9.17 where it is compared with the product distribution from fermentation without separation. Compared to conventional batch fermentation, succinic acid was more favorably produced in the IF-EDI-SB as fermentation progressed because the lower succinate concentration resulted in higher succinate productivity. The effect of minimized product inhibition *via* the *in situ*

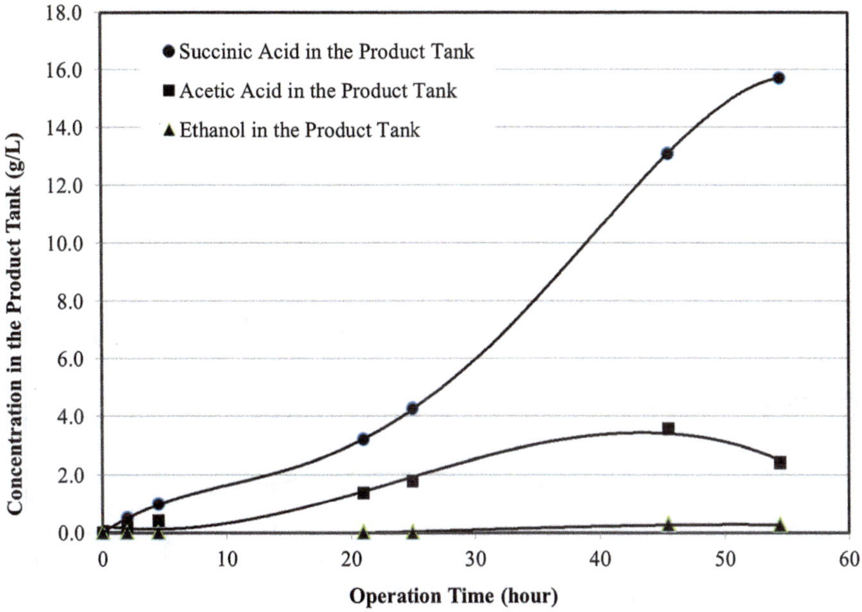

Figure 9.16 Co-product distribution in succinic acid fermentation using IF EDI-SB.

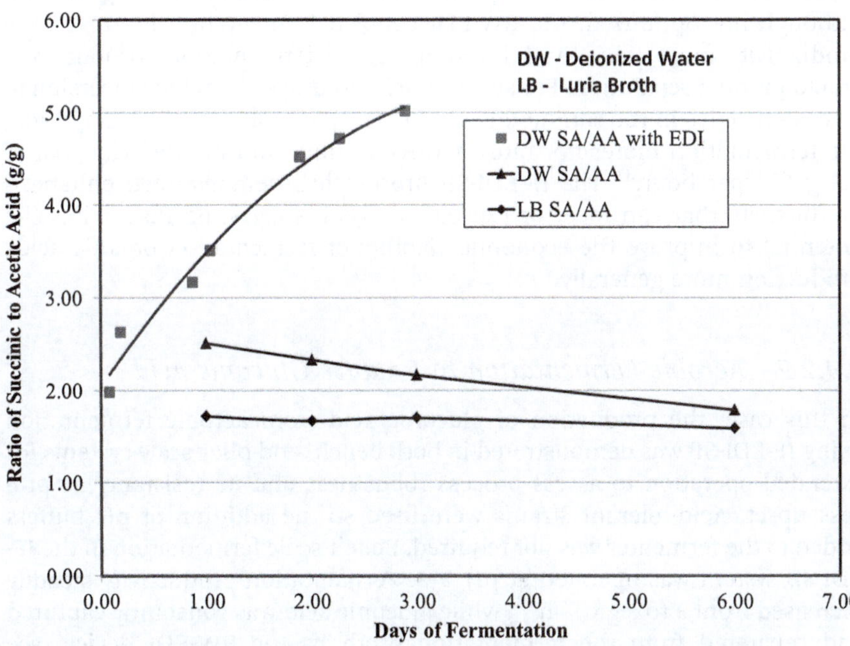

Figure 9.17 Succinic *vs.* acetic acid production ratios in different succinic fermenters.

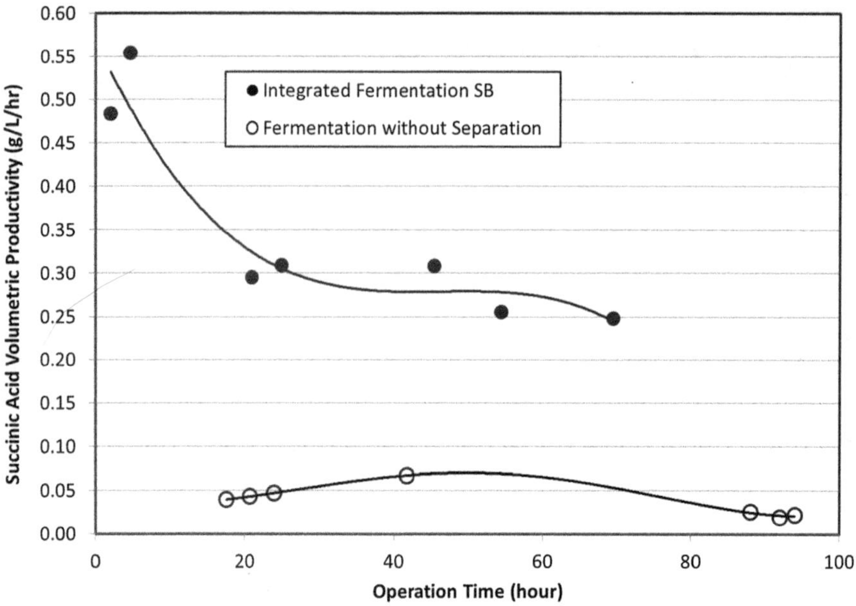

Figure 9.18 Fermentation productivity of succinic acid production with/without EDI-SB.

separation was shown to have a pronounced effect, as shown in Figure 9.18. Although not optimized, the RW-EDI coupled fermentation improved the productivity more than five-fold compared to fermentation without integrated product separation. For succinic acid produced from bioconversion to be cost-effective in the commodity market (*e.g.*, to replace maleic anhydride) the fermentation bioreactor must achieve a minimum productivity goal of 2.5 g L^{-1} per hour.[31] The IF-EDI-SB process has demonstrated enhanced productivity that can meet or exceed this goal, suggesting that it has the potential to improve the economic viability of fermentation organic acids production more generally.

9.4.2.2 *Aerobic Fermentation to Produce Gluconic Acid*

In this case, the production of gluconic acid from aerobic fermentation using IF-EDI-SB was demonstrated in both bench- and pilot-scale systems for extended operation to assess process robustness and its resistance to process upset. Acid-tolerant strains were used so the addition of pH buffers added to the fermenter was not required. Bench-scale fermentation in the IF-EDI-SB system was operated at pH 2–3. Fermentation productivity steadily increased from 3 to 5 g L^{-1} h^{-1}, while gluconic acid was constantly captured and separated from the fermentation broth by the RW-EDI device (see Figure 9.19). The acid capture rate also increased along with fermentation productivity. Over 2400 hours of operation, process performance from the

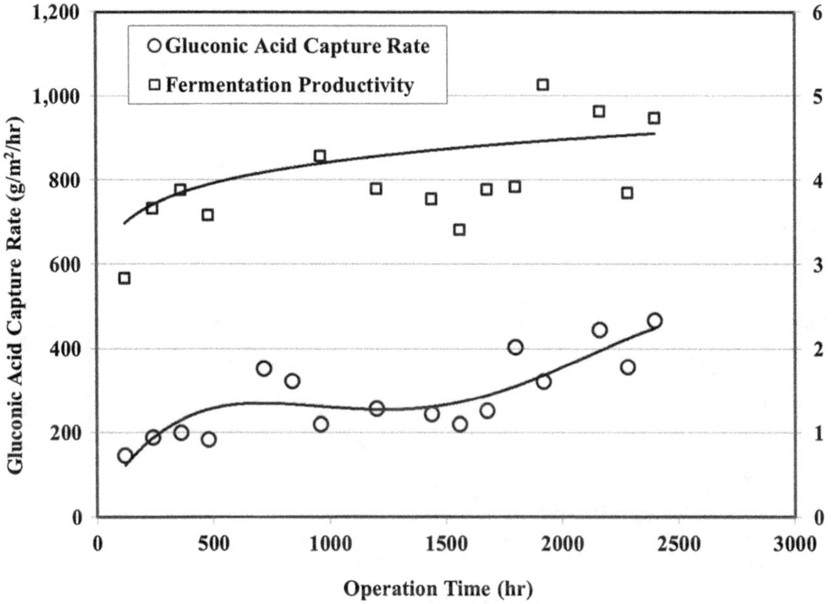

Figure 9.19 Gluconic acid production using bench-scale IF EDI-SB.

bench-scale IF-EDI-SB system was steady, demonstrating its robustness. Further assessment and validation of the technical and economic viability of the process was conducted in pilot-scale units, at a 30-fold scale-up in size from the bench-scale system. An initial batch-mode operation was conducted for 600 hours with the fermenter and RW-EDI operated independently. Subsequently, they were coupled together for more than 600 hours of continuous operation. Acid production and capture performance of the integrated pilot-scale system are shown in Figure 9.20. Fermentation productivity was significantly increased to greater than 20 g L^{-1} h^{-1} compared to 5 g L^{-1} h^{-1} observed with the bench-scale SB. Table 9.2 summarizes the process performance of gluconic acid production from aerobic fermentation in both bench-scale and pilot-scale tests.

In summary, the separative bioreactor platform provides the following advantages for producing organic acids in bioconversion processes:

- Overcomes the effects of product inhibition
- Eliminates or significantly reduces the need for neutralization chemicals
- Directly produces organic acids (rather than their salts)
- Significantly reduces the energy and processing costs for concentrating and purifying the organic acid products
- Can reduce fermenter size by several fold (owing to an increased fermentation productivity)
- Makes a continuous fermentation possible

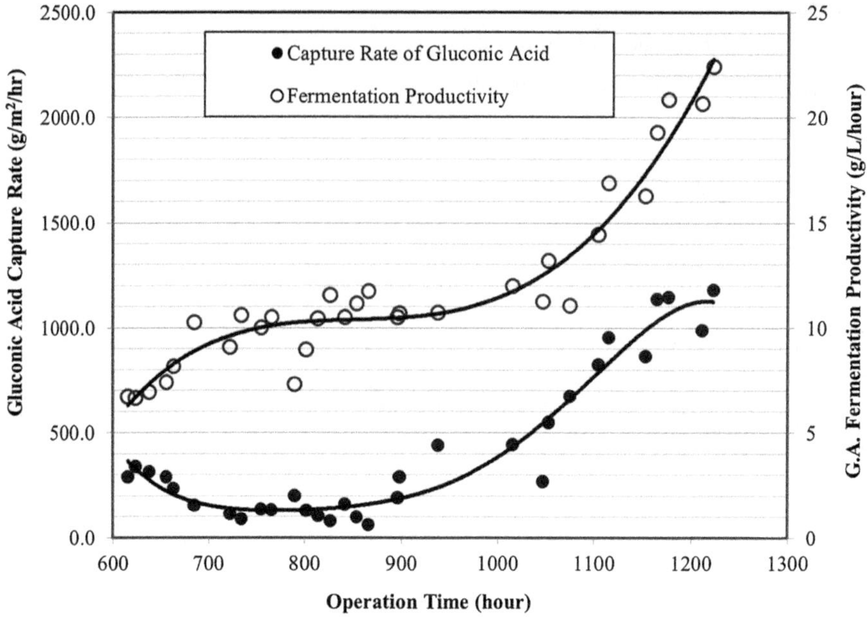

Figure 9.20 Gluconic acid capture rate and fermentation productivity in a pilot-scale IF EDI-SB.

Table 9.2 Process performance of gluconic acid production using IF EDI-SB.

Scale of fermenter and EDI	Total operation time (hours)	Captured gluconic acid concentration (wt%)	Gluconic acid capture rate (g m^{-2} h^{-1}	Power consumption (kWh lb^{-1})
Pilot-scale	1200	50	606	0.58
Bench-scale	2400	40	468	0.56

9.4.3 Economics of Organic Acid Separations by EDI-SB

Separation cost and overall bioprocessing cost for producing organic acids from biomass feedstocks depend strongly on the process efficiency, fermentation productivity and bioconversion yield. Figure 9.21 illustrates a process flow diagram of a separative bioreactor process to produce 85 wt% lactic acid from sugar fermentation. Based on the results of an ASPEN Plus simulation and cost estimates for capital installed equipment,[32] Figure 9.22 compares the production cost for 85 wt% lactic acid from conventional fermentation and SB processes. This example shows that SB can provide a significant reduction in the production cost, enabling bio-produced lactic acid to be cost-competitive in the commodity chemicals market. A number of different organic acids, including acetic, lactic, citric, gluconic, succinic, amino, propionic, 3-hydroxypropionic, and ascorbic acids, were evaluated for their capture from fermentation broth using EDI-SB. The electric energy consumption and processing cost of these organic

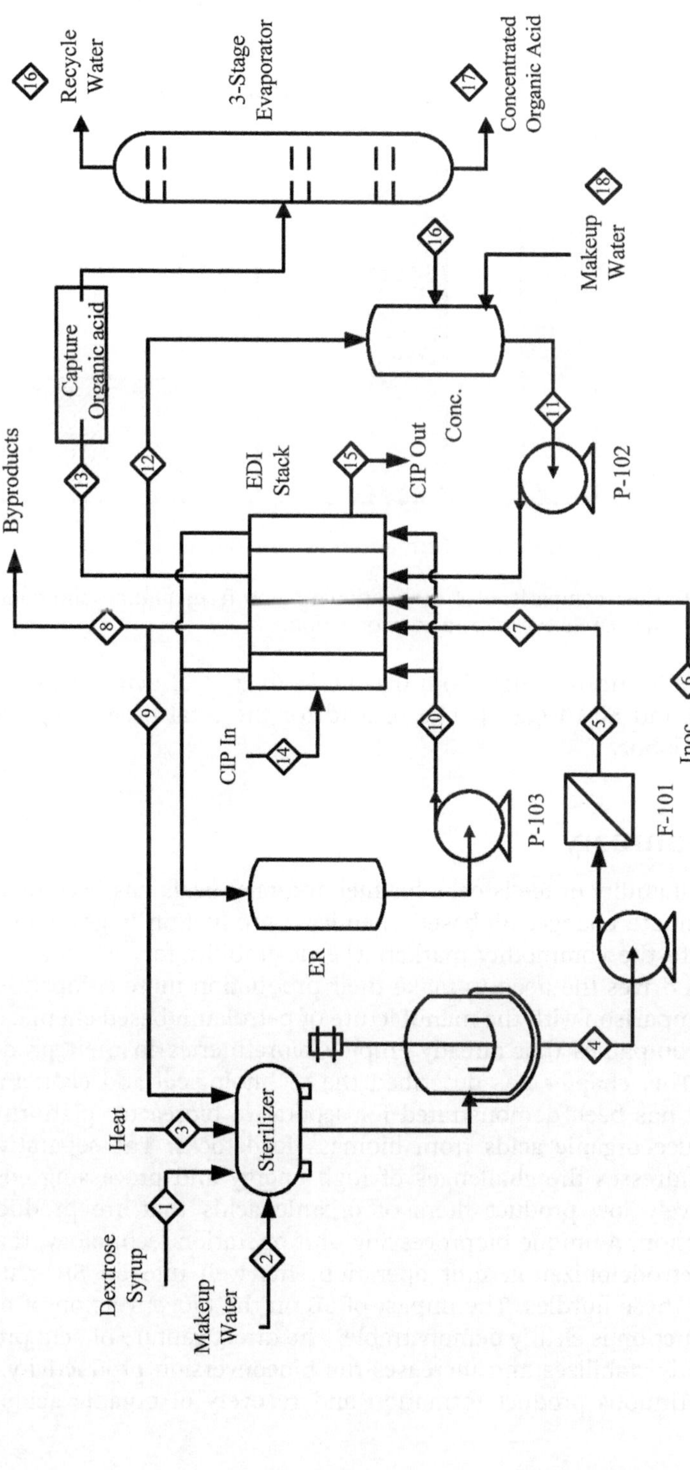

Figure 9.21 Process flow diagram of an IF EDI-SB used in an ASPEN Plus simulation.

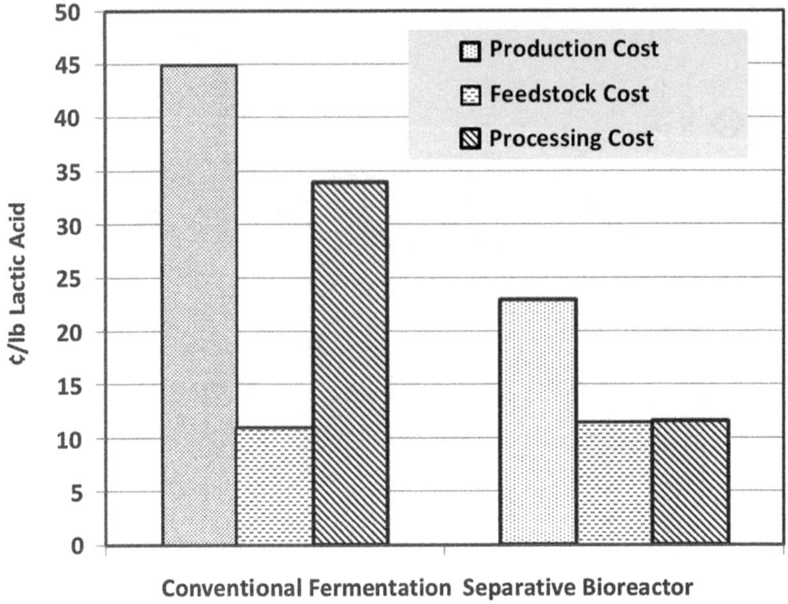

Figure 9.22 Cost comparison of conventional batch fermentation and continuous IF EDI-SB for lactic acid production.

acids were found to range from 0.4–3.0 kWh kg^{-1} of acid for power consumption and \$0.10–0.35 per kg of acid for the total processing cost, exclusive of labor.

9.5 Summary

The sustainability of feedstocks for fuel and chemicals has become critical due to climate change. Biobased chemicals production is gradually penetrating into the commodity market. The desirability for sustainable green chemicals drives the need to make their production more competitive, not just in comparison with the manufacture of petroleum-based chemicals, but also for companies that already employ biorefineries in their production strategy. This chapter has described the technological and economic viability that has been demonstrated for separative bioreactor platforms that can produce organic acids from biomass feedstocks. The separative bioreactor addresses the challenges of high energy and processing costs for the relatively low product titers of organic acids that are produced by bioconversion. A unique bioprocessing unit operation technology, the resin wafer electrodeionization unit operation, fits well into an SB system to overcome these hurdles. The impact of SB on the bioconversion of organic acid production is clearly demonstrable. The direct capture of acid products significantly stabilizes and increases the bioconversion productivity. It enables continuous product formation and recovery of organic acids while avoiding product inhibition. In addition, the separative bioreactor is capable of producing organic acids in the acid form, which avoids the need and cost for neutralization and subsequent acid regeneration. At both the bench-scale and pilot-scale, long-term operation of SB for organic acid production has demonstrated process performance advantages and the robustness to resist membrane fouling and contaminations in fermentation. The economic viability of using SB to capture organic acids was shown to be feasible in the range of \$0.1–0.35 per kg of organic acid.

References

1. J. A. Blinco and W. O. Doherty, in *Proceedings of the Conference of the Australian Society of Sugar Cane Technologists*, 2005.
2. C. S. López-Garzón and A. J. J. Straathof, Recovery of carboxylic acids produced by fermentation, *Biotechnol. Adv.*, 2014, **32**(5), 873.
3. T. Magarifuchi, K. Goto, Y. Iimura, M. Tadenuma and G. Tamura, *J. Ferment. Bioeng.*, 1995, **80**, 355.
4. L. P. Ling, H. F. Leow and M. R. Sarmidi, *J. Membr. Sci.*, 2002, **199**, 59.
5. P. Garcia-Fraile, L. R. Silva, S. Sanchez-Marquez, E. Velaquez and R. Rivas, *Food Chem.*, 2013, **139**, 31.
6. S. Yoshida and A. Yokoyama, *J. Biosci. Bioeng.*, 2012, **113**(5), 556.
7. A. Kutyla-Olesiuk, U. E. Wawrzyniak, P. Coisek and W. Wroblewski, *Anal. Chim. Acta*, 2014, **823**, 25.
8. S. Yu, D. Huang, J. Wen, S. Li, Y. Chen and X. Jia, *Bioresour. Technol.*, 2012, **104**, 610.
9. Z. Y. J. B. K. J. M. Zhang, *Biochem. Eng. J.*, 2007, **36**, 251.
10. B. Gu, P. Zheng, Q. Yan and W. Liu, *Sep. Purif. Technol.*, 2014, **138**, 47.
11. M. Matsumoto, K. Mochidduki, K. Fukunishi and K. Kondo, *Sep. Purif. Technol.*, 2004, **40**, 97.
12. F. S. Oliveira, J. M. Araujo, R. Ferreira, N. L. Rebelo Paulo and I. M. Marrucho, *Sep. Purif. Technol.*, 2012, **85**, 137.
13. T. Ali, D. Bylund, S. A. Essen and U. S. Lundstrom, *Soil Biol. Biochem.*, 2011, **43**, 2417.
14. D. G. Cherkasov and K. K. Il'in, *Russ. J. Appl. Chem.*, 2009, **82**(5), 920.
15. D. Wu, H. Chen, L. Jiang, J. X. Z. Cai and P. Cen, *Chin. J. Chem. Eng.*, 2010, **18**(4), 533.
16. J. E. Drewes, J. Mitterwallner, S. Gruenheid and C. A. Bellona, *Proceedings Water Quality Technology Conference in American,* 2002, p. 252.
17. J. M. K. Timmer, J. Kromkamp and T. Robbertsen, *J. Membr. Sci.*, 1994, **92**(2), 185.
18. J. Lemba, R. Karklins and I. L. Lapele, *Materialzinat. Lietiska Kim.*, 2000, **1**, 73.
19. S. M. R. Ullah, M. Takeuchi and P. K. Dasgupta, *Environ. Sci. Technol.*, 2006, **40**(3), 962.
20. J. Wisniewski, A. Rozanska and T. Winnicki, *Desalination*, 2005, **182**(1–3), 339.

21. Y. Wang, C. Huang and T. Xu, *J. Membr. Sci.*, 2011, **374**, 150.
22. A. Vertova, G. Aricci, S. Rondinini, R. Miglio, L. Carnelli and P. D'Olimpio, *J. Appl. Electrochem.*, 2009, **39**, 2051.
23. J. G. Zeikus, M. K. Jain and P. Elankovan, *Appl. Microbiol. Biotechnol.*, 1999, **51**, 545.
24. R. Datta and M. Henry, *J. Chem. Technol. Biotechnol.*, 2006, **81**, 1119.
25. R. P. John, K. M. Nampoothiri and A. Pandey, *Microbiol. Biotechnol.*, 2007, **74**, 524.
26. C. Frederick, A. J. Wilkins and G. C. Ganzi, *U. S. Pat.* 8182693 B2, 22 May 2012.
27. R. Datta, Y. Lin, B. Daniel and S.-p. Tsai, *U. S. Pat.* 6495014, 2003.
28. Y. J. Lin, M. P. Henry and S. W. Snyder, *U. S. Pat.* 7452920, 2008.
29. Y. J. Lin, M. B. Arora, J. A. Hestekin and E. J. St Martin, *U. S. Pat.* 6979140, 2004.
30. M. Donnelly, C. Millard and N. P. Nghiem, *U. S. Pat. No.* 6743610 B2, 2004.
31. T. Werpy and G. Petersen, Top Chemicals from Biomass, *USDOE: Washington, DC, Vol. I.*, 2004.
32. B. Atherton, Technical Memorandum, National Renewable Energy Laboratory, Denver, 2005.

CHAPTER 10

CO_2 Conversion to Chemicals with Emphasis on using Renewable Energy/Resources to Drive the Conversion

RICH MASEL,*[a] ZENGCAI LIU,[a] DI ZHAO,[a] QINGMEI CHEN,[a] DALE LUTZ[b] AND LAURA NERENG[b]

[a] Dioxide Materials, 3998 FAU Boulevard, Suite 300, Boca Raton, FL 33431-6429, USA; [b] 3M Company, 3M Center, 201-1S-13, St. Paul, MN 55144-1000, USA
*Email: rich.masel@dioxidematerials.com

10.1 Introduction

The critical driving factors for commercializing biobased products are decreased dependence on fossil sources for carbon and energy, diversification of feedstock choices, and reduced greenhouse gas emissions. CO_2 conversion to chemicals directly addresses dependence on fossil sources for carbon and diversification of feedstocks. Using renewable energy resources to drive conversion addresses fossil energy inputs and greenhouse gas reductions. Therefore this chapter addresses the primary drivers of this book.

CO_2 conversion to fuels and chemicals has had increasing interest due to a worldwide need to reduce CO_2 emissions, combined with the realization that CO_2 might be an inexpensive feedstock for fuel and chemical production. In the last several years, almost 2000 papers and many excellent reviews of the

RSC Green Chemistry No. 43
Commercializing Biobased Products: Opportunities, Challenges, Benefits, and Risks
Edited by Seth W. Snyder

Published by the Royal Society of Chemistry, www.rsc.org

subject have appeared.[1–35] Given this vast literature, and the page limit of this chapter, we could not review all of the excellent work that has appeared. Consequently, we decided to concentrate on approaches that discuss moving toward commercialization, and not laboratory studies. Also, we concentrate on only chemical, and not biological or photochemical, approaches, and emphasize topics that are only briefly covered in previous review articles.

Any CO_2 reduction technology requires energy input, and we will be discussing three possible energy inputs: fossil fuels, renewable electricity, and concentrated sunlight. We will consider primarily fossil fuels and renewable energy, with minimal coverage of sunlight as an energy source.

10.2 Reacting CO_2 with Fossil Fuels

Reactions of CO_2 with fossil fuels and other hydrocarbons have been used for years to make urea[36,37] and salicylic acid. Production of polycarbonates and polyurethanes from CO_2 has also reached pilot scale production.[38–40] CO_2 has also been proposed as a feedstock for the production of methanol and other hydrocarbons. Generally, reactions of CO_2 with hydrocarbons are less "green" than other approaches discussed here, but they have the advantage that the processes use well developed chemical process technology, so they are closer to commercialization.

10.2.1 Chemical Production from CO_2

10.2.1.1 Urea

Presently, urea manufacture is the largest application of CO_2 as a chemical feedstock. Most urea manufactured today is manufactured *via* the Bosch–Meiser process, where ammonia reacts with carbon dioxide in a two-step process yielding urea.

$$2NH_3 + CO_2 \rightarrow NH_2COONH_4 \text{ (ammonium carbamate)}$$

$$NH_2COONH_4 \rightarrow H_2O + NH_2CONH_2 \text{ (urea)}$$

The process was reported by Bosch and Meiser[41] and Krase and Gaddy[42] in 1922. There have been several process improvements since 1922. The latest is called the Avancore process, which incorporates improved heat and mass transfer, and materials that resist corrosion. Ulmann's Encyclopedia of Chemical Technology has a wonderful description of the process,[37] as does a recent paper by Meessen.[36] One is referred to these publications for further details.

One important point for the discussion here is that this process is a net source rather than sink of CO_2. One produces significant CO_2 during ammonia production and consumes only part of the CO_2 during the conversion to urea. Consequently, the overall process ends up being a source rather than a sink for CO_2.

10.2.1.2 Salicylic Acid

CO_2 is also commonly used for the production of salicylic acid. The process was first invented by Kolbe and Schmitt in 1860.[43,44] It is the main process today for the production of aspirin, the acetyl ester of salicylic acid. The overall reaction for the production of salicylic acid from phenol and CO_2 is shown in Scheme 10.1.

The reaction is believed to proceed by way of production of sodium phenoxide, followed by nucleophilic addition of the phenoxide to the carbon of CO_2, as described by Lindsey and Jeskey (see particularly section E, p. 599).[46]

Early publications described the reaction in Scheme 10.2 as being carried out at a temperature of 135 °C and a pressure of 4.3 atmospheres (4.4 bar).[47] However, one recent source[48] lists the current reaction conditions as 125 °C and 100 atmospheres.

The final aspirin product (acetylsalicylic acid) is then obtained by reacting salicylic acid with acetic anhydride in the presence of a small amount of a strong acid catalyst, such as sulfuric acid. This reaction produces an acetyl ester at the hydroxyl site of the salicylic acid, with acetic acid as a byproduct.

10.2.1.3 Polymer Production from CO_2

Production of polycarbonates from CO_2 has already been run on a pilot plant scale[38–40] and commercial scale plants are scheduled to start in mid-2016. Generally, one reacts an epoxide or other highly reactive molecule with CO_2 to form polycarbonate or a polyol (Scheme 10.3).

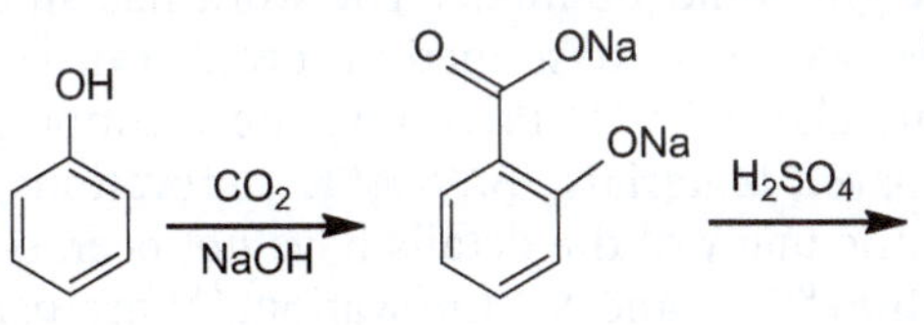

Scheme 10.1 Reaction diagram from "Kolbe-Schmitt" by Edgar.[45]

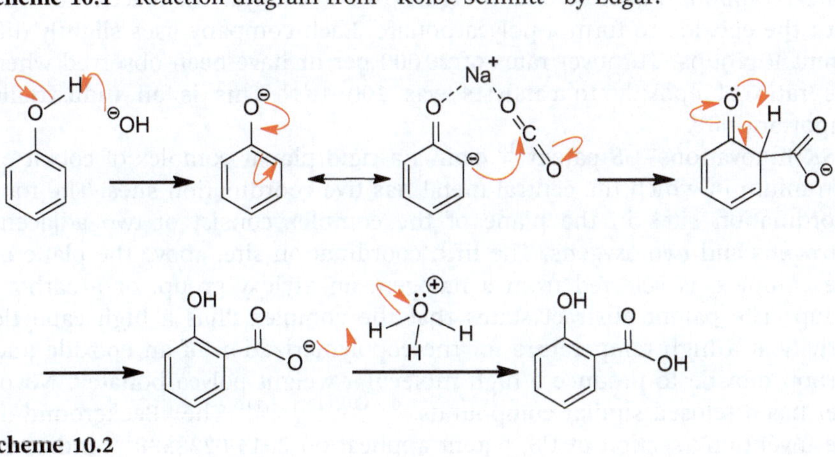

Scheme 10.2

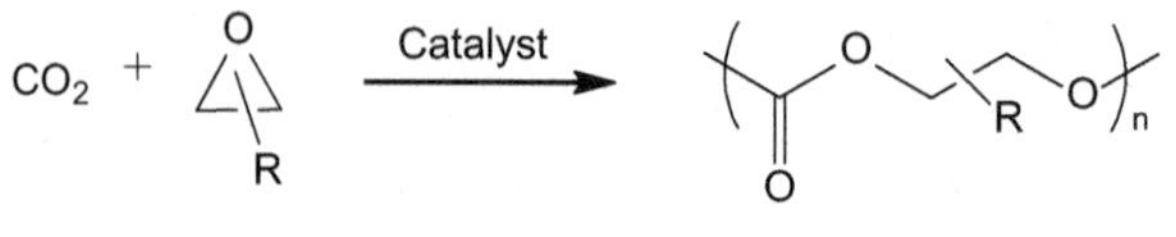

Scheme 10.3

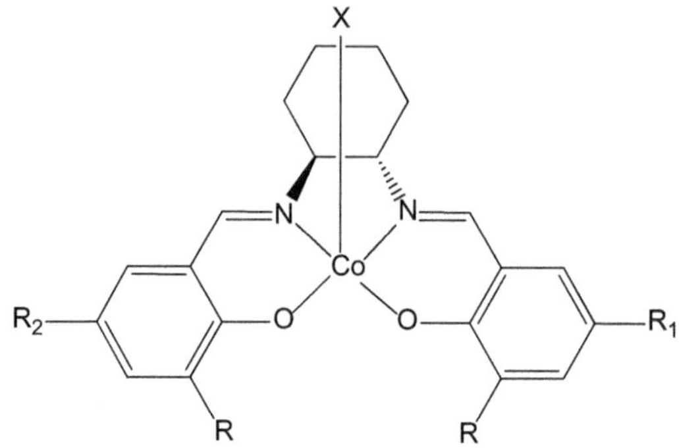

Figure 10.1 Structure of a salen-cobalt compound.

Reactions of CO_2 with an epoxide to form short chain polycarbonates were first reported by Inoue *et al.* in 1969.[49,50] The original work used zinc or aluminum organometallic catalysts that produce only low molecular weight materials. However, impressive work by Darensbourg,[51–54] Rieger,[55–71] Nozaki,[72–86] and Coates[87–100] improved the catalysis so that materials with a practical molecular weight could be made. The work has since been expanded by multiple investigators, as described in recent reviews,[101–106] and has started to be commercialized.[38–40] Presently, the technology is being pursued by Novomer, Bayer Materials, BASF, SK Innovations, Empower Materials and others. While many of the details have not been published, it appears that Novomer, Inc.[107–113] and SK Innovations[114] are using a salen cobalt compound like the one in Figure 10.1 to catalyze the reaction of CO_2 with the epoxide to form a polycarbonate. Each company uses slightly different R groups. Turnover rates of 20 000 per hr have been observed when the ratio of epoxide to catalysts was 100 : 1.[114] This is an industrially important rate.

SK Innovations' US patent[115] claims a rigid planar complex of cobalt or chromium in which the central metal has five coordination sites. The four coordination sites in the plane of the complex consist of two adjacent nitrogens and two oxygens. The fifth coordination site, above the plane of the complex, is selected from a halogen, an aryloxy group, or a carboxy group. The patent abstract states that the complex "has a high catalytic activity at a high temperature for the copolymerization of an epoxide and carbon dioxide to produce a high molecular weight polycarbonate". Novomer has disclosed similar compounds.[107–110,112,113,116] The "Background of the Invention" section of US. patent application 2014/0228538[116] indicates

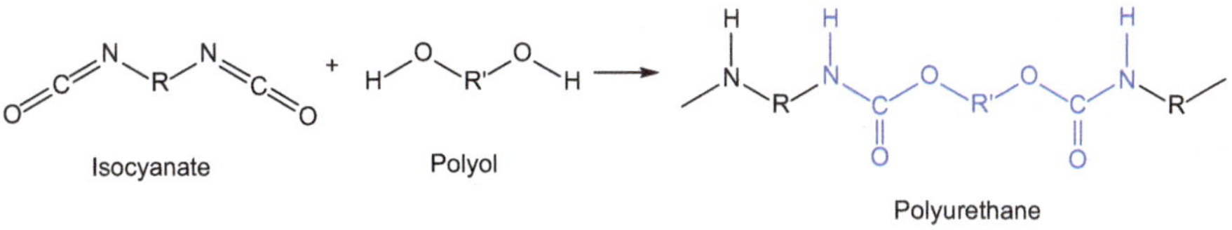

Scheme 10.4

that the synthesis and testing of the disclosed metal complexes was related to a search for better catalysts for producing polycarbonates from the reaction of epoxides and carbon dioxide. Example 2 of that application indicates that the complexes may be useful as catalysts for producing poly(propylene carbonate).

Bayer appears to be using a different catalyst to first form a small polyol that then reacts with isocyanate to form polyurethane (Scheme 10.4).

Bayer has announced that they expect to be producing industrial quantities of their polyurethane from CO_2 by mid-2016.

Previous investigators have already published reviews[101–106] of the conversion of CO_2 to polymers. One is referred to these previous papers for more details.

10.2.1.4 Reactions of CO_2 with Methane

Several investigators have leveraged the current abundance of CH_4 that has resulted from the use of hydraulic fracturing ("fracking") to obtain "shale gas" in the United States and elsewhere. Examples of such methane-based CO_2 conversion processes are described next. The same conversion technology could be made "greener" by using methane derived from biomass sources, such as anaerobic digesters, or by using hydrogen produced from the electrolysis of water, with carbon-neutral renewable energy sources such as wind or solar providing the necessary electricity.

Researchers at the University of Southern California, led by George A. Olah and G. K. Suryah Prakash, have produced an impressive body of work[117–144] supporting their concept of converting captured CO_2 from fossil fuel combustion into alternative transportation fuels and chemicals, primarily methanol and dimethyl ether. Much of this work has focused on combining high energy methane from natural gas with low energy CO_2 gas to produce liquid methanol, which is more easily transported and stored. Some of these processes involve the reverse water–gas shift reaction,

$$CO_2 + H_2 \rightleftharpoons CO + H_2O$$

in which CO_2 and hydrogen are converted to carbon monoxide and water. Carbon monoxide can then be combined with additional hydrogen to provide a "synthesis gas" (or syngas) mixture that serves as the feedstock for methanol production and other processes. A series of patents have been issued on various aspects of this work, the most recent of which focus on bi-reforming processes. For example, in one of the disclosed processes for producing methanol, "The molar ratio of hydrogen to carbon monoxide is at least two moles of hydrogen to one mole of carbon monoxide and the overall molar ratio between methane, water and carbon dioxide is about 3 : 2 : 1."[143]

The economy of the Canadian province of Alberta relies heavily on fossil fuel resources including natural gas, oil, and tar sands. However, other provinces, and Canada as a whole, have made various commitments to reducing their contribution to global CO_2 greenhouse gas emissions. As a result, the provincial government of Alberta established the Climate Change and Emissions Management Corporation (CCEMC) to administer a Grand Challenge competition to solicit and fund the development of novel technical solutions to reducing greenhouse gas emissions. A total of CAD$35 million was committed to the development of the selected technologies. Note that in Chapter 11, life cycle GHG emissions of several biobased products are evaluated. While the technologies could be developed anywhere, they had to be applicable to Alberta, and to be capable of being commercialized in the province.[145] The first stage awardees were announced in April 2014. One set of the selected technologies was designated as "Carbon Uses". Several of these projects were based on combining captured CO_2 emissions from Alberta's fossil fuel production, processing, or consumption with Alberta's natural gas, in order to produce useful chemicals. Example projects are described next.

Enerkem, Inc., was a first round winner of Alberta's CCEMC Grand Challenge. They proposed to catalytically convert CO_2 and methane to carbon monoxide (CO) and hydrogen, a mixture referred to as "synthesis gas," or syngas. The syngas can then be used in conventional processes to produce chemicals such as propanol, propionic acid, and acrylic acid.[146]

Another CCEMC awardee is Pioneer Energy, who plans to leverage earlier work by NASA and the US Department of Energy to produce butanol from CH_4 and CO_2. The butanol could potentially serve as an alternative fuel.[147]

University of California Riverside proposes to convert CH_4 and CO_2 to syngas using a thermochemical process and a new catalyst with superior thermal stability. The syngas would then be used to produce methanol. (Methanex, one of the world's largest producers of methanol, has a facility in Medicine Hat, Alberta, Canada.)[148]

In another variation on a theme, Quantiam Technologies proposes to produce hydrogen using "environmentally sustainable methods" (presumably biomass gasification, bio-methane reforming, or water electrolysis using renewable energy sources such as wind or solar). The hydrogen and CO_2 would then be converted to methanol, presumably using standard reforming technology.[149]

Robert Gordon University is developing a high temperature catalytic membrane reactor to combine CO_2, water, and oxygen from a flue gas stream with methane, in order to produce Fischer–Tropsch fuels, hydrogen, and ammonia/urea.[150]

The University of Alberta is developing a solid oxide fuel cell that will combine CH_4, CO_2, and air, resulting in the production of useful carbon monoxide, water, and electricity.[151]

The Gas Technology Institute (GTI) has developed a technology that combines CH_4 and CO_2 to produce acetic acid in a two-step isothermal catalytic process. This bypasses the need to use methanol as an intermediate, as in the conventional acetic acid synthesis, thus providing a more efficient and lower cost process.[152]

10.3 Reducing CO_2 using Renewable Energy/ Resources to Drive the Conversion

A variety of "green" approaches have been developed to convert CO_2 to higher energy forms of carbon without the use of fossil fuels.

10.3.1 Producing H_2 using Renewable Energy, Reacting H_2 with CO_2

One approach for "recycling" CO_2 is to produce hydrogen from a renewable energy source, then combine the hydrogen with the CO_2 to produce useful "synthetic" hydrocarbons that can displace hydrocarbons that are currently derived from fossil fuels. Three examples are provided.

Carbon Recycling International (CRI), founded in 2006 in Reykjavik, Iceland, uses Iceland's volcanic geothermal energy to drive steam turbines to generate electricity, then uses the electricity to electrolyze water to hydrogen and oxygen. The hydrogen is then combined with captured CO_2 from flue gas to produce methanol at their George Olah Renewable Methanol plant in Svartsengi, Iceland. Much of the renewable methanol is then blended into fuels for flex-fuel vehicles in Iceland and Europe. In early 2015, CRI announced a project to install their technology at a 500 MWe coal-fired power plant at the STEAG GmbH site in Lünen, Germany. This project is being funded 80% by the European Union "Horizon 2020" research program. Due to Germany's high penetration of intermittent renewable energy sources such as solar and wind, there are often times when more electricity is generated than is needed on the electrical grid. This surplus energy will be leveraged to convert captured CO_2 from flue gas into commercial products, such as transportation fuels. This general concept is referred to, particularly in Europe, as "valorizing CO_2".[153]

Automobile maker Audi has partnered with Etogas GmbH (Stuttgart, Germany) to build a large power-to-gas installation in Werlte, Germany, in which renewable electricity (from sources such as wind and solar) is used to electrolyze water for the production of hydrogen. The hydrogen may then be combined with CO_2 in the Sabatier methanation reaction:

$$CO_2 + 4H_2 \rightarrow CH_4 + 2H_2O + \text{energy} \quad \Delta H = -165.0 \text{ kJ mol}^{-1},$$

in order to produce methane, which is sometimes referred to as "synthetic natural gas" or SNG. This SNG can then be stored in the existing natural gas pipeline infrastructure, where it can then be distributed for use in heating, in industrial processes, or as fuel for Audi's compressed natural gas (CNG) hybrid vehicles, such as the Audi A3 g-tron.[154] Similarly, Audi announced in April 2015 the production of their first batch of diesel fuel made from CO_2 and H_2O, in cooperation with Sunfire and Climateworks. The plant converts CO_2 to CO in order to make a synthesis gas mixture with H_2 from the steam electrolysis of water, using renewable energy as the power source. The synthesis gas is then converted to a mixture of long chain aliphatic hydrocarbons that can be refined to produce a clean, transparent diesel fuel substitute.[155] These integrated systems provide long term storage of excess intermittent renewable energy, while reducing the net CO_2 greenhouse gas emissions of the transportation sector and reducing the need to import fossil fuels.

Dong Energy (Fredericia, Denmark) operates combined heat and power (CHP) units in Denmark, and is taking various steps to contribute to Copenhagen's target of making the city CO_2 neutral by 2025. In March, 2015, they announced that, by 2016, the last CHP unit at the Avedøre Power Station, Denmark's largest, will be converted to allow it to use biomass fuel (wood pellets) instead of coal.[156] Dong Energy has received US patent 8 109 095, "Method and Apparatus for Regulating Electrical Power," for a combined heat and power system that derives its energy from gasification of biomass.[157] According to the patent, at the times when more heat than electricity is required from the system, the excess electricity can be used to electrolyze water to produce H_2, which is then combined with CO_2 from the biomass gasification to produce hydrocarbon fuel, such as synthetic methane or methanol. The oxygen from the water electrolysis can also be used to improve the efficiency of the gasification and combustion processes of the CHP plant. In related work, the Avedøre Power Station is also involved with the BioCat Project, a biological catalyst power-to-gas project[158] in which hydrogen from water electrolysis is combined with CO_2 and methanogenic microorganisms to produce methane that can be injected into the natural gas infrastructure. The methane is then available for heating, electricity generation, or fueling CNG vehicles.[159]

10.3.2 Electrochemically Reducing CO_2 to Chemicals and Fuels

Electricity generated by renewable carbon neutral sources such as solar and wind energy will be the next generation of clean energy. However, effectively integrating renewable energy sources with both transportation and the power grid requires an extensive infrastructure of electric energy storage (EES) systems. Meanwhile, carbon dioxide (CO_2), a greenhouse gas, increases its concentration in the air rapidly because of the increased consumption of fossil fuels. Fortunately, CO_2 can be electrochemically reduced to chemicals and fuels using the electricity generated by carbon neutral sources. In this way, the renewable but intermittent energy can be stored in chemicals and fuels.

10.3.2.1 Fundamentals and Challenges

Carbon dioxide (CO_2) is the most stable carbon-based substance in nature, and consequently its reactivity is low. However, the equilibrium potential for CO_2 reduction is similar to that for hydrogen evolution from water. Table 10.1 lists the equilibrium potentials for some CO_2 reduction reactions.[160,161] Since the hydrogen evolution is dominant in acidic media, and molecular CO_2 does not exist in alkaline media, most carbon dioxide reduction occurs in near neutral media. As shown in Table 10.1, the equilibrium potential for formation of carbon monoxide (CO) and formate ($HCOO^-$) are only 100 mV and 20 mV negative with respect to the SHE, while the rest of the equilibrium potentials are even more positive relative to the SHE. In these cases, the carbon dioxide reduction should proceed with very little difficulty, perhaps even more easily than the hydrogen evolution.

In reality, carbon dioxide reduction does not occur until a much more negative potential is applied. It is generally accepted that carbon dioxide reduction involves the formation of intermediate species, *via* single electron transfer (eqn (10.1)). People call that species $CO_2^{\bullet -}$, although the exact structure of the species is unknown.[162]

$$CO_2 + e^- = CO_2^{\bullet -} \tag{10.1}$$

It costs energy to create the $CO_2^{\bullet -}$ species as indicated in the blue line in Figure 10.2. As a result, one needs to provide extra energy to get the reaction to occur.

Another issue is that the final products from carbon dioxide reduction and their selectivity depend on the electrode materials and electrolyte formulation. Hori and others have given excellent reviews of the electrochemical reduction of CO_2 on metal electrodes.[160,163] The metal electrodes are divided into four groups, based on the major product and the selectivity in an aqueous system.

Table 10.1 The equilibrium electrode potentials for CO_2 reduction to different products.

No.	CO_2 reduction reaction	Equilibrium potential/V *vs.* SHE	
		pH = 7	pH = 0
1	$CO_2 + H_2O + 2e^- = CO + 2OH^-$	−0.52	−0.10
2	$CO_2 + H_2O + 2e^- = HCOO^- + OH^-$	−0.43	−0.02
3	$2H_2O + 2e^- = H_2 + 2OH^-$	−0.41	0
4	$CO_2 + 5H_2O + 6e^- = CH_3OH + 6OH^-$	−0.39	0.02
5	$2CO_2 + 8H_2O + 12e^- = C_2H_4 + 12OH^-$	−0.34	0.07
6	$2CO_2 + 9H_2O + 12e^- = C_2H_5OH + 12OH^-$	−0.33	0.08
7	$2CO_2 + 13H_2O + 18e^- = C_3H_7OH + 18OH^-$	−0.32	0.09
8	$CO_2 + 6H_2O + 8e^- = CH_4 + 8OH^-$	−0.25	0.16

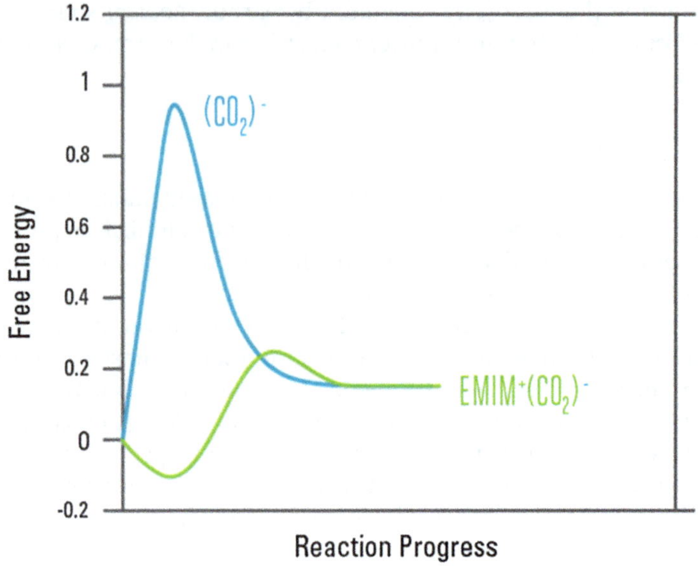

Figure 10.2 The reaction pathway for CO_2 reduction in the presence and absence of $EMIM^+$.

The 1st group includes Cd, In, Sn, Hg, Tl, Pb, Bi, *etc.*, which catalyze the reduction of CO_2 to formate. These metal electrodes exhibit very low coverage of $CO_2^{\bullet-}$. The free $CO_2^{\bullet-}$ accepts a proton (H^+) from water and a second electron, and finally is reduced to $HCOO^-$ with the H^+ bonding to the nucleophilic carbon atom (eqn (10.2) and (10.3)).

$$CO_2^{\bullet-} + H_2O \rightarrow HCOO^{\bullet} + OH^- \tag{10.2}$$

$$HCOO^{\bullet} + e^- \rightarrow HCOO^- \tag{10.3}$$

The electrode potentials (at 5 mA cm^{-2}) are −1.6 V or more negative on Cd, Hg, Tl and Pb electrodes. Since $CO_2^{\bullet-}$ could be weakly adsorbed on In and Sn electrodes, they can achieve the same current density at much less negative potentials.

The 2nd group consists of Au, Ag, Pd, Ga, Zn, *etc.*, and the main product is CO. CO formation occurs at a less negative potential than $HCOO^-$ formation. The decrease in the overpotential results from the stabilization of adsorbed $CO_2^{\bullet-}$ on these metals. The electrode potential correlates well with the heat of fusion of the electrode materials: the higher the heat of fusion, the lower (less negative) the electrode potential. Unlike $HCOO^-$ formation, H^+ will not take part in the CO formation, but H_2O will react with the nucleophilic O atoms of adsorbed $CO_2^{\bullet-}$, leading to the formation of CO and OH^- (eqn (10.4)).

$$CO_2^{\bullet-} + H_2O + e^- \rightarrow CO + 2OH^- \tag{10.4}$$

Cu is unique for carbon dioxide reduction and belongs to the 3rd group. The CO_2 can be reduced to CO and $HCOO^-$ at a Cu electrode at lower overpotentials, but with very low faradaic efficiency. As the overpotential increases, the intermediately formed CO is further reduced to CH_4, C_2H_4 and alcohols, probably due to the moderate CO heat of adsorption on Cu. Au has a much lower heat of adsorption, and therefore CO desorbs easily before it is reduced. Pt and Ni have much higher heats of adsorption and adsorb CO strongly, so that CO cannot be reduced.

The 4th group of metals essentially promote hydrogen evolution, and they are Pt, Ni, Fe, Ti, *etc.* As discussed earlier, CO is stabilized on Pt and Ni electrode surfaces, which prevents further reduction of CO_2, while Fe and Ti do not reduce CO_2 at ambient conditions.

In addition to the aqueous electrolyte, non-aqueous electrolyte solutions have been evaluated for carbon dioxide reduction due to the high solubility of CO_2. Propylene carbonate (PC), acetonitrile (AN), dimethylformamide (DMF), and dimethyl sulfoxide (DMSO) are aprotic solvents, and therefore hydrogen evolution is greatly suppressed in them. The major products from carbon dioxide reduction are different in a nonaqueous electrolyte than in an aqueous electrolyte. The metal electrodes are categorized into three groups. Pb, Hg, and Tl produce mainly oxalate ($C_2O_4^{2-}$) from carbon dioxide reduction. Ni, Cu, Zn, Ag, In, Sn, Pt and Au give CO as the main product. Cr, Fe, Co, Pd, and Cd yield CO and $C_2O_4^{2-}$ at a comparable rate. The addition of a small amount of water generally switches the main product to $HCOO^-$. Generally, the applied potential is much more negative in a nonaqueous electrolyte than in an aqueous electrolyte. Methanol is also a nonaqueous solvent, but it is a protic solvent like water. The major products in methanol are similar to those in aqueous systems.

In order for CO_2 to be reduced to industrially viable products, (i) the overpotential for carbon dioxide reduction should be low enough to be energy efficient; (ii) the faradaic efficiency towards the target product (product selectivity) should be quite high, in order to save energy and to avoid costly separation processes; (iii) the rate of carbon dioxide reduction (determined as current) should be high enough to be economically viable; and (iv) the electrochemical cell should be scaled up to match other industrial processes.

10.3.2.2 Cell Configuration

To date, tremendous progress has been made to increase the current density and to lower the overpotential for carbon dioxide reduction.[164,165] Most of the results were obtained from half-cell measurements using traditional three-electrode systems in a glass cell. The half-cell configuration is very useful to study the fundamentals and mechanisms of carbon dioxide reductions. However, it is impossible to convert CO_2 to products in a continuous and economic way using this configuration. A full electrolysis cell (CO_2 electrolyzer) is therefore needed. A CO_2 electrolyzer consists of a cathode, an anode and the electrolyte(s) in between, as shown in Figure 10.3.

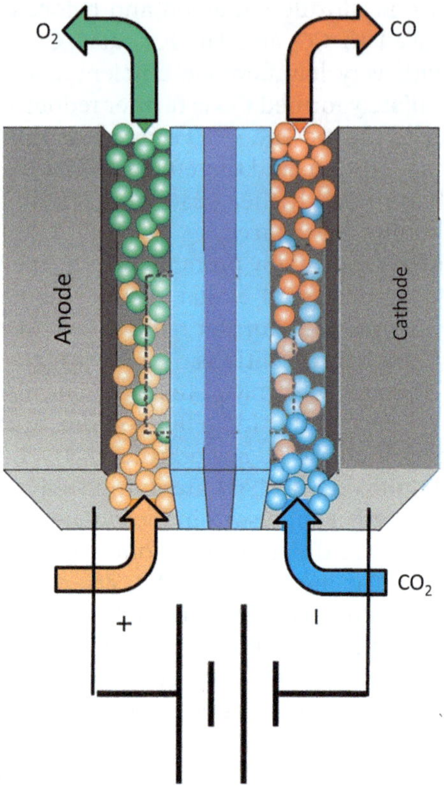

Figure 10.3 Schematic of a CO_2 electrolyzer.

CO_2 is reduced to chemicals such as CO, $HCOO^-$, *etc.* at the cathode. Oxygen evolution and/or oxidation of other compounds may be involved at the anode. The electrolyte options include liquid electrolytes,[166,167] solid polymer electrolytes[168,170,171] and solid oxide electrolytes.[172] The cell configuration varies from one electrolyte to another and depends on the final products as well.[159,164,166–168] Although there are many different products from CO_2 electrolysis, here we will limit the discussion to those products that show the most promise to be industrially viable.

10.3.2.3 Formic Acid or Formate

Formate or formic acid is one of the first products identified from carbon dioxide reduction on the 1st group metal electrodes, which have very high hydrogen evolution overpotentials.[173] A valuable example is Alvarez-Guerra *et al.*'s (University of Cantabria, Spain) work on the reduction of CO_2 to formate on Sn and Pb plates in a filter-press or parallel-plate type electrochemical cell.[174,175] Sn electrodes exhibited higher faradaic efficiency and current for formate production.

Oloman *et al.* (University of British Columbia, UBC, Canada) developed an interesting flow cell with continuous two-phase flow containing a catholyte solvent and carbon dioxide.[176–179] The mixed flow of the gas/liquid phase provides dynamic conditions that favor CO_2 reduction to formate. The reactor was scaled up with a geometric cathode area of over 300 cm^2. The output current was 20 to 100 A at a cell potential of 2.7 to 4 V. The faradaic efficiency varied from 90 to 60% with an increase in the current. With a single pass, the reactor is able to produce up to 1 M formate (4.6 wt%) in the catholyte. The faradaic efficiency decreased from 65% to 48% in 6 hours of operation, and was recovered either by chemical treatment of the cathode or reversal of polarization. Mantra Energy Alternative Ltd.,[180,181] a subsidiary of Mantra Venture Group Ltd., acquired the intellectual property from UBC in 2008, and now reports to be on the path towards commercialization.

Sridhar *et al.* from Det Norske Veritas (DNV, US) disclosed a lovely method and apparatus for the electrochemical reduction of CO_2 to formate.[180] They claimed that a Sn alloy with a few percent of Zn exhibited higher faradaic efficiency at lower overpotentials. The faradaic efficiency was stable at about 80% for more than 20 hours in a gas–liquid–solid half-cell, and then dropped to 20%. They demonstrated a stable faradaic efficiency of 60–70% and current density of about 40 mA cm^{-2} with an electrodeposited Sn cathode for four days.

Oloman *et al.* and Sridhar *et al.*'s systems ran at relatively high overpotential, but Masel *et al.* discovered that they could lower the overpotential by adding imidazolium moieties to the reaction mixture (Figure 10.2). Masel *et al.* have reported[182] initial results on CO_2 conversion to formic acid in a half-cell. They observed currents up to 100 mA cm^{-2} at high (>80%) faradaic efficiencies.

Bocarsly's group has also demonstrated a wonderful lowering of the overpotential for formic acid production using pyridine rather than an imidazolium species[183,184] although he also finds that imidazoliums lower the potential more than pyridine.[185,186]

10.3.2.4 Carbon Monoxide and Syngas

Carbon monoxide (CO) offers broad options for the downstream processing of syngas using Fischer–Tropsch chemistry. CO can be produced from CO_2 reduction on the 2nd group metals. However, these metals exhibit lower overpotential for the hydrogen evolution than 1st group metals. Therefore, hydrogen evolution is the main side reaction that lowers the faradaic efficiency towards CO_2 reduction and lowers the CO selectivity. There are two ways to solve this problem. The first way is to suppress the hydrogen evolution, either by using a co-catalyst in addition to the metal electrode or by incorporating a novel structured metal electrode.

Dioxide Materials Inc. has been using imidazolium-based ionic liquids as co-catalysts to suppress hydrogen evolution and to significantly lower the overpotential for CO_2 reduction.[169,182,187–202] For example, on an Ag catalyst, they achieved 100 mA cm^{-2} at cathode overpotentials less than 0.6 V, with selectivities over 95%. The current and selectivity were stable in a 250 hour run.[203] At the metal loadings in the experiment, this corresponds to over 1 billion turnovers and a turnover frequency over 1000 per sec. They also discovered that the rate is strongly dependent on particle size[195] and pH in the catalyst layer.

Sun *et al.* (Brown University, USA) reported selective carbon dioxide reduction to CO on monodispersed Au nanoparticles in 0.5 M $KHCO_3$. The 8 nm Au nanoparticles showed the maximum faradaic efficiency of 90% at −0.67 V *vs.* RHE (reversible hydrogen electrode).[204] When combined with butyl-3-methylimidazolium hexafluorophosphate ($BMIMPF_6$), the faradaic efficiency was boosted up to 97% at −0.52 V. Density functional theory (DFT) calculations suggested that edge sites are active for carbon dioxide reduction, while corner sites are active for hydrogen evolution. The nanoparticles with optimal edge sites exhibited the highest activity for carbon dioxide reduction. They further synthesized ultrathin Au nanowires and achieved a selectivity of 94% at −0.35 V due to the increased reactive edge sites and the weak CO binding on these sites.[205] Wang *et al.* observed the size dependent selectivity for reduction of CO_2 to CO on Pd nanoparticles,[206] repeating the earlier work of Salehi-Khojin *et al.*[195] The faradaic efficiency increased significantly from 6% to 91% when the particle size decreased from 10.3 nm to 2.4 nm at −0.89 V (*vs.* RHE). This is due to easier adsorption of CO_2 and formation of COOH* on edge and corner sites than on terrace sites of Pd nanoparticles.

Jiao *et al.* (University of Delaware, USA) did excellent work in which they reported a nanoporous Ag catalyst with about 92% faradaic efficiency for carbon dioxide reduction.[165] The current was over 3000 times higher than polycrystalline Ag at a moderate overpotential of <0.5 V. They attributed the high activity to the increased electrochemical active area and the intrinsic high activity due to the greater stabilization of the $CO_2^{\bullet -}$ intermediate on the highly curved surface.

Salehi-Khojin (University of Illinois at Chicago, USA) led first-class work in which his group reported non-noble metals and metal-free carbon nanofibers as alternative catalysts to Ag for carbon dioxide reduction in the ionic liquid system.[207,208] Molybdenum disulfide (MoS_2) showed superior performance for carbon dioxide reduction over noble metals.[207] The overpotential was as low as 54 mV and the CO selectivity increased significantly from 3% to 98% with the increase in the overpotential. Metal-free carbon nanofibers (CNFs) exhibited the same overpotential (0.17 V) as Ag did in the presence of $EMIMBF_4$, with a selectivity of 98%. Increases in rate when water was added was observed for carbon dioxide reduction on both MoS_2 and CNFs as well as Ag.[209]

Kenis *et al.* (University of Illinois at Urbana-Champaign, USA) did wonderful work in which they found that the larger cations in the aqueous electrolyte can suppress hydrogen evolution and favor carbon dioxide reduction on an Ag electrode.[166] It is believed that stronger cation hydration reduces the cation adsorption on the electrode, and therefore increases the product selectivity and the reaction rate as well. They also improved the CO selectivity from 80% to 95% because of improved Ag coverage on the gas diffusion layer by using air-brushed rather than hand-painted Ag electrodes.[167] TiO_2 is more effective than carbon as a catalyst support to improve the utilization of Ag without sacrificing the CO selectivity.[159] Tornow *et al.* synthesized carbon supported nitrogen-based organometallic Ag catalysts and achieved more than 90% faradaic efficiency for carbon dioxide reduction, suggesting a similar co-catalyst mechanism.[164,210–213]

A number of investigators have used homogeneous catalysts[18,214–240] for the same reaction. While the faradaic efficiencies have been reasonable, the reaction rates and catalyst lifetimes have been tiny. For example, a recent paper quoted "High turnover nos. (*ca.* 54 000) and turnover frequencies (*ca.* 1.5 s^{-1})".[216] By comparison, the catalysts used by Dioxide Materials and Liquid Light run at turnover frequencies between 1000 and 3000 s^{-1} and achieve turnover numbers in the billions.

The second way to address the lower overpotential for hydrogen evolution from water electrolysis on 2nd group metals is to utilize the side reaction of the hydrogen evolution in order to produce CO and H_2 simultaneously, thus generating a syngas mixture. That mixture can be fed directly into a downstream reactor to generate a wide range of hydrocarbons *via*, for example, Fischer–Tropsch heterogeneous catalysis. As another example, a mixture with an H_2–CO ratio of 2 : 1 would be desirable to produce methanol. Olah and Prakash (University of Southern California, USA) disclosed electrochemical cells producing syngas.[241,242] In one cell, an optimal mixture of CO and H_2 was produced at −1.5 or −4.0 V *vs.* Ag/AgCl on an Au electrode in an aqueous methanol solution containing tetrabutylammonium bromide. In another cell, the same optimal mixture was generated at −3.2 V *vs.* Ag/AgCl on Au in a 0.1 M $KHCO_3$ solution. They also demonstrated the feasibility of producing syngas using a membrane electrode assembly (MEA) with an anion exchange membrane.[242] The MEA was prepared by sandwiching the polymer membrane electrolyte between an Au coated cathode and a Pt coated anode. Newman *et al.* compared the different electrochemical cell configurations with polymer electrolytes for the production of syngas.[168] They found that carbon dioxide reduction was not favorable when an Ag electrode was in direct contact with Nafion (a proton exchange membrane from DuPont). By inserting a pH-buffer layer (aqueous $KHCO_3$) between the Ag electrode and the Nafion membrane, the syngas with a CO–H_2 ratio of $\frac{1}{2}$ was generated at −2 V *vs.* a saturated calomel electrode (SCE). However, the CO selectivity decreased with time. A polyethersulfone-based anion exchange membrane was also tested, but only 3% CO was detected, with the rest being H_2. They attributed the low CO efficiency to the limited CO_2 transport.

Kanan *et al.* (Stanford University, USA) authored an impressive series of papers in which they found that using nanocrystalline metal films (Cu, Au, Pb and Sn) as catalysts can significantly suppress hydrogen evolution and favor carbon dioxide reduction.[243–250] The nanocrystalline catalysts prepared by reducing their oxide precursors are believed to have relatively high grain boundary surface density, which can promote the formation of a very thin oxide/hydroxide layer over the electrode surface and act as the active surface for CO_2 reduction. For example, the oxide-derived nanocrystalline Au (OD-Au) catalyst had lower specific (*i.e.*, surface-area-normalized) activity for H^+ reduction and enhanced specific activity for CO reduction compared to other Au materials at overpotentials as low as 140 mV.

Syngas can also be produced by high temperature co-electrolysis of steam and CO_2 using solid oxide electrolysis cells (SOECs).[251–258] The SOECs, which are fundamentally solid oxide fuel cells (SOFCs) operated in a reverse manner, consist of three parts: yttria-stabilized zirconia (YSZ) as an electrolyte, perovskites (*e.g.* strontium-doped lanthanum manganite or LSM, strontium-doped lanthanum cobaltite or LSC) as an anode, and nickel/YSZ cermet as a cathode. During co-electrolysis, H_2O and CO_2 both simultaneously undergo decomposition, the produced O^{2-} ions transport through the electrolyte layer and form O_2 on the anode. As a result, only CO and H_2 mixtures are left on the cathode side. Different from the other syngas production technologies discussed previously, the CO_2-steam co-electrolysis is a high temperature (typically 600–1000 °C) process, which involves not just steam and CO_2 electrolysis but also the water gas shift reaction and the reverse water gas shift reaction. The overall reaction for the co-electrolysis of steam and CO_2 can be written as:

$$H_2O + CO_2 \rightarrow H_2 + CO + O_2 \tag{10.5}$$

This process is composed of the three partial reactions:
on the cathode side:

$$H_2O + 2e^- \rightarrow H_2 + O^{2-} \tag{10.6}$$

$$CO_2 + 2e^- \rightarrow CO + O^{2-} \tag{10.7}$$

and on the anode side:

$$2O^{2-} \rightarrow O_2 + 4e^- \tag{10.8}$$

The water gas shift reaction and its reverse reaction can be seen as:

$$CO + H_2O \underset{\text{heat}}{\Longleftrightarrow} CO_2 + H_2 \tag{10.9}$$

Due to its fast startup and shutdown ramp rates, the high temperature CO_2-steam co-electrolysis technology is easily coupled with intermittent energy sources (wind, solar, *etc.*) to store the generated heat and power in the form of chemical energies. The Idaho National Laboratory (USA) in collaboration with Ceramatec Inc. (Salt Lake City, USA) has been studying the use of SOEC stacks to electrolyze steam for large-scale H_2 production and co-electrolyze CO_2-steam mixtures for syngas production.[259] They found that by using a cobalt ferrite oxygen electrode and bond layer, the 10-cell stacks showed good stability during a 2500-hour test with a degradation rate of just 8.15% per 1000 hours.[260] Haldor Topsoe (Denmark) also claims that co-electrolysis using SOECs is a more efficient solution for CO_2 reduction. The content of CO in the output from the SOEC stack can be as high as 80 wt%.[261] Liu *et al.* reported a tubular solid oxide electrolysis cell with yttrium-stabilized-zirconia (YSZ) as the electrolyte, strontium-doped lanthanum manganite (LSM) as the anode and cermet of Ag-GDC (gadolinium-doped-ceria) as the cathode, which shows a minimum electrolyzing voltage of 0.70 V and a maximum faradaic efficiency of 98.5%. The stability test shows that the SOEC has good CO_2 electrolysis performance for 18 hours at constant voltage of 1.59 V at 800 °C.[262]

10.3.2.5 Higher Order Products

Production of CO and of formate from carbon dioxide reduction have shown quite high faradaic efficiency, and they are potentially feedstock chemicals for other industrial processes. These technologies are on the route to commercialization. However, there is growing interest in other chemicals that can be produced directly from carbon dioxide reduction. For example, methanol (methyl alcohol) can be synthesized from syngas over a catalyst comprising a mixture of copper, zinc oxide, and alumina using a method developed by Imperial Chemical Industries (ICI) in 1966.[263] The syngas feedstock could be provided by CO from CO_2 reduction and H_2 from water splitting, either separately from two electrolyzers or simultaneously from one electrolyzer. Methanol can also be produced directly from CO_2 electrolysis.

Early work related to energy efficient CO_2 conversion chemistry using renewable energy, and eventually also "artificial photosynthesis," by Bocarsly's group[19,184–186,264–292] at Princeton University produced a superb paper titled, "A new homogeneous electrocatalyst for the reduction of carbon dioxide to methanol at low overpotential".[283] Scheme 1 of Ref. 283 shows pyridine being reversibly protonated to a positively charged pyridinium ion, which then takes on an electron, followed by the transfer of a (neutral) hydrogen atom to the target molecule, thus regenerating the original pyridine. See also related patents US8313634, US8663447, and related applications, titled "Conversion of carbon dioxide to organic products".[293,294]

In 2012, this technology was spun out to the startup company Liquid Light Corporation (Monmouth Junction, NJ, USA). The company continues to collaborate with Bocarsly's group at Princeton.[295]

Continued research has shown that a variety of low molecular weight multi-carbon compounds such as isopropanol and butanol can be produced by varying the electrode material, organic catalyst, and reaction conditions. In April 2014, Liquid Light was selected as one of the awardees of the Alberta CCEMC Grand Challenge, to design and build a pilot plant to convert CO_2 into chemicals using clean energy sources. Their first commercial target is ethylene glycol, with plans to extend the possible products to include propylene, isopropanol, methyl-methacrylate, and acetic acid.[296,297] Liquid

Light has filed over 20 patent families related to various aspects of this technology.[183,184,268,272,273,286,298–317]

10.3.2.6 Other Electrochemical Processes

The University of British Columbia (UBC, Canada) developed a coupled CO_2 and wastewater treatment process to create high value gas/oil field chemicals.[318] This novel technology mineralizes CO_2 using the salts found in industrial wastewater in an electrochemical cell. The product could be high-value oil-field chemicals such as acids and carbonate salts. This process could simultaneously remove CO_2 and desalinate the wastewater or brine. The concentration of salts in the wastewater determines the amount of CO_2 that can be converted.

Dioxide Materials has reported a process in which formic acid is formed from CO_2 by electrolysis or other means, and then the formic acid further reacts to form other products such as formaldehyde.[200]

10.4 Concentrated Solar for CO_2 Conversion

Several high temperature thermochemical reactions of CO_2 have been explored, in which the required thermal energy is provided by a concentrated solar energy source. Two examples are provided.

The Sandia National Laboratory "Sunshine to Petrol" team is developing a process to use concentrated solar energy and metal oxides to split CO_2 into CO and O_2, and to also split water into H_2 and O_2. The CO and H_2 can then be combined to form a syngas mixture, which can then be converted by known processes to synthetic fuels that can replace their current petroleum-based counterparts.[319] A 2007 prototype device was named the Counter-Rotating Ring Receiver Reactor Recuperator (or CR5).[320] In this device, one side of a pair of counter-rotating rings of metal oxide was exposed to concentrated solar radiation and to one or both of CO_2 and H_2O. The metal oxides were raised to a temperature at which they could extract oxygen atoms from the gases, leaving CO and H_2 that could be used as syngas for fuel production. As the metal oxide rings rotated out of the concentrated sunlight, they cooled enough to then release the O_2 in the rear chamber of the device.

Professor Jane Davidson and her group at the University of Minnesota Solar Energy Laboratory are developing ways to "efficiently 'reenergize' CO_2 and H_2O back into hydrocarbon form using solar energy as the heat source to drive chemical reactions".[321] They are using the metal oxide ceria to split CO_2 and H_2O to form syngas for fuel production, much like the Sandia Laboratory project described earlier. A second project involves splitting CO_2 and H_2O by means of a solar thermochemical Zn/ZnO redox cycle.[322] A third project, nicknamed "SUNgas", uses concentrated solar thermochemical gasification of biomass to generate syngas for fuel production. The current SUNgas system uses a molten salt as a catalyst and heat transfer medium, as well as for energy storage.[323]

10.5 Summary

In summary then, CO_2 utilization is a growing field. There are many different approaches. Here we have summarized two main approaches: first, one in which hydrocarbons react with CO_2 to produce useful products; and second, one in which electricity from renewable energy is used to convert CO_2 into useful products. There is a huge amount of work that has been done already. The big picture is that CO_2 is becoming a useful feedstock for the production of chemicals. CO_2 could also be a useful feedstock for the production of fuels, assuming that the economics of the processes improve.

Conflict of Interest Statement

Dioxide Materials and 3M are working toward commercializing certain processes and materials described in this chapter. The authors have a financial interest in the outcome.

Acknowledgements

This work was supported in part by the US Department of Energy ARPA-E project under contract DE-AR 0000345 and the US Department of Energy under contract DE-SC0004453.

References

1. A. M. Appel, J. E. Bercaw, A. B. Bocarsly, H. Dobbek, D. L. DuBois, M. Dupuis, J. G. Ferry, E. Fujita, R. Hille, P. J. A. Kenis *et al.*, Frontiers, Opportunities, and Challenges in Biochemical and Chemical Catalysis of CO2 Fixation, *Chem. Rev.*, 2013, **113**, 6621–6658, DOI: 10.1021/cr300463y.
2. M. Aresta, *Carbon Dioxide: Utilization Options to Reduce its Accumulation in the Atmosphere*, Wiley-VCH Verlag GmbH & Co. KGaA, 2010, pp. 1–13.
3. M. Aresta and A. Dibenedetto, Utilisation of CO2 as a chemical feedstock: opportunities and challenges, *Dalton Trans.*, 2007, 2975–2992, DOI: 10.1039/b700658f.
4. M. Aresta, A. Dibenedetto and A. Angelini, Catalysis for the Valorization of Exhaust Carbon: from CO2 to Chemicals, Materials, and Fuels. Technological Use of CO_2, *Chem. Rev.*, 2014, **114**, 1709–1742, DOI: 10.1021/cr4002758.
5. M. Bevilacqua, J. Filippi, H. A. Miller and F. Vizza, Recent Technological Progress in CO2 Electroreduction to Fuels and Energy Carriers in Aqueous Environments, *Energy Technol.*, 2015, **3**, 197–210, DOI: 10.1002/ente.201402166.
6. A. Boddien, F. Gaertner, C. Federsel, I. Piras, H. Junge, R. Jackstell and M. Beller, *Catalytic Utilization of Carbon Dioxide: Actual Status and Perspectives*, Wiley-VCH Verlag GmbH & Co. KgaA, 2012, pp. 685–724.

7. G. Centi, G. Iaquaniello and S. Perathoner, Can We Afford to Waste Carbon Dioxide? Carbon Dioxide as a Valuable Source of Carbon for the Production of Light Olefins, *ChemSusChem*, 2011, **4**, 1265–1273, DOI: 10.1002/cssc.201100313.
8. G. Centi, E. A. Quadrelli and S. Perathoner, Catalysis for CO_2 conversion: a key technology for rapid introduction of renewable energy in the value chain of chemical industries, *Energy Environ. Sci.*, 2013, **6**, 1711–1731, DOI: 10.1039/c3ee00056g.
9. J. J. Concepcion, R. L. House, J. M. Papanikolas and T. J. Meyer, Chemical approaches to artificial photosynthesis, *Proc. Natl. Acad. Sci. U. S. A.*, 2012, **109**, 15560–15564, DOI: 10.1073/pnas.1212254109.
10. C. Costentin, M. Robert and J.-M. Saveant, Catalysis of the electrochemical reduction of carbon dioxide, *Chem. Soc. Rev.*, 2013, **42**, 2423–2436, DOI: 10.1039/c2cs35360a.
11. A. Deronzier and S. Chardon-Noblat, Electrocatalytic reduction of carbon dioxide: a future solution for sustainable chemistry?, *Actual. Chim.*, 2013, **371–372**, 84–88.
12. A. Dibenedetto, A. Angelini and P. Stufano, Use of carbon dioxide as feedstock for chemicals and fuels: homogeneous and heterogeneous catalysis, *J. Chem. Technol. Biotechnol.*, 2014, **89**, 334–353, DOI: 10.1002/jctb.4229.
13. D. L. DuBois, Carbon, Electrochemical reactions of carbon dioxide, *Encycl. Electrochem.*, 2006, **7a**, 202–225.
14. I. Ganesh, Conversion of carbon dioxide into methanol – a potential liquid fuel: Fundamental challenges and opportunities (a review), *Renewable Sustainable Energy Rev.*, 2014, **31**, 221–257, DOI: 10.1016/j.rser.2013.11.045.
15. M. Gattrell, N. Gupta and A. Co, A review of the aqueous electrochemical reduction of CO2 to hydrocarbons at copper, *J. Electroanal. Chem.*, 2006, **594**, 1–19, DOI: 10.1016/j.jelechem.2006.05.013.
16. A. Goeppert, M. Czaun, G. K. Surya Prakash and G. A. Olah, Air as the renewable carbon source of the future: an overview of CO2 capture from the atmosphere, *Energy Environ. Sci.*, 2012, **5**, 7833–7853, DOI: 10.1039/c2ee21586a.
17. Y. Hori, Electrochemical CO_2 reduction on metal electrodes, *Mod. Aspects Electrochem.*, 2008, **42**, 89–189.
18. P. Kang, Z. Chen, M. Brookhart and T. J. Meyer, Electrocatalytic Reduction of Carbon Dioxide: Let the Molecules Do the Work, *Top. Catal.*, 2015, **58**, 30–45, DOI: 10.1007/s11244-014-0344-y.
19. K. A. Keets, E. B. Cole, A. J. Morris, N. Sivasankar, K. Teamey, P. S. Lakkaraju and A. B. Bocarsly, Analysis of pyridinium catalyzed electrochemical and photoelectrochemical reduction of CO_2: chemistry and economic impact, *Indian J. Chem., Sect. A: Inorg., Bio-inorg., Phys., Theor. Anal. Chem.*, 2012, **51A**, 1284–1297.
20. E. V. Kondratenko, G. Mul, J. Baltrusaitis, G. O. Larrazabal and J. Perez-Ramirez, Status and perspectives of CO_2 conversion into fuels

and chemicals by catalytic, photocatalytic and electrocatalytic processes, *Energy Environ. Sci.*, 2013, **6**, 3112–3135, DOI: 10.1039/c3ee41272e.
21. Y.-N. Li, R. Ma, L.-N. He and Z.-F. Diao, Homogeneous hydrogenation of carbon dioxide to methanol, *Catal. Sci. Technol.*, 2014, **4**, 1498–1512, DOI: 10.1039/c3cy00564j.
22. Q. Lu, J. Rosen and F. Jiao, Nanostructured metallic electrocatalysts for carbon dioxide reduction, *ChemCatChem*, 2015, **7**, 38–47, DOI: 10.1002/cctc.201402669.
23. X. Lu, D. Y. C. Leung, H. Wang, M. K. H. Leung and J. Xuan, Electrochemical Reduction of Carbon Dioxide to Formic Acid, *ChemElectroChem*, 2014, **1**, 836–849, DOI: 10.1002/celc.201300206.
24. D. G. Nocera, The Artificial Leaf, *Acc. Chem. Res.*, 2012, **45**, 767–776, DOI: 10.1021/ar2003013.
25. I. Omae, Recent developments in carbon dioxide utilization for the production of organic chemicals, *Coord. Chem. Rev.*, 2012, **256**, 1384–1405, DOI: 10.1016/j.ccr.2012.03.017.
26. S. Perathoner and G. Centi, CO_2 Recycling: A Key Strategy to Introduce Green Energy in the Chemical Production Chain, *ChemSusChem*, 2014, **7**, 1274–1282, DOI: 10.1002/cssc.201300926.
27. J. Qiao, Y. Liu, F. Hong and J. Zhang, A review of catalysts for the electroreduction of carbon dioxide to produce low-carbon fuels, *Chem. Soc. Rev.*, 2014, **43**, 631–675, DOI: 10.1039/c3cs60323g.
28. Y. Qin, X. Sheng, S. Liu, G. Ren, X. Wang and F. Wang, Recent advances in carbon dioxide based copolymers, *J. CO2 Util.*, 2014, DOI: 10.1016/j.jcou.2014.10.003.
29. N. S. Spinner, J. A. Vega and W. E. Mustain, Recent progress in the electrochemical conversion and utilization of CO_2, *Catal. Sci. Technol.*, 2012, **2**, 19–28, DOI: 10.1039/c1cy00314c.
30. D. Srinivas, Carbon dioxide to chemicals and fuels: green synthetic approach, *Chem. Ind. Dig.*, 2011, **24**, 68–73.
31. J. M. Thomas, Heterogeneous Catalysis and the Challenges of Powering the Planet, Securing Chemicals for Civilised Life, and Clean Efficient Utilization of Renewable Feedstocks, *ChemSusChem*, 2014, **7**, 1801–1832, DOI: 10.1002/cssc.201301202.
32. W. Wang, S. Wang, X. Ma and J. Gong, Recent advances in catalytic hydrogenation of carbon dioxide, *Chem. Soc. Rev.*, 2011, **40**, 3703–3727, DOI: 10.1039/c1cs15008a.
33. C. D. Windle and R. N. Perutz, Advances in molecular photocatalytic and electrocatalytic CO_2 reduction, *Coord. Chem. Rev.*, 2012, **256**, 2562–2570, DOI: 10.1016/j.ccr.2012.03.010.
34. C. S. Yeung, Recent catalytic approaches to chemical synthesis from carbon feedstocks, *Pure Appl. Chem.*, 2013, **85**, 941–956, DOI: 10.1351/pac-con-12-09-14.
35. Q. Liu, L. Wu, R. Jackstell and M. Beller, Using carbon dioxide as a building block in organic synthesis, *Nat. Commun.*, 2015, **6**, 6933, DOI: 10.1038/ncomms6933.

36. J. Meessen, Urea synthesis, *Chem. Ing. Tech.*, 2014, **86**, 2180–2189, DOI: 10.1002/cite.201400064.
37. J. H. Meessen, Urea, in *Ullmann's Encyclopedia of Industrial Chemistry*, Wiley-VCH Verlag GmbH & Co. KGaA, 2000.
38. Empower Materials, http://www.empowermaterials.com/. Accessed April 10 2015.
39. Green Polymer Made of CO2 from Exhaust Gases, http://www.siemens.com/innovation/en/news/2012/e_inno_1213_2.htm. Accessed April 10, 2015.
40. Dream Production – CO2-based polyols: Sustainability and Future Research Perspectives, http://www.polyurethanes.bayer.com/News-and-Events/Events/UTECH/Dr-Christoph-Guertler.aspx. Accessed April 11, 2015.
41. C. Bosch and W. Meiser, Process of manufacturing urea Patent or Patent Application US1429483 A, 1922.
42. N. W. Krase and V. L. Gaddy, Synthesis of Urea from Ammonia and Carbon Dioxide, *J. Ind. Eng. Chem.*, 1922, **14**, 611–615, DOI: 10.1021/ie50151a009.
43. H. Kolbe, Ueber Synthese der Salicylsäure, *Justus Liebigs Ann. Chem.*, 1860, **113**, 125–127, DOI: 10.1002/jlac.18601130120.
44. R. Schmitt, Beitrag zur Kenntniss der Kolbe'schen Salicylsäure Synthese, *J. Prakt. Chem.*, 1885, **31**, 397–411, DOI: 10.1002/prac.18850310130.
45. Edgar181, http://commons.wikimedia.org/wiki/File:Kolbe-Schmitt.png#/media/File:Kolbe-Schmitt.png. Accessed May 4, 2015.
46. A. S. Lindsey and H. Jeskey, The Kolbe-Schmitt Reaction, *Chem. Rev.*, 1957, **57**, 583–620, DOI: 10.1021/cr50016a001.
47. N. Kuhnert, One Hundred Years of Aspirin, *Chem. Unserer Zeit*, 1999, **33**, 213–220.
48. Kolbe-Schmitt reaction, http://en.wikipedia.org/wiki/Kolbe%E2%80%93Schmitt_reaction. Accessed May 4, 2015.
49. S. Inoue, H. Koinuma and T. Tsuruta, Copolymerization of carbon dioxide and epoxide, *J. Polym. Sci., Part B: Polym. Lett.*, 1969, **7**, 287–292, DOI: 10.1002/pol.1969.110070408.
50. S. Inoue, H. Koinuma and T. Tsuruta, Copolymerization of carbon dioxide and epoxide with organometallic compounds, *Makromol. Chem.*, 1969, **130**, 210–220.
51. D. J. Darensbourg, Making Plastics from Carbon Dioxide: Salen Metal Complexes as Catalysts for the Production of Polycarbonates from Epoxides and CO_2, *Chem. Rev.*, 2007, **107**, 2388–2410, DOI: 10.1021/cr068363q.
52. D. J. Darensbourg, Personal adventures in the synthesis of copolymers from carbon dioxide and cyclic ethers, *Adv. Inorg. Chem.*, 2014, **66**, 1–23, DOI: 10.1016/b978-0-12-420221-4.00001-9.
53. D. J. Darensbourg, W.-C. hung, A. D. Yeung and M. Luna, Dramatic Behavioral Differences of the Copolymerization Reactions of 1,4-Cyclohexadiene and 1,3-Cyclohexadiene Oxides with Carbon Dioxide,

Macromolecules, 2015, **48**, 1679–1687, DOI: 10.1021/acs.macromol.5b00172.

54. D. J. Darensbourg and Y. Wang, Terpolymerization of propylene oxide and vinyl oxides with CO_2: copolymer cross-linking and surface modification *via* thiol-ene click chemistry, *Polym. Chem.*, 2015, **6**, 1768–1776, DOI: 10.1039/c4py01612b.
55. M. Cokoja, C. Bruckmeier, B. Rieger, W. A. Herrmann and F. E. Kuehn, Transformation of Carbon Dioxide with Homogeneous Transition-Metal Catalysts: A Molecular Solution to a Global Challenge?, *Angew. Chem., Int. Ed.*, 2011, **50**, 8510–8537, DOI: 10.1002/anie.201102010.
56. R. Eberhardt, M. Allmendinger, G. Luinstra and B. Rieger, New β-diiminate zinc catalysts with ethylsulfinate initiating groups for the alternating copolymerization of CO_2 and epoxides, *PMSE Prepr.*, 2002, **87**, 215–216.
57. R. Eberhardt, M. Allmendinger, G. A. Luinstra and B. Rieger, The Ethylsulfinate Ligand: A Highly Efficient Initiating Group for the Zinc β-Diiminate Catalyzed Copolymerization Reaction of CO2 and Epoxides, *Organometallics*, 2003, **22**, 211–214, DOI: 10.1021/om020734f.
58. R. Eberhardt, M. Allmendinger and B. Rieger, DMAP [4-(N,N-dimethylamino)pyridine]/Cr(III) catalyst ratio: The decisive factor for poly(propylene carbonate) formation in the coupling of CO_2 and propylene oxide, *Macromol. Rapid Commun.*, 2003, **24**, 194–196, DOI: 10.1002/marc.200390022.
59. R. Eberhardt, M. Allmendinger, M. Zintl, C. Troll, G. A. Luinstra and B. Rieger, New zinc dicarboxylate catalysts for the CO_2/propylene oxide copolymerization reaction: Activity enhancement through Zn(II)-ethylsulfinate initiating groups, *Macromol. Chem. Phys.*, 2004, **205**, 42–47, DOI: 10.1002/macp.200350081.
60. S. Kissling, P. T. Altenbuchner and B. Rieger, Materials for the 21st century: can the carbon come from CO2, *DGMK Tagungsber.*, 2012, **2012-3**, 91–96.
61. S. Kissling, M. W. Lehenmeier, P. T. Altenbuchner, A. Kronast, M. Reiter, P. Deglmann, U. B. Seemann and B. Rieger, Dinuclear zinc catalysts with unprecedented activities for the copolymerization of cyclohexene oxide and CO_2, *Chem. Commun.*, 2015, **51**, 4579–4582, DOI: 10.1039/c5cc00784d.
62. S. Klaus, M. W. Lehenmeier, C. E. Anderson and B. Rieger, Recent advances in CO_2/epoxide copolymerization-New strategies and co-operative mechanisms, *Coord. Chem. Rev.*, 2011, **255**, 1460–1479, DOI: 10.1016/j.ccr.2010.12.002.
63. S. Klaus, M. W. Lehenmeier, E. Herdtweck, P. Deglmann, A. K. Ott and B. Rieger, Mechanistic Insights into Heterogeneous Zinc Dicarboxylates and Theoretical Considerations for CO_2-Epoxide Copolymerization, *J. Am. Chem. Soc.*, 2011, **133**, 13151–13161, DOI: 10.1021/ja204481w.
64. S. Klaus, S. I. Vagin, M. W. Lehenmeier, P. Deglmann, A. K. Brym and B. Rieger, Kinetic and Mechanistic Investigation of Mononuclear and

Flexibly Linked Dinuclear Complexes for Copolymerization of CO_2 and Epoxides, *Macromolecules*, 2011, **44**, 9508–9516, DOI: 10.1021/ma201387f.

65. M. W. Lehenmeier, C. Bruckmeier, S. Klaus, J. E. Dengler, P. Deglmann, A.-K. Ott and B. Rieger, Differences in Reactivity of Epoxides in the Copolymerisation with Carbon Dioxide by Zinc-Based Catalysts: Propylene Oxide *versus* Cyclohexene Oxide, *Chem. – Eur. J.*, 2011, **17**, 8858–8869, DOI: 10.1002/chem.201100578, 8851–8827.
66. M. W. Lehenmeier, S. Kissling, P. T. Altenbuchner, C. Bruckmeier, P. Deglmann, A.-K. Brym and B. Rieger, Flexibly Tethered Dinuclear Zinc Complexes: A Solution to the Entropy Problem in CO_2/Epoxide Copolymerization Catalysis?, *Angew. Chem., Int. Ed.*, 2013, **52**, 9821–9826, DOI: 10.1002/anie.201302157.
67. G. A. Luinstra, G. R. Haas, F. Molnar, V. Bernhart, R. Eberhardt and B. Rieger, On the formation of aliphatic polycarbonates from epoxides with chromium(III) and aluminum(III) metal-salen complexes, *Chem. – Eur. J.*, 2005, **11**, 6298–6314, DOI: 10.1002/chem.200500356.
68. M. Reiter, P. T. Altenbuchner, S. Kissling, E. Herdtweck and B. Rieger, Amine-bis(phenolato)cobalt(II) Catalysts for the Formation of Organic Carbonates from Carbon Dioxide and Epoxides, *Eur. J. Inorg. Chem.*, 2015, **2015**, 1766–1774, DOI: 10.1002/ejic.201403087.
69. K. A. Salmeia, S. Vagin, C. E. Anderson and B. Rieger, Poly(propylene carbonate): Insight into the Microstructure and Enantioselective Ring-Opening Mechanism, *Macromolecules*, 2012, **45**, 8604–8613, DOI: 10.1021/ma301916r.
70. W. Xia, K. A. Salmeia, S. I. Vagin and B. Rieger, Concerning the deactivation of cobalt(III)-based porphyrin and salen catalysts in epoxide/CO_2 copolymerization, *Chem. – Eur. J.*, 2015, **21**, 4384–4390, DOI: 10.1002/chem.201406258.
71. W. Xia, S. I. Vagin and B. Rieger, Regarding Initial Ring Opening of Propylene Oxide in its Copolymerization with CO_2 Catalyzed by a Cobalt(III) Porphyrin Complex, *Chem. – Eur. J.*, 2014, **20**, 15499–15504, DOI: 10.1002/chem.201404147.
72. K. Nakano, R. Fujie, R. Shintani and K. Nozaki, Efficient catalyst removal and recycling in copolymerization of epoxides with carbon dioxide *via* simple liquid-liquid phase separation, *Chem. Commun.*, 2013, **49**, 9332–9334, DOI: 10.1039/c3cc45622f.
73. K. Nakano, S. Hashimoto and K. Nozaki, Bimetallic mechanism operating in the copolymerization of propylene oxide with carbon dioxide catalyzed by cobalt-salen complexes, *Chem. Sci.*, 2010, **1**, 369–373, DOI: 10.1039/c0sc00220h.
74. K. Nakano, T. Kamada and K. Nozaki, Selective formation of polycarbonate over cyclic carbonate: copolymerization of epoxides with carbon dioxide catalyzed by a cobalt(III) complex with a piperidinium end-capping arm, *Angew. Chem., Int. Ed.*, 2006, **45**, 7274–7277, DOI: 10.1002/anie.200603132.

75. K. Nakano, K. Kobayashi and K. Nozaki, Tetravalent Metal Complexes as a New Family of Catalysts for Copolymerization of Epoxides with Carbon Dioxide, *J. Am. Chem. Soc.*, 2011, **133**, 10720–10723, DOI: 10.1021/ja203382q.
76. K. Nakano, K. Kobayashi, T. Ohkawara, H. Imoto and K. Nozaki, Copolymerization of Epoxides with Carbon Dioxide Catalyzed by Iron-Corrole Complexes: Synthesis of a Crystalline Copolymer, *J. Am. Chem. Soc.*, 2013, **135**, 8456–8459, DOI: 10.1021/ja4028633.
77. K. Nakano, N. Kosaka, T. Hiyama and K. Nozaki, Metal-catalyzed synthesis of stereoregular polyketones, polyesters, and polycarbonates, *Dalton Trans.*, 2003, 4039–4050, DOI: 10.1039/b304690g.
78. K. Nakano, M. Nakamura and K. Nozaki, Alternating Copolymerization of Cyclohexene Oxide with Carbon Dioxide Catalyzed by (salalen)CrCl Complexes, *Macromolecules*, 2009, **42**, 6972–6980, DOI: 10.1021/ma9012626.
79. K. Nakano, K. Nozaki and T. Hiyama, Asymmetric Alternating Copolymerization of Cyclohexene Oxide and CO_2 with Dimeric Zinc Complexes, *J. Am. Chem. Soc.*, 2003, **125**, 5501–5510, DOI: 10.1021/ja028666b.
80. K. Nozaki, Asymmetric polymerization using C1 resources, *J. Polym. Sci., Part A: Polym. Chem.*, 2003, **42**, 215–221, DOI: 10.1002/pola.11025.
81. K. Nozaki, Asymmetric catalytic synthesis of polyketones and polycarbonates, *Pure Appl. Chem.*, 2004, **76**, 541–546, DOI: 10.1351/pac200476030541.
82. K. Nozaki, K. Nakano and T. Hiyama, Optically Active Polycarbonates: Asymmetric Alternating Copolymerization of Cyclohexene Oxide and Carbon Dioxide, *J. Am. Chem. Soc.*, 1999, **121**, 11008–11009, DOI: 10.1021/ja992433b.
83. K. Nozaki, K. Nakano and T. Hiyama, Asymmetric alternating copolymerization of cyclohexene oxide and carbon dioxide, *PMSE Prepr.*, 2002, **87**, 88.
84. T. Ohkawara, K. Suzuki, K. Nakano, S. Mori and K. Nozaki, Facile Estimation of Catalytic Activity and Selectivities in Copolymerization of Propylene Oxide with Carbon Dioxide Mediated by Metal Complexes with Planar Tetradentate Ligand, *J. Am. Chem. Soc.*, 2014, **136**, 10728–10735, DOI: 10.1021/ja5046814.
85. A. Okada, S. Kikuchi, K. Nakano, K. Nishioka, K. Nozaki and T. Yamada, New class of catalysts for alternating copolymerization of alkylene oxide and carbon dioxide, *Chem. Lett.*, 2010, **39**, 1066–1068, DOI: 10.1246/cl.2010.1066.
86. C. Robert, T. Ohkawara and K. Nozaki, Manganese-Corrole Complexes as Versatile Catalysts for the Ring-Opening Homo- and Co-Polymerization of Epoxide, *Chem. – Eur. J.*, 2014, **20**, 4789–4795, DOI: 10.1002/chem.201303703.
87. G. W. Coates and R. C. Jeske, *Homogeneous Catalyst Design for the Synthesis of Aliphatic Polycarbonates and Polyesters*, Wiley-VCH Verlag GmbH & Co. KgaA, 2009, pp. 343–373.

88. S. D. Allen, C. M. Byrne and G. W. Coates, Carbon dioxide as a renewable C1 feedstock: synthesis and characterization of polycarbonates from the alternating copolymerization of epoxides and CO2, *ACS Symp. Ser.*, 2006, **921**, 116–129, DOI: 10.1021/bk-2006-0921.ch009.
89. S. D. Allen, D. R. Moore, E. B. Lobkovsky and G. W. Coates, High-Activity, Single-Site Catalysts for the Alternating Copolymerization of CO2 and Propylene Oxide, *J. Am. Chem. Soc.*, 2002, **124**, 14284–14285, DOI: 10.1021/ja028071g.
90. M. I. Childers, J. M. Longo, N. J. Van Zee, A. M. LaPointe and G. W. Coates, Stereoselective Epoxide Polymerization and Copolymerization, *Chem. Rev.*, 2014, **114**, 8129–8152, DOI: 10.1021/cr400725x.
91. G. W. Coates and D. R. Moore, Discrete metal-based catalysts for the copolymerization of CO2 and epoxides: Discovery, reactivity, optimization, and mechanism, *Angew. Chem., Int. Ed.*, 2004, **43**, 6618–6639, DOI: 10.1002/anie.200460442.
92. C. T. Cohen, T. Chu and G. W. Coates, Cobalt catalysts for the alternating copolymerization of propylene oxide and carbon dioxide: combining high activity and selectivity, *J. Am. Chem. Soc.*, 2005, **127**, 10869–10878, DOI: 10.1021/ja051744l.
93. C. T. Cohen and G. W. Coates, Alternating copolymerization of propylene oxide and carbon dioxide with highly efficient and selective (salen)Co(III) catalysts: effect of ligand and cocatalyst variation, *J. Polym. Sci., Part A: Polym. Chem.*, 2006, **44**, 5182–5191, DOI: 10.1002/pola.21606.
94. C. T. Cohen, C. M. Thomas, K. L. Peretti, E. B. Lobkovsky and G. W. Coates, Copolymerization of cyclohexene oxide and carbon dioxide using (salen)Co(III) complexes: synthesis and characterization of syndiotactic poly(cyclohexene carbonate), *Dalton Trans.*, 2005, 237–249, DOI: 10.1039/b513107c.
95. W. C. Ellis, Y. Jung, M. Mulzer, R. Di Girolamo, E. B. Lobkovsky and G. W. Coates, Copolymerization of CO2 and meso epoxides using enantioselective β-diiminate catalysts: a route to highly isotactic polycarbonates, *Chem. Sci.*, 2014, **5**, 4004–4011, DOI: 10.1039/c4sc01686f.
96. D. R. Moore, S. D. Allen and G. W. Coates, Alternating copolymerization of epoxides and carbon dioxide: Advances in beta-diiminate zinc catalysts, *Polym. Prepr. (Am. Chem. Soc., Div. Polym. Chem.)*, 2003, **44**, 735–736.
97. D. R. Moore, M. Cheng, E. B. Lobkovsky and G. W. Coates, Electronic and steric effects on catalysts for CO2/epoxide polymerization: subtle modifications resulting in superior activities, *Angew. Chem., Int. Ed.*, 2002, **41**, 2599–2602, DOI: 10.1002/1521-3773(20020715)41:14 < 2599::aid-anie2599 > 3.0.co;2-n.
98. D. R. Moore, M. Cheng, E. B. Lobkovsky and G. W. Coates, Mechanism of the alternating copolymerization of epoxides and CO2 using β-diiminate zinc catalysts: Evidence for a bimetallic epoxide enchainment, *J. Am. Chem. Soc.*, 2003, **125**, 11911–11924, DOI: 10.1021/ja030085e.

99. Z. Qin, C. M. Thomas, S. Lee and G. W. Coates, Cobalt-based complexes for the copolymerization of propylene oxide and CO2: Active and for polycarbonate synthesis, *Angew. Chem., Int. Ed.*, 2003, **42**, 5484–5487, DOI: 10.1002/anie.200352605).
100. N. J. Robertson, Z. Qin, G. C. Dallinger, E. B. Lobkovsky, S. Lee and G. W. Coates, Two-dimensional double metal cyanide complexes: highly active catalysts for the homopolymerization of propylene oxide and copolymerization of propylene oxide and carbon dioxide, *Dalton Trans.*, 2006, 5390–5395, DOI: 10.1039/b607963f.
101. J. W. Comerford, I. D. V. Ingram, M. North and X. Wu, Sustainable metal-based catalysts for the synthesis of cyclic carbonates containing five-membered rings, *Green Chem.*, 2015, **17**, 1966–1987, DOI: 10.1039/c4gc01719f.
102. D. J. Darensbourg and S. J. Wilson, Recent advances in polycarbonate production, *Green Chem.*, 2012, **14**, 2665–2671, DOI: 10.1039/c2gc35928f.
103. X.-B. Lu, W.-M. Ren and G.-P. Wu, CO2 Copolymers from Epoxides: Catalyst Activity, Product Selectivity, and Stereochemistry Control, *Acc. Chem. Res.*, 2012, **45**, 1721–1735, DOI: 10.1021/ar300035z.
104. I. Omae, Recent developments in carbon dioxide utilization for the production of organic chemicals, *Coord. Chem. Rev.*, 2012, **256**, 1384–1405, DOI: 10.1016/j.ccr.2012.03.017.
105. M. Taherimehr and P. P. Pescarmona, Green polycarbonates prepared by the copolymerization of CO2 with epoxides, *J. Appl. Polym. Sci.*, 2014, **131**, 41141–41117, DOI: 10.1002/app.41141.
106. H. Zhang, H.-B. Liu and J.-M. Yue, Organic Carbonates from Natural Sources, *Chem. Rev.*, 2014, **114**, 883–898, DOI: 10.1021/cr300430e.
107. S. D. Allen, A. E. Cherian, C. A. Simoneau, J. J. Farmer, G. W. Coates, A. Gridnev and R. E. Lapointe, Catalysts and methods for synthesis of carbon dioxide-epoxide copolymers Patent or Patent Application WO2010022388A2, 2010.
108. S. D. Allen, G. W. Coates, A. E. Cherian, C. A. Simoneau, A. A. Gridnev and J. J. Farmer, Preparation of polycarbonate polyols from epoxides and carbon dioxide Patent or Patent Application WO2010028362A1, 2010.
109. S. D. Allen, G. Job and J. J. Farmer, Metal complexes for the copolymerization of carbon dioxide and epoxides Patent or Patent Application WO2013067460A1, 2013.
110. S. D. Allen, C. A. Simoneau and J. J. Farmer, Carbon dioxide-epoxide copolymer compositions and methods Patent or Patent Application WO2012154849A1, 2012.
111. J. J. Farmer, Catalysts for synthesis of polymers from epoxides and carbon dioxide Patent or Patent Application WO2012037282A2, 2012.
112. G. Job, S. D. Allen, C. Simoneau, R. Valente and J. J. Farmer, Metal complexes of cationic tetradentate diamine-salicylaldehyde Schiff base ligands as catalysts for copolymerization of carbon dioxide with epoxides Patent or Patent Application WO2014031811A1, 2014.

113. R. E. Lapointe, S. D. Allen and J. R. Conuel, Carbon dioxide-based polycarbonate compositions with tunable decomposition properties Patent or Patent Application WO2010075232A1, 2010.
114. B. Y. Lee, S. Sujith, E. K. Noh and J. K. Min, Process for producing polycarbonates and a coordination complex used therefor Patent or Patent Application US8829156 B2, 2012.
115. B. Y. Lee, S. Sujith, E. K. Noh and J. K. Min, Process for producing polycarbonates and a coordination complex used therefor Patent or Patent Application US20140249323 A1, 2014.
116. S. D. Allen and J. J. Farmer, Metal complexes Patent or Patent Application US20140228538 A1, 2014.
117. A. Goeppert, M. Czaun, J.-P. Jones, G. K. Surya Prakash and G. A. Olah, Recycling of carbon dioxide to methanol and derived products - closing the loop, *Chem. Soc. Rev.*, 2014, **43**, 7995–8048, DOI: 10.1039/c4cs00122b.
118. A. Goeppert, G. K. S. Prakash and G. A. Olah, Carbon dioxide recycling to methanol, dimethyl ether and derived products for a sustainable future, *Actual. Chim.*, 2013, **371–372**, 78–83.
119. J.-P. Jones, G. K. S. Prakash and G. A. Olah, Electrochemical CO_2 Reduction: Recent Advances and Current Trends, *Isr. J. Chem.*, 2014, **54**, 1451–1466, DOI: 10.1002/ijch.201400081.
120. G. A. Olah, After Oil and Gas: Methanol Economy, *Catal. Lett.*, 2004, **93**, 1–2, DOI: 10.1023/B:CATL.0000017043.93210.9c.
121. G. A. Olah, Beyond oil and gas: The methanol economy, *Angew. Chem., Int. Ed.*, 2005, **44**, 2636–2639, DOI: 10.1002/anie.200462121.
122. G. A. Olah, The Role of Catalysis in Replacing Oil by Renewable Methanol Using Carbon Dioxide Capture and Recycling (CCR), *Catal. Lett.*, 2013, **143**, 983–987, DOI: 10.1007/s10562-013-1096-1.
123. G. A. Olah, Towards Oil Independence Through Renewable Methanol Chemistry, *Angew. Chem., Int. Ed.*, 2013, **52**, 104–107, DOI: 10.1002/anie.201204995.
124. G. A. Olah, A. Goeppert, M. Czaun and G. K. S. Prakash, Bi-reforming of Methane from Any Source with Steam and Carbon Dioxide Exclusively to Metgas (CO-2H2) for Methanol and Hydrocarbon Synthesis, *J. Am. Chem. Soc.*, 2013, **135**, 648–650, DOI: 10.1021/ja311796n.
125. G. A. Olah, A. Goeppert and G. K. S. Prakash, Chemical Recycling of Carbon Dioxide to Methanol and Dimethyl Ether: From Greenhouse Gas to Renewable, Environmentally Carbon Neutral Fuels and Synthetic Hydrocarbons, *J. Org. Chem.*, 2009, **74**, 487–498, DOI: 10.1021/jo801260f.
126. G. A. Olah, G. K. S. Prakash and A. Goeppert, Anthropogenic Chemical Carbon Cycle for a Sustainable Future, *J. Am. Chem. Soc.*, 2011, **133**, 12881–12898, DOI: 10.1021/ja202642y.
127. G. K. S. Prakash, G. A. Olah and A. Goeppert, Beyond oil and gas: the methanol economy, *ECS Trans.*, 2011, **35**, 31–40, DOI: 10.1149/1.3645178.

128. G. K. S. Prakash, M. C. Smart, G. A. Olah, S. R. Narayanan, W. Chun, S. Surampudi and G. Halpert, Performance of dimethoxymethane and trimethoxymethane in liquid-feed direct oxidation fuel cells, *J. Power Sources*, 2007, **173**, 102–109, DOI: 10.1016/j.jpowsour.2007.05.003.
129. G. A. Olah and R. Aniszfeld, Preparation of methanol, dimethyl ether, derived synthetic hydrocarbons from carbon dioxide and water using air as sole source Patent or Patent Application US20080039538A1, 2008.
130. G. A. Olah, A. Goepert, S. Meth and G. K. S. Prakash, Nanostructured solid regenerable absorbents for reversible absorption of carbon dioxide from gases, including air Patent or Patent Application WO2008021700A1, 2008.
131. G. A. Olah and G. k. S. Prakash, Efficient and selective chemical recycling of carbon dioxide to methanol, dimethyl ether and derived products Patent or Patent Application US20070254969A1, 2007.
132. G. A. Olah and G. K. S. Prakash, Conversion of carbon dioxide to methanol and/or dimethyl ether using bi-reforming of methane or natural gas Patent or Patent Application US20080319093A1, 2008.
133. G. A. Olah and G. K. S. Prakash, Environmentally friendly ternary transportation flex-fuel of gasoline, methanol and bioethanol Patent or Patent Application US20090172997A1, 2009.
134. G. A. Olah and G. k. S. Prakash, Conversion of carbon dioxide to dimethyl ether using bi-reforming of methane or natural gas Patent or Patent Application US20090030240A1, 2009.
135. G. A. Olah and G. K. S. Prakash, Electrolysis of carbon dioxide in aqueous media to carbon monoxide and hydrogen for production of methanol Patent or Patent Application US20090014336A1, 2009.
136. G. A. Olah and G. k. S. Prakash, Rendering petroleum oil as an environmentally carbon dioxide neutral source material for fuels, derived products and as a regenerative carbon source Patent or Patent Application WO2010118137A1, 2010.
137. G. A. Olah and G. k. S. Prakash, Rendering natural gas as an environmentally carbon dioxide neutral fuel and a regenerative carbon source Patent or Patent Application WO2010118133A1, 2010.
138. G. A. Olah and G. k. S. Prakash, Rendering coal as an environmentally carbon dioxide neutral fuel and a regenerative carbon source Patent or Patent Application WO2010118127A1, 2010.
139. G. A. Olah and G. k. S. Prakash, Electrolysis of carbon dioxide in aqueous media to carbon monoxide and hydrogen for production of methanol Patent or Patent Application US20100193370A1, 2010.
140. G. A. Olah and G. K. S. Prakash, Producing methanol and its products exclusively from geothermal sources and their energy Patent or Patent Application WO2010011504A2, 2010.
141. G. A. Olah and G. k. S. Prakash, Mitigating or eliminating the carbon footprint of human activities Patent or Patent Application US20110086928A1, 2011.

142. G. A. Olah and G. k. S. Prakash, Recycling carbon dioxide *via* capture and temporary storage to produce renewable fuels and derived products Patent or Patent Application WO2012047443A2, 2012.
143. G. A. Olah and G. K. S. Prakash, Efficient, self sufficient production of methanol from a methane source *via* oxidative bi-reforming Patent or Patent Application US20140100391A1, 2014.
144. G. A. Olah and S. G. K. Prakash, Efficient and selective conversion of carbon dioxide into methanol or dimethyl ether and derived products Patent or Patent Application WO2006113293A1, 2006.
145. The CCEMC Grand Challenge: Innovative Carbon Uses http://ccemc.redaffect.com/grand-challenge/ Accessed May 4, 2015.
146. Valorizing Industrially Produced CO2: A Reliable and Cost-Effective Solution for Carbon Capture and its Conversion to Marketable Products, http://ccemc.ca/project/valorizing-industrially-produced-co2/. Accessed May 4, 2015.
147. High-Value Synthetic Chemicals and Gasoline Drop-In Liquid Fuels from Canada's CO2 and Flare Gas Emissions, http://ccemc.ca/project/high-value-synthetic-chemicals-gasoline-drop-liquid-fuels/. Accessed May 4, 2015.
148. CO2 Conversion to Methanol through Bi-reforming, http://ccemc.ca/project/co2-conversion-methanol-bi-reforming/. Accessed May 4, 2015.
149. Green Methanol From Carbon Dioxide and Renewable Hydrogen (Methanol +) http://ccemc.ca/project/green-methanol-carbon-dioxide-renewable-hydrogen-methanol/ Accessed May 4, 2015.
150. Integration of Advanced Hybrid Inorganic Membranes for Carbon Dioxide Conversion, http://ccemc.ca/project/integration-advanced-hybrid-inorganic-membranes-carbon-dioxide-conversion/. Accessed May 4, 2015.
151. Novel Internal Dry Reforming Solid Oxide Fuel Cell Technology for CO2 Utilization, http://ccemc.ca/project/dry-reforming-solid-oxide-fuel-cell-technology-for-co2-utilization/. Accessed May 4, 2015.
152. Direct Catalytic Synthesis of Acetic Acid from CO2 AND CH4, http://ccemc.ca/project/direct-catalytic-synthesis-acetic-acid-co2-ch4/. Accessed May 4, 2015.
153. Emission to Liquid http://www.carbonrecycling.is/ Accessed May 4, 2015.
154. ETOGAS Delivers World's Largest Methane Production Plant To Audi, http://www.ngvaeurope.eu/etogas-delivers-worlds-largest-methane-production-plant-to-audi. Accessed May 4, 2015.
155. Fuel of the future: Research facility in Dresden produces first batch of Audi e-diesel, https://www.audi-mediaservices.com/publish/ms/content/en/public/pressemitteilungen/2015/04/21/fuel_of_the_future.html. Accessed May 9, 2015.
156. DONG Energy is converting coal-fired power station to green energy, http://www.dongenergy.com/en/media/newsroom/news/articles/dong-energy-is-converting-coal-fired-power-station-to-green-energy. Accessed May 4, 2015.

157. J. Blum, J. Sonntag and M. Strohmaier, Method and apparatus for regulating electrical power output of a fuel cell system Patent or Patent Application US7422809 B2, 2008.
158. BioCat Project, http://biocat-project.com/about-the-project/technology-components/. Accessed May 4, 2015.
159. S. Ma, Y. Lan, G. M. J. Perez, S. Moniri and P. J. A. Kenis, Silver Supported on Titania as an Active Catalyst for Electrochemical Carbon Dioxide Reduction, *ChemSusChem*, 2014, 7, 866–874, DOI: 0.1002/cssc.201300934.
160. Y. Hori, Electrochemical CO_2 Reduction on Metal Electrodes, in *Modern Aspects of Electrochemistry*, ed. C. Vayenas, R. White and M. Gamboa-Aldeco, Springer, New York, 2008, pp. 89–189.
161. K. W. Frese and S. Leach, Electrochemical Reduction of Carbon Dioxide to Methane, Methanol, and CO on Ru Electrodes, *J. Electrochem. Soc.*, 1985, **132**, 259–260, DOI: 10.1149/1.2113780.
162. K. Chandrasekaran and L. O. M. Bockris, In-situ spectroscopic investigation of adsorbed intermediate radicals in electrochemical reactions: CO_2- on platinum, *Surf. Sci.*, 1987, **185**, 495–514, DOI: 10.1016/S0039-6028(87)80173-5.
163. L. Wenzhen, Electrocatalytic Reduction of CO_2 to Small Organic Molecule Fuels on Metal Catalysts, in Advances in *CO_2* Conversion and Utilization, American Chemical Society, 2010, pp. 55–76.
164. B. A. Rosen, A. Salehi-Khojin, M. R. Thorson, W. Zhu, D. T. Whipple, P. J. A. Kenis and R. I. Masel, Ionic Liquid–Mediated Selective Conversion of CO_2 to CO at Low Overpotentials, *Science*, 2011, **334**, 643–644, DOI: 10.1126/science.1209786.
165. Q. Lu, J. Rosen, Y. Zhou, G. S. Hutchings, Y. C. Kimmel, J. G. Chen and F. Jiao, A selective and efficient electrocatalyst for carbon dioxide reduction, *Nat. Commun.*, 2014, **5**, DOI: 10.1038/ncomms4242.
166. M. R. Thorson, K. I. Siil and P. J. A. Kenis, Effect of Cations on the Electrochemical Conversion of CO_2 to CO, *J. Electrochem. Soc.*, 2013, **160**, F69–F74, DOI: 10.1149/2.052301jes.
167. H.-R. M. Jhong, F. R. Brushett and P. J. A. Kenis, The Effects of Catalyst Layer Deposition Methodology on Electrode Performance, *Adv. Energy Mater.*, 2013, **3**, 589–599, DOI: 10.1002/aenm.201200759.
168. C. Delacourt, P. L. Ridgway, J. B. Kerr and J. Newman, Design of an Electrochemical Cell Making Syngas ($CO+H_2$) from CO_2 and H_2O Reduction at Room Temperature, *J. Electrochem. Soc.*, 2008, **155**, B42–B49, DOI: 10.1149/1.2801871.
169. R. I. Masel, Q. Chen, Z. Liu and R. Kutz, Electrolyzers and Membranes Patent or Patent Application PCT/US2015/26507, 2015 and PCT/US15/14328, 2015.
170. L. M. Aeshala, S. U. Rahman and A. Verma, Effect of solid polymer electrolyte on electrochemical reduction of CO_2, *Sep. Purif. Technol.*, 2012, **94**, 131–137, DOI: 10.1016/j.seppur.2011.12.030.
171. L. M. Aeshala, R. Uppaluri and A. Verma, Electrochemical conversion of CO_2 to fuels: tuning of the reaction zone using suitable functional

groups in a solid polymer electrolyte, *Phys. Chem. Chem. Phys.*, 2014, **16**, 17588–17594, DOI: 10.1039/c4cp02389g.

172. C. Gaudillere, L. Navarrete and J. M. Serra, Syngas production at intermediate temperature through H_2O and CO_2 electrolysis with a Cu-based solid oxide electrolyzer cell, *Int. J. Hydrogen Energy*, 2014, **39**, 3047–3054, DOI: 10.1016/j.ijhydene.2013.12.045.
173. T. E. Teeter and P. Van Rysselberghe, Reduction of Carbon Dioxide on Mercury Cathodes, *J. Chem. Phys.*, 1954, **22**, 759–760, DOI: 10.1063/1.1740178.
174. M. Alvarez-Guerra, A. Del Castillo and A. Irabien, Continuous electrochemical reduction of carbon dioxide into formate using a tin cathode: Comparison with lead cathode, *Chem. Eng. Res. Des.*, 2014, **92**, 692–701, DOI: 10.1016/j.cherd.2013.11.002.
175. M. Alvarez-Guerra, S. Quintanilla and A. Irabien, Conversion of carbon dioxide into formate using a continuous electrochemical reduction process in a lead cathode, *Chem. Eng. J.*, 2012, **207–208**, 278–284, DOI: 10.1016/j.cej.2012.06.099.
176. H. Li and C. Oloman, The Electro-Reduction of Carbon Dioxide in a Continuous Reactor, *J. Appl. Electrochem.*, 2005, **35**, 955–965.
177. H. Li and C. Oloman, Development of a continuous reactor for the electro-reduction of carbon dioxide to formate – Part 1: Process variables, *J. Appl. Electrochem.*, 2006, **36**, 1105–1115, DOI: 10.1007/s10800-006-9194-z.
178. H. Li and C. Oloman, Development of a continuous reactor for the electro-reduction of carbon dioxide to formate – Part 2: Scale-up, *J. Appl. Electrochem.*, 2007, **37**, 1107–1117, DOI: 10.1007/s10800-007-9371-8.
179. C. Oloman and H. Li, Electrochemical processing of carbon dioxide, *ChemSusChem*, 2008, **1**, 385–391.
180. A. S. Agarwal, Y. Zhai, D. Hill and N. Sridhar, The Electrochemical Reduction of Carbon Dioxide to Formate/Formic Acid: Engineering and Economic Feasibility, *ChemSusChem*, 2011, **4**, 1301–1310, DOI: 10.1002/cssc.201100220.
181. H. L. Colin Oloman, Continuous co-current electrochemical reduction of carbon dioxide Patent or Patent Application US20080223727 A1, 2008.
182. R. I. Masel, Novel catalyst mixtures for electrochemical conversion of carbon dioxide to formic acid Patent or Patent Application WO2012006240A1, 2012.
183. E. B. Cole, K. Teamey, A. B. Bocarsly and N. Sivasankar, Reduction of Carbon Dioxide to Carboxylic Acids, Glycols, and Carboxylates Patent or Patent Application US20140027303, 2014.
184. A. B. Bocarsly and Z. M. Detweiler, Reducing carbon dioxide to products with an indium oxide electrode Patent or Patent Application WO2014032000A1, 2014.
185. J. D. Watkins and A. B. Bocarsly, Direct Reduction of Carbon Dioxide to Formate in High-Gas-Capacity Ionic Liquids at Post-Transition-Metal

Electrodes, *ChemSusChem*, 2014, 7, 284–290, DOI: 10.1002/cssc.201300659.

186. K. Liao, M. Askerka, E. L. Zeitler, A. B. Bocarsly and V. S. Batista, Electrochemical Reduction of Aqueous Imidazolium on Pt(111) by Proton Coupled Electron Transfer, *Top. Catal.*, 2015, **58**, 23–29, DOI: 10.1007/s11244-014-0340-2.
187. R. I. Masel and C. Rice, Fuel Cell Membranes and Crossover Prevention Patent or Patent Application US7569297 B2, 2009.
188. R. I. Masel, Novel catalyst mixtures for electrochemical conversion of carbon dioxide to formic acid Patent or Patent Application US20110237830A1, 2011.
189. R. I. Masel and B. Rosen, Novel catalyst mixtures for CO2 electrochemical conversion Patent or Patent Application WO2011120021A1, 2011.
190. B. A. Rosen, A. Salehi-Khojin, M. R. Thorson, W. Zhu, D. T. Whipple, P. J. A. Kenis and R. I. Masel, Ionic Liquid-Mediated Selective Conversion of CO_2 to CO at Low Overpotentials, *Science*, 2011, **334**, 643–644, DOI: 10.1126/science.1209786.
191. R. I. Masel and B. Rosen, Low cost carbon dioxide sensors Patent or Patent Application WO2012177952A2, 2012.
192. M. Sayyah, M. Rostam-Abadi, Y. Lu, R. I. Masel and K. S. Suslick, Magnesia-stabilized calcium oxide adsorbents for high temperature CO_2 capture *via* mechanical milling, *Prepr. Symp. - Am. Chem. Soc., Div. Fuel Chem.*, 2012, **57**, 828–829.
193. R. I. Masel and B. A. Rosen, On demand carbon monoxide generator for therapeutic and other applications Patent or Patent Application US20130075273A1, 2013.
194. B. A. Rosen, W. Zhu, G. Kaul, A. Salehi-Khojin and R. I. Masel, Water enhancement of CO_2 conversion on silver in 1-ethyl-3-methylimidazolium tetrafluoroborate, *J. Electrochem. Soc.*, 2013, **160**, H138–H141, DOI: 10.1149/2.004303jes.
195. A. Salehi-Khojin, H.-R. M. Jhong, B. A. Rosen, W. Zhu, S. Ma, P. J. A. Kenis and R. I. Masel, Nanoparticle Silver Catalysts That Show Enhanced Activity for Carbon Dioxide Electrolysis, *J. Phys. Chem. C*, 2013, **117**, 1627–1632, DOI: 10.1021/jp310509z.
196. M. Sayyah, Y. Lu, R. I. Masel and K. S. Suslick, Mechanical Activation of CaO-Based Adsorbents for CO_2 Capture, *ChemSusChem*, 2013, **6**, 193–198, DOI: 10.1002/cssc.201200454.
197. W. Zhu, B. A. Rosen, A. Salehi-Khojin and R. I. Masel, Monolayers of choline chloride can enhance desired electrochemical reactions and inhibit undesirable ones, *Electrochim. Acta*, 2013, **96**, 18–22, DOI: 10.1016/j.electacta.2013.02.061.
198. R. Masel, R. Ni, Z. Liu, Q. Chen, R. Kutz, L. Nereng, D. Lutz and K. Lewinski, Unlocking the Potential of CO_2 Conversion to Fuels and Chemicals as an Economically Viable Route to CCR, *Energy Procedia*, 2014, **63**, 7959–7962, DOI: 10.1016/j.egypro.2014.11.832.

199. R. I. Masel, Z. R. Ni, Q. Chen and B. A. Rosen, Carbon dioxide conversion to fuels and chemicals Patent or Patent Application US20140239231A1, 2014.
200. R. I. Masel, B. A. Rosen and W. Zhu, Devices and processes for carbon dioxide conversion into useful fuels and chemicals Patent or Patent Application WO2014047661A2, 2014.
201. R. Masel and A. Salehi-Khojin, Electrocatalysts for carbon dioxide conversion Patent or Patent Application US20130157174 A1, 2015.
202. T. S. Matthews, M. Kaplun, Z. Liu, Q. Chen, R. Kutz, S. Luopa, K. Lewinski and R. I. Masel, On the route to commercialization of a CO_2 electrolyzer: lessons learned from an industry effort to fight climate change, *Prepr. - Am. Chem. Soc., Div. Energy Fuels*, 2015, **60**, 222–223.
203. R. I. Masel, Z. Liu, Q. Chen, R. Kutz, H. Yang, K. Lewinski, T. Matthews and M. Kaplun submitted Electrochemical Conversion of Carbon Dioxide to High Volume Products: Current Progress and Future Research Needs Faraday Discussions.
204. W. Zhu, R. Michalsky, Ö. Metin, H. Lv, S. Guo, C. J. Wright, X. Sun, A. A. Peterson and S. Sun, Monodisperse Au Nanoparticles for Selective Electrocatalytic Reduction of CO_2 to CO, *J. Am. Chem. Soc.*, 2013, **135**, 16833–16836, DOI: 10.1021/ja409445p.
205. W. Zhu, Y.-J. Zhang, H. Zhang, H. Lv, Q. Li, R. Michalsky, A. A. Peterson and S. Sun, Active and Selective Conversion of CO_2 to CO on Ultrathin Au Nanowires, *J. Am. Chem. Soc.*, 2014, **136**, 16132–16135, DOI: 10.1021/ja5095099.
206. D. Gao, H. Zhou, J. Wang, S. Miao, F. Yang, G. Wang, J. Wang and X. Bao, Size-Dependent Electrocatalytic Reduction of CO_2 over Pd Nanoparticles, *J. Am. Chem. Soc.*, 2015, **137**, 4288–4291, DOI: 10.1021/jacs.5b00046.
207. B. Kumar, M. Asadi, D. Pisasale, S. Sinha-Ray, B. A. Rosen, R. Haasch, J. Abiade, A. L. Yarin and A. Salehi-Khojin, Renewable and metal-free carbon nanofibre catalysts for carbon dioxide reduction, *Nat. Commun.*, 2013, **4**, DOI: 10.1038/ncomms3819.
208. M. Asadi, B. Kumar, A. Behranginia, B. A. Rosen, A. Baskin, N. Repnin, D. Pisasale, P. Phillips, W. Zhu, R. Haasch *et al.*, Robust carbon dioxide reduction on molybdenum disulphide edges, *Nat. Commun.*, 2014, **5**, DOI: 10.1038/ncomms5470.
209. B. A. Rosen, W. Zhu, G. Kaul, A. Salehi-Khojin and R. I. Masel, Water Enhancement of CO_2 Conversion on Silver in 1-Ethyl-3-Methylimidazolium Tetrafluoroborate, *J. Electrochem. Soc.*, 2013, **160**, H138–H141, DOI: 10.1149/2.004303jes.
210. C. E. Tornow, M. R. Thorson, S. Ma, A. A. Gewirth and P. J. A. Kenis, Nitrogen-Based Catalysts for the Electrochemical Reduction of CO_2 to CO, *J. Am. Chem. Soc.*, 2012, **134**, 19520–19523, DOI: 10.1021/ja308217w.
211. E. Barton Cole, P. S. Lakkaraju, D. M. Rampulla, A. J. Morris, E. Abelev and A. B. Bocarsly, Using a One-Electron Shuttle for the Multielectron

Reduction of CO_2 to Methanol: Kinetic, Mechanistic, and Structural Insights, *J. Am. Chem. Soc.*, 2010, **132**, 11539–11551, DOI: 10.1021/ja1023496.
212. A. J. Morris, R. T. McGibbon and A. B. Bocarsly, Electrocatalytic Carbon Dioxide Activation: The Rate-Determining Step of Pyridinium-Catalyzed CO_2 Reduction, *ChemSusChem*, 2011, **4**, 191–196, DOI: 10.1002/cssc.201000379.
213. G. Seshadri, C. Lin and A. B. Bocarsly, A new homogeneous electrocatalyst for the reduction of carbon dioxide to methanol at low overpotential, *J. Electroanal. Chem.*, 1994, **372**, 145–150.
214. S. Zhang, P. Kang, S. Ubnoske, M. K. Brennaman, N. Song, R. L. House, J. T. Glass and T. J. Meyer, Polyethylenimine-Enhanced Electrocatalytic Reduction of CO_2 to Formate at Nitrogen-Doped Carbon Nanomaterials, *J. Am. Chem. Soc.*, 2014, **136**, 7845–7848, DOI: 10.1021/ja5031529.
215. S. Zhang, P. Kang and T. J. Meyer, Nanostructured Tin Catalysts for Selective Electrochemical Reduction of Carbon Dioxide to Formate, *J. Am. Chem. Soc.*, 2014, **136**, 1734–1737, DOI: 10.1021/ja4113885.
216. P. Kang, S. Zhang, T. J. Meyer and M. Brookhart, Rapid selective electrocatalytic reduction of carbon dioxide to formate by an iridium pincer catalyst immobilized on carbon nanotube electrodes, *Angew. Chem., Int. Ed.*, 2014, **53**, 8709–8713, DOI: 10.1002/anie.201310722.
217. Z. Chen, P. Kang, M.-T. Zhang and T. J. Meyer, Making syngas electrocatalytically using a polypyridyl ruthenium catalyst, *Chem. Commun.*, 2014, **50**, 335–337, DOI: 10.1039/c3cc47251e.
218. A. K. Vannucci, L. Alibabaei, M. D. Losego, J. J. Concepcion, B. Kalanyan, G. N. Parsons and T. J. Meyer, Crossing the divide between homogeneous and heterogeneous catalysis in water oxidation, *Proc. Natl. Acad. Sci. U. S. A.*, 2013, **110**, 20918–20922, DOI: 10.1073/pnas.1319832110, 20911–20914.
219. A. M. Lapides, D. L. Ashford, K. Hanson, D. A. Torelli, J. L. Templeton and T. J. Meyer, Stabilization of a Ruthenium(II) Polypyridyl Dye on Nanocrystalline TiO2 by an Electropolymerized Overlayer, *J. Am. Chem. Soc.*, 2013, **135**, 15450–15458, DOI: 10.1021/ja4055977.
220. P. Kang, T. J. Meyer and M. Brookhart, Selective electrocatalytic reduction of carbon dioxide to formate by a water-soluble iridium pincer catalyst, *Chem. Sci.*, 2013, **4**, 3497–3502, DOI: 10.1039/c3sc51339d.
221. Z. Chen, J. J. Concepcion, M. K. Brennaman, P. Kang, M. R. Norris, P. G. Hoertz and T. J. Meyer, Splitting CO_2 into CO and O_2 by a single catalyst, *Proc. Natl. Acad. Sci. U. S. A.*, 2012, **109**, 15606–15611, DOI: 10.1073/pnas.1203122109, 15601–15602.
222. M. D. Sampson, A. D. Nguyen, K. A. Grice, C. E. Moore, A. L. Rheingold and C. P. Kubiak, Manganese catalysts with bulky bipyridine ligands for the electrocatalytic reduction of carbon dioxide: Eliminating dimerization and altering catalysis, *J. Am. Chem. Soc.*, 2014, **136**, 5460–5471, DOI: 10.1021/ja501252f.

223. C. W. Machan, M. D. Sampson, S. A. Chabolla, T. Dang and C. P. Kubiak, Developing a Mechanistic Understanding of Molecular Electrocatalysts for CO_2 Reduction using Infrared Spectroelectrochemistry, *Organometallics*, 2014, **33**, 4550–4559, DOI: 10.1021/om500044a.
224. K. A. Grice and C. P. Kubiak, Recent studies of rhenium and manganese bipyridine carbonyl catalysts for the electrochemical reduction of CO_2, *Adv. Inorg. Chem.*, 2014, **66**, 163–188, DOI: 10.1016/b978-0-12-420221-4.00005-6.
225. M. Garcia, M. J. Aguirre, G. Canzi, C. P. Kubiak, M. Ohlbaum and M. Isaacs, Electro and photoelectrochemical reduction of carbon dioxide on multimetallic porphyrins/polyoxotungstate modified electrodes, *Electrochim. Acta*, 2014, **115**, 146–154, DOI: 10.1016/j.electacta.2013.10.142.
226. M. L. Clark, K. A. Grice, C. E. Moore, A. L. Rheingold and C. P. Kubiak, Electrocatalytic CO_2 reduction by M(bpy-R)(CO)4 (M = Mo, W; R = H, tBu) complexes. Electrochemical, spectroscopic, and computational studies and comparison with group 7 catalysts, *Chem. Sci.*, 2014, **5**, 1894–1900, DOI: 10.1039/c3sc53470g.
227. C. S. Seu, D. Ung, M. D. Doud, C. E. Moore, A. L. Rheingold and C. P. Kubiak, Synthesis, Structural, and Electrocatalytic Reduction Studies of [Pd(P2N2)2]2 + Complexes, *Organometallics*, 2013, **32**, 4556–4563, DOI: 10.1021/om400472s.
228. J. A. Keith, K. A. Grice, C. P. Kubiak and E. A. Carter, Elucidation of the selectivity of proton-dependent electrocatalytic CO_2 reduction by fac-Re(bpy)(CO)3Cl, *J. Am. Chem. Soc.*, 2013, **135**, 15823–15829, DOI: 10.1021/ja406456g.
229. K. A. Grice, N. X. Gu, M. D. Sampson and C. P. Kubiak, Carbon monoxide release catalysed by electron transfer: electrochemical and spectroscopic investigations of [Re(bpy-R)(CO)4](OTf) complexes relevant to CO_2 reduction, *Dalton Trans.*, 2013, **42**, 8498–8503, DOI: 10.1039/c3dt50612f.
230. E. E. Benson, M. D. Sampson, K. A. Grice, J. M. Smieja, J. D. Froehlich, D. Friebel, J. A. Keith, E. A. Carter, A. Nilsson and C. P. Kubiak, The Electronic States of Rhenium Bipyridyl Electrocatalysts for CO_2 Reduction as Revealed by X-ray Absorption Spectroscopy and Computational Quantum Chemistry, *Angew. Chem., Int. Ed.*, 2013, **52**, 4841–4844, DOI: 10.1002/anie.201209911.
231. D. L. DuBois, Development of Molecular Electrocatalysts for Energy Storage, *Inorg. Chem.*, 2014, **53**, 3935–3960, DOI: 10.1021/ic4026969.
232. B. J. Boro, A. M. Appel, B. R. Galan, J. C. Linehan and D. L. DuBois, *Preparation of Electrocataysts for the Reduction of CO_2 to CO*, American Chemical Society, 2011, pp. INOR-430.
233. M. Rakowski Dubois and D. L. Dubois, Development of Molecular Electrocatalysts for CO_2 Reduction and H_2 Production/Oxidation, *Acc. Chem. Res.*, 2009, **42**, 1974–1982, DOI: 10.1021/ar900110c.

234. K. Kobayashi and K. Tanaka, Approach to multi-electron reduction beyond two-electron reduction of CO_2, *Phys. Chem. Chem. Phys.*, 2014, **16**, 2240–2250, DOI: 10.1039/c3cp52635f.
235. M. Bourrez, M. Orio, F. Molton, H. Vezin, C. Duboc, A. Deronzier and S. Chardon-Noblat, Pulsed-EPR evidence of a manganese(II) hydroxycarbonyl intermediate in the electrocatalytic reduction of carbon dioxide by a manganese bipyridyl derivative, *Angew. Chem., Int. Ed.*, 2014, **53**, 240–243, DOI: 10.1002/anie.201306750.
236. M. Bourrez, F. Molton, S. Chardon-Noblat and A. Deronzier, [Mn(bipyridyl)CO3Br]: An Abundant Metal Carbonyl Complex as Efficient Electrocatalyst for CO_2 Reduction, *Angew. Chem., Int. Ed.*, 2011, **50**, 9903–9906, DOI: 10.1002/anie.201103616, 9901–9909.
237. A. Deronzier and S. Chardon-Noblat, Electrocatalytic reduction of carbon dioxide: a future solution for sustainable chemistry?, *Actual. Chim.*, 2013, **371–372**, 84–88.
238. C. Costentin, S. Drouet, G. Passard, M. Robert and J.-M. Saveant, Proton-Coupled Electron Transfer Cleavage of Heavy-Atom Bonds in Electrocatalytic Processes. Cleavage of a C-O Bond in the Catalyzed Electrochemical Reduction of CO_2, *J. Am. Chem. Soc.*, 2013, **135**, 9023–9031, DOI: 10.1021/ja4030148.
239. C. Costentin, S. Drouet, M. Robert and J.-M. Saveant, Turnover Numbers, Turnover Frequencies, and Overpotential in Molecular Catalysis of Electrochemical Reactions. Cyclic Voltammetry and Preparative-Scale Electrolysis, *J. Am. Chem. Soc.*, 2012, **134**, 11235–11242, DOI: 10.1021/ja303560c.
240. C. Costentin, S. Drouet, M. Robert and J.-M. Saveant, A Local Proton Source Enhances CO_2 Electroreduction to CO by a Molecular Fe Catalyst, *Science*, 2012, **338**, 90–94, DOI: 10.1126/science.1224581.
241. G. A. Olah and G. K. S. Prakash, Electrolysis of Carbon Dioxide in Aqueous Media to Carbon Monoxide and Hydrogen for Production of Methanol Patent or Patent Application US 7,704,369 B2, 2010.
242. G. A. Olah and G. K. S. Prakash, Electrolysis of Carbon Dioxide in Aqueous Media to Carbon Monoxide and Hydrogen for Production of Methanol Patent or Patent Application US 8138380 B2, 2012.
243. C. W. Li and M. W. Kanan, CO_2 Reduction at Low Overpotential on Cu Electrodes Resulting from the Reduction of Thick Cu2O Films, *J. Am. Chem. Soc.*, 2012, **134**, 7231–7234, DOI: 10.1021/ja3010978.
244. M. W. Kanan, E. Beh and C. Gorin, Catalysis controlled by interfacial electric fields Patent or Patent Application WO2013028911A1, 2013.
245. Y. Chen, C. Li and M. W. Kanan, Catalysts for low temperature electrolytic CO2 or CO reduction Patent or Patent Application WO2014018091A1, 2014.
246. C. W. Li, J. Ciston and M. W. Kanan, Electroreduction of carbon monoxide to liquid fuel on oxide-derived nanocrystalline copper, *Nature*, 2014, **508**, 504–507, DOI: 10.1038/nature13249.

247. X. Min, Y. Chen and M. W. Kanan, Alkaline O_2 reduction on oxide-derived Au: high activity and 4e- selectivity without (100) facets, *Phys. Chem. Chem. Phys.*, 2014, **16**, 13601–13604, DOI: 10.1039/c4cp01337a.
248. X. Feng, K. Jiang, S. Fan and M. W. Kanan, Grain-Boundary-Dependent CO_2 Electroreduction Activity, *J. Am. Chem. Soc.*, 2015, **137**, 4606–4609, DOI: 10.1021/ja5130513.
249. C. H. Lee and M. W. Kanan, Controlling H+ *vs.* CO_2 Reduction Selectivity on Pb Electrodes, *ACS Catal.*, 2015, **5**, 465–469, DOI: 10.1021/cs5017672.
250. X. Min and M. W. Kanan, Pd-Catalyzed Electrohydrogenation of Carbon Dioxide to Formate: High Mass Activity at Low Overpotential and Identification of the Deactivation Pathway, *J. Am. Chem. Soc.*, 2015, **137**, 4701–4708, DOI: 10.1021/ja511890h.
251. S. D. Ebbesen, S. H. Jensen, A. Hauch and M. B. Mogensen, High temperature electrolysis in alkaline cells, solid proton conducting cells, and solid oxide cells, *Chem. Rev.*, 2014, **114**, 10697–10734, DOI: 10.1021/cr5000865.
252. C. Cheng, G. Kelsall and L. Kleiminger, Reduction of CO_2 to CO at Cu–ceria-gadolinia (CGO) cathode in solid oxide electrolyser, *J. Appl. Electrochem.*, 2013, **43**, 1131–1144.
253. N. Minh, M. C. Williams and B. Washom, Hydrogen/Oxygen Production by Solid Oxide Electrolysis, *ECS Trans.*, 2015, **65**, 93–97.
254. J. Yan, H. Chen, E. Dogdibegovic, J. W. Stevenson, M. Cheng and X.-D. Zhou, High-efficiency intermediate temperature solid oxide electrolyzer cells for the conversion of carbon dioxide to fuels, *J. Power Sources*, 2014, **252**, 79–84.
255. W. Becker, R. Braun, M. Penev and M. Melaina, Production of Fischer–Tropsch liquid fuels from high temperature solid oxide co-electrolysis units, *Energy*, 2012, **47**, 99–115.
256. S. D. Ebbesen, R. Knibbe and M. Mogensen, Co-electrolysis of steam and carbon dioxide in solid oxide cells, *J. Electrochem. Soc.*, 2012, **159**, F482–F489.
257. M. Ni, 2D thermal modeling of a solid oxide electrolyzer cell (SOEC) for syngas production by H 2 O/CO 2 co-electrolysis, *Int. J. Hydrogen Energy*, 2012, **37**, 6389–6399.
258. M. Ni, An electrochemical model for syngas production by co-electrolysis of H 2 O and CO 2, *J. Power Sources*, 2012, **202**, 209–216.
259. C. M. Stoots, J. O'Brien., G. Hawkes, J. Herring and J. Hartvigsen, High-temperature co-electrolysis of H_2O and CO_2 for syngas production, Idaho National Laboratory, 2006.
260. C. Stoots, J. O'brien, J. Herring, K. Condie, L. Moore-McAteer, J. J. Hartvigsen and D. Larsen, High Temperature Solid-Oxide Electrolyzer 2500 Hour Test Results at the Idaho National Laboratory, in *American Society of Chemical Engineers Annual Meeting*, 2009, pp. 8–13.

261. N. B. Jakobsson, Friis Pederson, Bogild Claus Hansen, A process for producing CO from CO2 in a solid oxide electrolysis cell Patent or Patent Application WO2014154253, 2014.
262. X. Chen, C. Guan, X. Guoping, X. Du and J.-Q. Wang, FD Electrolysis: Syngas Production by High Temperature Steam/ CO2 Coelectrolysis Using Solid Oxide Electrolysis Cells. *Faraday Discussions*, 2015.
263. K. C. Waugh, Methanol Synthesis, *Catal. Today*, 1992, **15**, 51–75, DOI: 10.1016/0920-5861(92)80122-4.
264. E. Barton Cole, P. S. Lakkaraju, D. M. Rampulla, A. J. Morris, E. Abelev and A. B. Bocarsly, Using a One-Electron Shuttle for the Multielectron Reduction of CO_2 to Methanol: Kinetic, Mechanistic, and Structural Insights, *J. Am. Chem. Soc.*, 2010, **132**, 11539–11551, DOI: 10.1021/ja1023496.
265. E. E. Barton Cole, M. F. Baruch, R. P. L'Esperance, M. T. Kelly, P. S. Lakkaraju, E. L. Zeitler and A. B. Bocarsly, Substituent Effects in the Pyridinium Catalyzed Reduction of CO_2 to Methanol: Further Mechanistic Insights, *Top. Catal.*, 2015, **58**, 15–22, DOI: 10.1007/s11244-014-0343-z.
266. A. Bocarsly, G. Seshadri and L. Chao, The efficient reduction of carbon dioxide to methanol at platinum and palladium electrodes using a pyridinium catalyst, *Proc. - Electrochem. Soc.*, 1993, **93–18**, 30–37.
267. A. B. Bocarsly and E. B. Cole, Conversion of carbon dioxide to organic products Patent or Patent Application US20100187123A1, 2010.
268. A. B. Bocarsly, A. Morris, P. Lakkaraju and R. Dominey, Advanced aromatic amine heterocyclic catalysts for carbon dioxide reduction Patent or Patent Application US20120175269A1, 2012.
269. A. B. Bocarsly, A. Morris, P. Lakkaraju and R. Dominey, Advanced aromatic amine heterocyclic catalysts for carbon dioxide reduction Patent or Patent Application WO2013102092A1, 2013.
270. A. B. Bocarsly, E. Zeitler and A. Mills, Gas phase electrochemical reduction of carbon dioxide Patent or Patent Application US20120171583A1, 2012.
271. A. B. Bocarsly, E. Zeitler and A. Mills, Gas phase electrochemical reduction of carbon dioxide Patent or Patent Application WO2013102086A1, 2013.
272. E. B. Cole and A. Bocarsly, Heterocycle catalyzed electrochemical reduction of a carbonyl compound Patent or Patent Application US20110226632A1, 2011.
273. E. B. Cole and A. B. Bocarsly, Heterocycle catalyzed electrochemical reduction of a carbonyl compound Patent or Patent Application WO2012015909A2, 2012.
274. E. B. Cole, N. Sivasankar, A. Bocarsly, K. Teamey and N. Krishna, Reducing carbon dioxide to products Patent or Patent Application US20110114502A1, 2011.

275. E. B. Cole, N. Sivasankar, A. B. Bocarsly, K. Teamey and N. Krishna, Reducing carbon dioxide to products Patent or Patent Application WO2012015905A1, 2012.
276. E. B. Cole, K. Teamey, A. B. Bocarsly and N. Sivasankar, Electrochemical reduction of carbon dioxide to carboxylic acids, glycols, and carboxylates Patent or Patent Application US20120277465A1, 2012.
277. E. B. Cole, K. Teamey, A. B. Bocarsly and N. Sivasankar, Electrochemical production of butanol from carbon dioxide and water Patent or Patent Application US20120132538A1, 2012.
278. E. B. Cole, K. Teamey, A. B. Bocarsly and N. Sivasankar, Electrochemical production of butanol from carbon dioxide and water Patent or Patent Application WO2013082492A1, 2013.
279. E. B. Cole, K. Teamey, A. B. Bocarsly and N. Sivasankar, Reduction of carbon dioxide to carboxylic acids, glycols, and carboxylates Patent or Patent Application WO2013006711A1, 2013.
280. Z. M. Detweiler, J. L. White, S. L. Bernasek and A. B. Bocarsly, Anodized Indium Metal Electrodes for Enhanced Carbon Dioxide Reduction in Aqueous Electrolyte, *Langmuir*, 2014, **30**, 7593–7600, DOI: 10.1021/la501245p.
281. K. A. Keets, E. B. Cole, A. J. Morris, N. Sivasankar, K. Teamey, P. S. Lakkaraju and A. B. Bocarsly, Analysis of pyridinium catalyzed electrochemical and photoelectrochemical reduction of CO_2: chemistry and economic impact, *Indian J. Chem., Sect. A: Inorg., Bio-inorg., Phys., Theor. Anal. Chem.*, 2012, **51A**, 1284–1297.
282. A. J. Morris, R. T. McGibbon and A. B. Bocarsly, Electrocatalytic carbon dioxide activation: The rate-determining step of pyridinium-catalyzed CO_2 reduction, *ChemSusChem*, 2011, **4**, 191–196, DOI: 10.1002/cssc.201000379.
283. G. Seshadri, C. Lin and A. B. Bocarsly, A new homogeneous electrocatalyst for the reduction of carbon dioxide to methanol at low overpotential, *J. Electroanal. Chem.*, 1994, **372**, 145–150, DOI: 10.1016/0022-0728(94)03300-5.
284. N. Sivasankar, E. Cole, K. Teamey and A. Bocarsly, Electrochemical production of urea from NOx and CO2 Patent or Patent Application US20110114503A1, 2011.
285. N. Sivasankar, E. B. Cole, R. Parajuli and A. B. Bocarsly, Heterocycle catalyzed electrochemical reduction of a carbonyl compound Patent or Patent Application WO2014043648A2, 2014.
286. N. Sivasankar, E. B. Cole, R. Parajuli and A. B. Bocarsly, Heterocycle catalyzed electrochemical process Patent or Patent Application US20140021060A1, 2014.
287. K. Teamey, E. B. Cole, N. Sivasankar and A. Bocarsly, Purification of carbon dioxide from a mixture of gases Patent or Patent Application US20110114501A1, 2011.
288. K. Teamey, J. J. Kaczur, E. B. Cole, P. Majsztrik, N. Sivasankar and A. B. Bocarsly, Electrochemical reduction of CO2 with co-oxidation of an alcohol Patent or Patent Application US20130140187A1, 2013.

289. K. Teamey, J. J. Kaczur, E. B. Cole, P. Majsztrik, N. Sivasankar and A. B. Bocarsly, Electrochemical reduction of CO2 with co-oxidation of an alcohol Patent or Patent Application WO2014046798A2, 2014.
290. J. L. White and A. B. Bocarsly, Electroreduction of carbon dioxide on indium-based nanoparticles, *Prepr. - Am. Chem. Soc., Div. Energy Fuels*, 2014, **59**, 408–409.
291. J. L. White, J. T. Herb, J. J. Kaczur, P. W. Majsztrik and A. B. Bocarsly, Photons to formate: Efficient electrochemical solar energy conversion *via* reduction of carbon dioxide, *J. CO2 Util.*, 2014, 7, 1–5, DOI: 10.1016/j.jcou.2014.05.002.
292. Y. Yan, E. L. Zeitler, J. Gu, Y. Hu and A. B. Bocarsly, Electrochemistry of Aqueous Pyridinium: Exploration of a Key Aspect of Electrocatalytic Reduction of CO_2 to Methanol, *J. Am. Chem. Soc.*, 2013, **135**, 14020–14023, DOI: 10.1021/ja4064052.
293. A. B. Bocarsly, E. Barton Cole, Conversion of carbon dioxide to organic products Patent or Patent Application US8313634 B2, 2012.
294. A. B. Bocarsly, E. Barton Cole, Conversion of carbon dioxide to organic products Patent or Patent Application US8663447 B2, 2014.
295. Startup born in Princeton lab turns carbon dioxide into fuels, http://www.princeton.edu/main/news/archive/S33/95/96G16/index.xml?section = featured. Accessed May 4, 2015.
296. Converting Carbon Dioxide into Chemicals and Fuels Using Clean, Domestic Sources of Energy in Alberta, http://ccemc.ca/project/converting-carbon-dioxide-using-clean-domestic-energy/. Accessed May 4, 2015.
297. Liquid Light wins Grand Challenge grant for innovative carbon uses from Climate Change and Emission Management Corporation, http://llchemical.com/news/15April14pressrelease. Accessed May 4, 2015.
298. N. Sivasankar, E. B. Cole and K. Teamey, Electrochemical production of synthesis gas from carbon dioxide Patent or Patent Application US 8721866 B2, 2014.
299. N. Sivasankar, E. B. Cole and K. Teamey, Electrochemical Production of Synthesis Gas from Carbon Dioxide Patent or Patent Application, 2014.
300. E. B. Cole, N. Sivasankar, A. Bocarsly, K. Teamey and N. Krishna, Reducing Carbon Dioxide To Products Patent or Patent Application, 2011.
301. N. Sivasankar, E. Cole, K. Teamey and A. Bocarsly, Electrochemical production of urea from NOx and carbon dioxide Patent or Patent Application US 8524066 B2, 2013.
302. N. Sivasankar, E. B. Cole, K. Teamey and A. B. Bocarsly, Electrochemical Production of Urea from NOx and Carbon Dioxide Patent or Patent Application US20140021042 A1, 2014.
303. E. B. Cole, K. Teamey, A. B. Bocarsly and N. Sivasankar, Electrochemical production of butanol from carbon dioxide and water Patent or Patent Application US20120132538, 2012.
304. A. B. Bocarsly, E. Zeitler and A. Mills, Gas Phase Electrochemical Reduction Of Carbon Dioxide Patent or Patent Application, 2012.

305. N. Sivasankar, I. Sullivan, E. B. Cole, K. Teamey and K. Keyshar, Process for making formic acid Patent or Patent Application US8562811 B2, 2013.
306. N. Sivasankar, I. Sullivan, E. B. Cole, K. Teamey and K. Keyshar, System and Process for Making Formic Acid Patent or Patent Application US20140021059 A1, 2014.
307. P. Lakkaraju and K. Teamey, Carbon dioxide capture and conversion to organic products Patent or Patent Application US8658016 B2, 2014.
308. P. Lakkaraju and K. Teamey, Carbon Dioxide Capture and Conversion to Organic Products Patent or Patent Application, 2014.
309. E. B. Cole, K. Teamey, A. B. Bocarsly and N. Sivasankar, Reduction of carbon dioxide to carboxylic acids, glycols, and carboxylates Patent or Patent Application, 2013.
310. K. Teamey, J. J. Kaczur and E. B. Cole, Electrochemical co-production of a glycol and an alkene employing recycled halide Patent or Patent Application US8444844 B1, 2013.
311. J. J. Kaczur, T. J. Kramer, K. Keyshar, P. Majsztrik and Z. Twardowski, Process and high surface area electrodes for the electrochemical reduction of carbon dioxide Patent or Patent Application US8858777 B2, 2014.
312. N. Sivasankar, J. J. Kaczur and E. B. Cole, Multiphase electrochemical reduction of CO2 Patent or Patent Application US8641885 B2, 2014.
313. N. Sivasankar, J. J. Kaczur and E. B. Cole, Multiphase Electrochemical Reduction of CO2 Patent or Patent Application, 2014.
314. J. J. Kaczur, Integrated Process for Producing Carboxylic Acids from Carbon Dioxide Patent or Patent Application US20130292257 A1, 2013.
315. E. B. Cole, N. Sivasankar, R. Parajuli and K. A. Keets, Reducing carbon dioxide to products Patent or Patent Application US8845878, 2014.
316. Cole Emily Barton, SivasankarI Narayanappa, Parajuli Rishi, A. Keets Kate and K. Kunttal, Reducing carbon dioxide to products Patent or Patent Application WO2013134418, 2013.
317. N. Sivasankar, E. B. Cole, R. Parajuli and A. B. Bocarsly, Heterocycle Catalyzed Electrochemical Process Patent or Patent Application WO2014043648, 2014.
318. A Coupled CO2 And Wastewater Treatment Process To Create High Value Gas/Oil Field Chemicals, http://ccemc.ca/project/coupled-co2-wastewater-treatment-to-create-high-value-field-chemicals/. Accessed May 4, 2015.
319. Sunshine to Petrol, http://energy.sandia.gov/energy/renewable-energy/solar-energy/sunshine-to-petrol/. Accessed May 4, 2015.
320. Sandia's Sunshine to Petrol project seeks fuel from thin air, https://share.sandia.gov/news/resources/releases/2007/sunshine.html. Accessed May 4, 2015.

321. Solar fuels, http://www.me.umn.edu/labs/solar/research/solar_fuels.shtml. Accessed May 9 2015.
322. L. J. Venstrom and J. H. Davidson, Splitting water and carbon dioxide *via* the heterogeneous oxidation of zinc vapor: Thermodynamic considerations, *J. Sol. Energy Eng.*, 2011, **133**, 011017.
323. SUNgas: Thermochemical gasification of biomass using concentrated solar energy, http://www.me.umn.edu/labs/solar/research/biomass.shtml. Accessed May 4, 2015.

CHAPTER 11

Methodological Considerations, Drivers and Trends in the Life Cycle Analysis of Bioproducts

JENNIFER B. DUNN,* FELIX K. ADOM, NORMAN F. SATHER AND JEONGWOO HAN

Argonne National Laboratory, 9700 S Cass Ave, Argonne, IL 60439, USA
*Email: jdunn@anl.gov

11.1 Introduction

The motivation to produce chemicals and products such as plastics from biomass is two-fold. First, consumers have become interested in purchasing products that could be considered "green," or produced with a lower environmental impact than conventional products. One technique of reducing a product's environmental impact could be to use biomass as a feedstock rather than a fossil-based feedstock such as crude oil or natural gas. For example, Coca-Cola has sought market differentiation and an enhanced relationship with customers through investment in "PlantBottle" technology, which uses bio-derived polyethylene terephthalate (PET).[1,2] The corporation has invested in three renewable fuels and chemicals companies, Gevo, Virent, and Avantium, to further develop bio-based routes to PET and is aiming to expand the technology's footprint through licensing agreements with H. J. Heinz and other companies. Interestingly, Virent's original business plan focused on production of gasoline, diesel, and jet fuels from renewable feedstocks, yet one of the company's key strategies currently is to

RSC Green Chemistry No. 43
Commercializing Biobased Products: Opportunities, Challenges, Benefits, and Risks
Edited by Seth W. Snyder

Published by the Royal Society of Chemistry, www.rsc.org

expand their renewable PET business. This strategy highlights the second reason bioproducts are receiving increased emphasis: they offer a bridge for companies to maintain profitable operation as the technology to produce biofuels continues to mature.

11.1.1 Life Cycle Analysis of Bioproducts

Given that one motivation to produce bioproducts is consumers' interest in products with reduced environmental footprints, it is important to assess the relative environmental performance of bioproducts compared to their conventional (fossil-based) counterparts and to identify opportunities to improve the environmental footprint of these products even further. One method of quantifying the environmental footprint of a bioproduct is life cycle analysis (LCA). In LCA, material and energy flows over each bioproduct life cycle step are aggregated to produce an estimate of life-cycle energy consumption, greenhouse gas (GHG) emissions, and metrics for other environmental impacts.

Figure 11.1 outlines the steps, inputs, and outputs of the life cycle of a bioproduct. The first life-cycle stage is feedstock production. This step includes production of fertilizers, herbicides, and pesticides that are

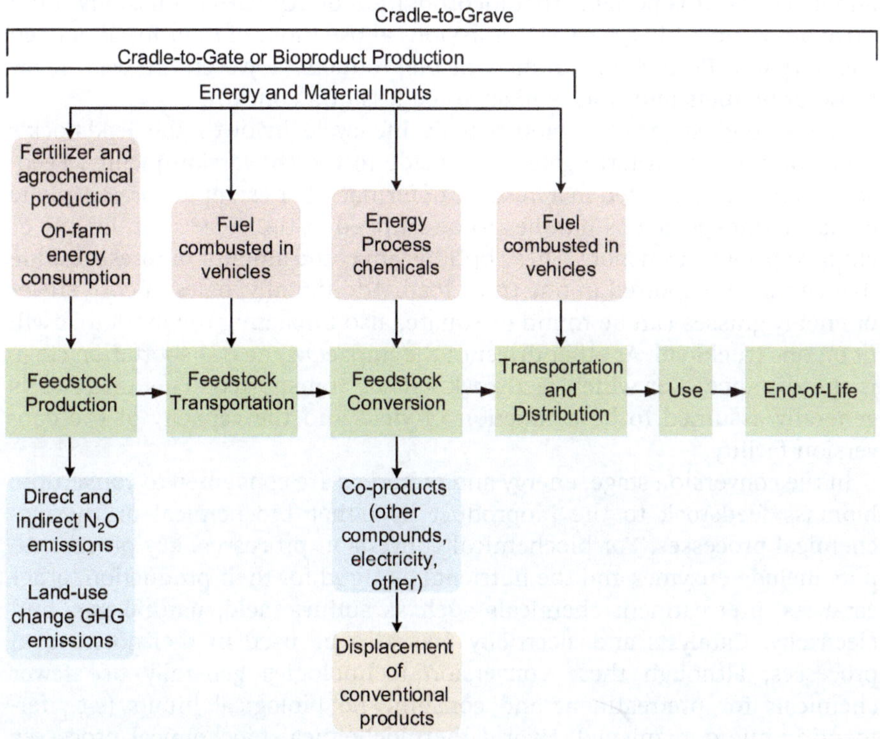

Figure 11.1 Bioproduct life cycle stages.

consumed during farming operations. It also includes consumption of fuel to operate farming equipment such as tractors. One potentially "hidden" source of emissions in this stage is N_2O emissions from nitrification–denitrification of nitrogen-containing fertilizers. Given that N_2O has a global warming potential 298 times that of CO_2, these emissions are very important in the life cycle of bioproducts and biofuels. Additionally, land-use change (LUC) occurs during feedstock production when land is converted to bioproduct feedstock production from other uses or states. For example, prior to conversion, land could be used for non-feedstock agriculture or be a forest or grassland. This type of LUC, which can be called direct LUC, has been observed in the United States.[3] The resulting change in crop production levels (*e.g.*, an uptick in corn production may cause a decline in soybean production) and exports may alter land uses domestically and internationally through economic linkages. This second type of LUC can be called indirect LUC and is typically estimated through the use of economic models.[4,5] Estimates of iLUC are uncertain and can cause calculated life-cycle GHG emissions for some biofuels to exceed life-cycle GHG emissions of petroleum fuels.[6] To our knowledge, indirect LUC emissions have not been estimated for bioproducts. When biofuel LUC GHG estimates are high, such as for biodiesel produced from oil crops in the European Union, the use of these fuels becomes controversial, leading to uncertainty for biofuel producers and investors. It is possible that incorporation of LUC GHG emissions could drive bioproduct life-cycle GHG emissions above those of their fossil-derived counterparts. Production volumes of bioproducts, however, are well below those of biofuels and LUC is likely to be less of a concern.

The second stage of a bioproduct's life cycle involves the feedstock's transportation from farm gate or roadside to the conversion facility. Feedstock can be transported in a number of formats. For example, forest residue could be transported as bundles to be chipped at the conversion facility or chipped prior to transport. These options affect the amount of forest residue that can be transported in one truck load. Additionally, bales of corn stover or energy grasses can be round or square, also impacting the mass that will fit on one truckload. Another driver of the impact of the transportation stage is the distance over which feedstock will be transported. This distance is generally assumed to be a function of yield and the capacity of the conversion facility.[7]

In the conversion stage, energy and materials are consumed to convert the biomass feedstock to the bioproduct by either biochemical or thermochemical processes. For biochemical conversion processes, key process inputs include enzymes and the nutrients required for their production, other catalysts, pretreatment chemicals such as sulfuric acid, natural gas, and electricity. Catalysts and electricity may also be used in thermochemical processes, although these conversion technologies generally use fewer chemicals for pretreatment and consume no biological inputs (*e.g.*, fermenting microorganisms). Hybrid thermochemical–biochemical processes are also feasible. After conversion, the bioproduct must be delivered to the point of use. For the most part, the use phase of bioproducts is considered to be burden-free. Treatment of bioproduct end-of-life can have a significant influence on LCA results as discussed in Section 11.2.

Because two main drivers of interest in biofuels and bioproducts are reducing GHG emissions and developing robust domestic biobased energy and raw materials resources, energy consumption and GHG emissions are often the focus of LCAs of biofuels and bioproducts. LCA is a key element of both the United States Environmental Protection Agency's (US EPA) Renewable Fuel Standard (RFS2)[8] and the California Air Resources Board's (CARB) Low Carbon Fuel Standard (LCFS),[9] which both aim to reduce GHG emissions from the transportation sector. RFS2 sets GHG reduction thresholds for biofuels as compared to a petroleum-derived fuel baseline. These thresholds must be met for the biofuel pathway (for example, corn-to-ethanol) to qualify under the standard. There is no set GHG emissions reduction threshold for biofuels under the LCFS, but overall the fuel mix in California must meet a 10% reduction in life-cycle GHG emissions by 2020 as compared to 2010. In the European Union, the Renewable Energy Directive (RED) is similar to the RFS2 in that qualifying biofuels must meet GHG emissions reduction thresholds.[10] No US legislation currently dictates GHG emissions levels that bioproducts must achieve but the LCA framework established for biofuels under the RFS2 and LCFS will likely shape approaches to LCA for bioproducts. For a discussion of US policy see Chapter 13.

In addition to GHG emissions and energy consumption, other relevant LCA outcomes for bioproducts may include water consumption, water quality, and emissions of other air pollutants such as particulate matter or volatile organic compounds. Water consumption, for example, must be considered, given the increased competition for water resources in some areas of the US. Water consumption during the production of biofuels has been the subject of intense study[11–14] and a key conclusion is that consumption during the feedstock production phase dominates overall life-cycle water consumption.[15] The use of water during conversion, however, is also key because conversion facilities may have difficulty being sited in regions with low water availability if the conversion process is water-intensive.

11.1.2 Bioproduct LCA Results in the Literature

Given the importance of the environmental performance of bioproducts, many companies that produce them have released estimates of GHG reduction of a plant-derived product as compared to its fossil-derived counterpart. Table 11.1 summarizes selected LCA results available in the public domain from bioproduct producers. It is important to note that these values are often provided without information about the system boundary of the study, key assumptions, or sensitivity analysis, likely due to concerns about protecting confidential business information. It is quite possible therefore that these results should not be directly compared to results from other

Table 11.1 Selected LCA results as reported by bioproduct producers.

Company	Bioproduct	Feedstock	Life-cycle GHG reductions compared to petroleum-derived counterpart	Notes
Myriant[17]	Succinic acid	Cellulosic	93/94%	Compared to fossil-derived adipic/succinic acid
BioAmber[18]	Succinic acid	Corn	102%	Compared to petroleum-derived succinic acid
Genomatica[19]	1,4-Butanediol	Sugarcane	117%	Sugarcane bagasse burned to power the facility
DuPont[20]	Propanediol	Corn	56%	
OPXBio/Dow[21]	Acrylic acid	Corn or sugarcane	$>70\%$	
Braskem[16]	High-density polyethylene (HDPE)	Sugarcane	$\sim 80\%$	Reported cradle-to-gate LCA results converted to cradle-to-grave results assuming full degradation of both fossil and bio-derived HDPE at end-of-life. Report documents assumptions about co-product treatment.

Table 11.2 Selected literature bioproduct LCA results.

		Fossil energy consumption: cradle-to-gate		Greenhouse gas emissions: cradle-to-grave	
Bioproduct	Feedstock	MJ kg^{-1}	% Reduction[e]	kg $CO_2{}^e$ kg^{-1}	% Reduction[e]
1,3-PDO[22,23]	Corn	41–50	55–68%	2.7–3.5[a]	46–71%
	Corn	22–58	37–55%	1.2–2.9	37–55%
	Sugarcane	−2.2–4.4	78–103%	−0.6–1.2	62–115%
Propylene glycol[24]	Soybeans and canola	NR		3.24	61%
Acrylic acid[22]	Corn starch	34	34%	2	43%
	Corn stover	18	66%	1.2	66%
	Sugarcane	4.4	91%	0.7	80%
Succinic acid[25]	Corn starch	33	74%[f]	0.88[b]	90%[f]
		49	60%[f]	1.7[c]	81%[f]
		45	64%[f]	1.47[d]	83%[f]

[a]Adjusted to include end-of-life carbon emissions.
[b]Direct crystallization.
[c]Electrodialysis.
[d]Acidulation and ammonium sulfate co-product formation.
[e]Reduction as reported in individual studies.
[f]Assuming biosuccinic acid displaces adipic acid in the market place.

studies because, as will be discussed in Section 11.2, assumptions and co-product treatment can greatly influence LCA results. Of the results in Table 11.1, only the Braskem study[16] clearly indicated that the results were on a cradle-to-gate basis and did not include end-of-life contributions. Nonetheless, a key point from Table 11.1 is that each company reports a significant GHG reduction for the bioproduct as compared to the conventional product. Additionally, aside from Myriant, the companies all assume either corn or sugarcane as a feedstock. The ramifications of this assumption will be discussed further in the next section.

Bioproduct LCA research has also been increasing in the academic literature (Table 11.2). Again, sugar and starch are the dominant feedstocks. Additionally, in every instance, the bioproduct exhibited lower life-cycle GHG emissions and energy consumption than its fossil counterpart.

11.1.3 Feedstock Choice

As biochemical production develops, a key technology hurdle is increasing production from cellulosic feedstocks or algae rather than starch- or sugar-based feedstocks. Reasons to look to these second and third generation feedstocks include avoiding competition with food markets and limiting land-use change (LUC) associated with feedstock production.

Despite these issues, a review of publicly available information, summarized in Table 11.3, reveals that corn remains a key feedstock for bioproducts. Although technologies to produce fuels from cellulosic feedstocks have improved in recent years, the economics of producing these

Table 11.3 Feedstocks used in the bioproducts industry.

Company	Product[a]	Current feedstock	Development plans
Myriant[29]	Succinic acid	Feedstock flexible among cellulosic feedstocks. May use grain sorghum.	Not specified
Bioamber[30]	Succinic acid, 1,3-propanediol	Dextrose 95, which is engineered for industrial use, from corn.	Expand production from non-genetically modified, non-food biomass including corn stover
DuPont[31,32]	1,3-Propanediol	Corn. "Renewably-sourced" materials contain at least 20% renewably-sourced materials by weight.	Soybeans, sugarcane, wheat, and others
OPXBio[21]	Acrylic acid	Corn or sugarcane	
Gevo[33,34]	Isobutanol	Corn	Cellulosic feedstocks

[a]Companies may produce bioproducts in addition to those listed.

high-volume, low profit margin fuels remain challenging. One interesting factor in the economics of biofuels and bioproducts is the role of the price of natural gas in the United States.[26,27] As shale gas swells the supply of domestic natural gas, its price has declined. On one hand, the low cost of natural gas has led some policy makers to see its potential as a feedstock for liquid fuel production to be greater than that of higher cost biomass.[27] On the other hand, ethylene producers have shifted away from using naphtha, a petroleum-derived product, as a feedstock to natural gas-derived feedstocks. As a result, the output of C_3–C_5 and pygas co-products has declined, which limits feedstock supply for a set of chemical intermediates and polymers such as synthetic rubber and nylon-6,6. This shortage has led some chemical producers to look to bio-derived feedstocks that offer price stability.[28] To this point, algae has remained a feedstock only for specialty chemicals production by companies such as Blue Marble and Aurora Algae, although it would be possible to convert glycerol from algae into a number of other bioproducts.

11.1.4 Study Background and Motivation

Recently, we have undertaken a study to examine the life-cycle fossil energy consumption and GHG emissions of eight selected bioproducts with two motivations. First, we aimed to assess whether these bioproducts could offer fossil energy consumption and GHG emissions reductions compared to their fossil counterparts. Second, we sought to identify the key factors that influence bioproduct LCA results and whether these factors were the same or different among the bioproducts. The process we used to select the eight bioproducts is described in an earlier report and was based on production

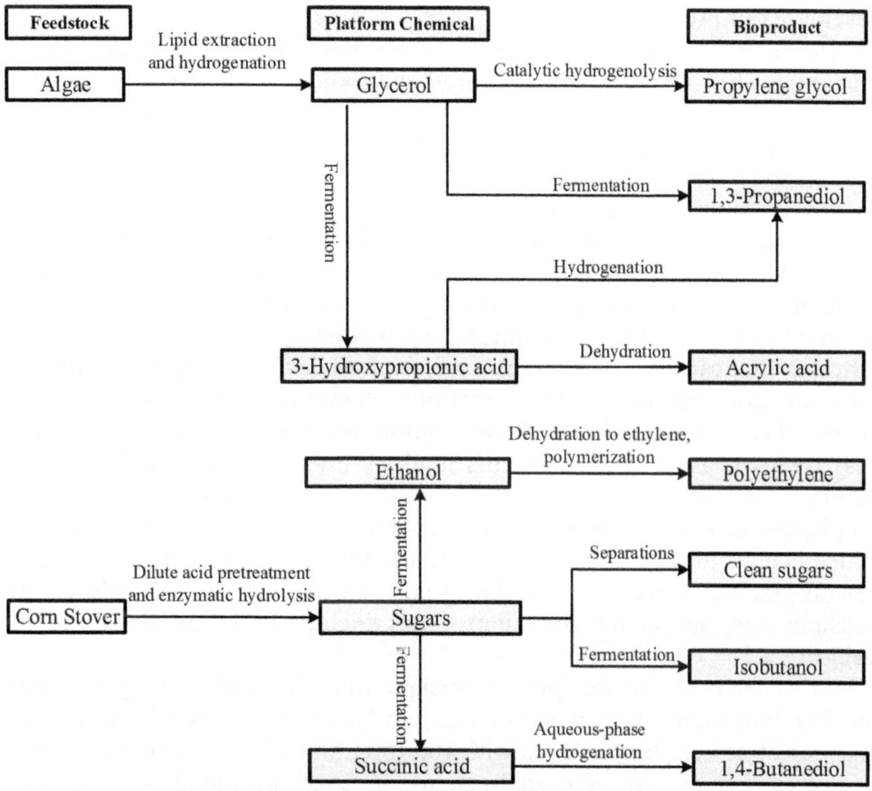

Figure 11.2 Bioproducts incorporated in the analysis.

technology readiness and market potential.[35] The products we selected based on this limited market assessment are presented in Figure 11.2.

Notably, we chose to use one second-generation feedstock, corn stover, and one third-generation feedstock, algae, to assess the GHG and fossil energy consumption of producing bioproducts from feedstocks other than corn and sugarcane, which, as Table 11.1 indicates, dominate current bioproduct production. Although conversion of new lands to produce corn stover or algae is unlikely,[4] the removal of corn stover from corn fields does decrease the carbon stored in soils compared to a case in which no stover is removed.[36] In our analysis, we assume that the soil organic carbon (SOC) is maintained with the adoption of good management practices and the removal of only sustainable amounts of stover.[37–39] US EPA, in its handling of SOC changes attributable to corn stover harvest, did take into account the reduction in SOC from stover harvest. The agency also considered reduced tillage in a stover harvest scenario that preserves SOC as compared to corn agriculture with conventional tillage.[40] They report substantial GHG reductions (67–129%) from biofuel pathways that use corn stover as a feedstock, despite the influence of residue removal on corn stover stocks, as do several of our previous analyses.[40–42] US government agencies, including the US Department of Energy (DOE), are using their proposal solicitations to encourage the growing bio-based products industry to use non-food feedstocks, which is another motivation for our selecting these two feedstocks. For example, DOE funding opportunity announcements frequently mandate that proposed research that uses feedstocks that could be used for food (*e.g.*, corn) are not eligible for funding.

The two selected feedstocks are converted to a group of bioproducts we term "platform chemicals". The platform chemical approach mimics the structure of the petrochemical industry, in which crude oil feedstocks are converted to a handful of chemical intermediates such as propylene and ethylene that can be converted to a range of other chemicals depending on demand and economic considerations. Subsequently, these platform chemicals are converted to end-use bioproducts. Several of the bioproducts that we treat as endpoints in this analysis are, however, possible intermediates on the route to other products. 1,3-Propanediol (PDO), for example, is often consumed to produce polytrimethylene terephthalate (PTT), which is used in carpet production.[43] In the US, a significant amount of 1,3-PDO is also used as a raw material in polyurethane production. Additional possible end points for the bioproducts included in this study are in Table 11.4.

In this chapter, we will provide insight into the methodology for conducting bioproduct LCA (Section 11.2) and will present results from our analysis of several bioproducts with the intention of examining key methodological issues. For our analysis, we inserted information about the

Table 11.4 Selected potential uses for bioproducts included in this analysis.

Bioproduct	Selected potential uses
Propylene glycol	Unsaturated polyester resins, antifreeze and deicing fluids, personal care products and cosmetics, food, detergents, paintings and coatings[44]
1,3-Propanediol	Deicing fluid, personal care products and cosmetics, food, coolants, polyurethanes, heat transfer fluids,[45] carpets[46]
3-Hydroxypropanoic acid	1,3-Propanediol, propylene glycol, acrylic acid,[46] acrylonitrile (precursor to carbon fiber)[47]
Acrylic acid	Superabsorbent polymers[48]
Polyethylene	Blown film extrusion or injection molding to plastic products
Isobutanol	Coatings, solvents, paraxylene[49]
Succinic acid	1,4-Butanediol, tetrahydrofuran, γ-butyrolactone,[50] polyester polyols, coatings, polymers and esters, plasticizers[17]
1,4-Butanediol	Tetrahydrofuran, γ-butyrolactone, polybutylene terephthalate resins, polyurethanes,[51] polybutylene terephthalate (PBT) which is used in engineered plastics for the automotive and electronics industries.

material and energy flows for the production of the bioproducts in Figure 11.2 into Argonne National Laboratory's GREET™ (greenhouse gases, regulated emissions, and energy use in transportation) model. The energy and material flow data were developed by creating process models for each bioproduct in Aspen Plus® as described in an earlier report.[35]

11.2 Methodology

The first step in conducting an LCA of a bioproduct is identifying the feedstock and developing material and energy flow data for the production of that feedstock. Two approaches to data collection are possible. In the case of a private company conducting a bioproduct LCA, a specific feedstock provider may be under contract. That provider may provide the requisite data. Alternatively, data from the literature or life cycle models such as GREET[52] could be used. Typical agricultural inputs include fertilizers that add nitrogen, phosphorus, and potassium to the soil. Pesticides and herbicides may also be consumed. Water consumption is becoming increasingly important and should be determined if possible. Additionally, fuel is consumed during feedstock production, harvest and transportation to the conversion facility. Figure 11.3 shows GREET estimates of the fossil energy and GHG intensity of producing several feedstocks.[7,42,53]

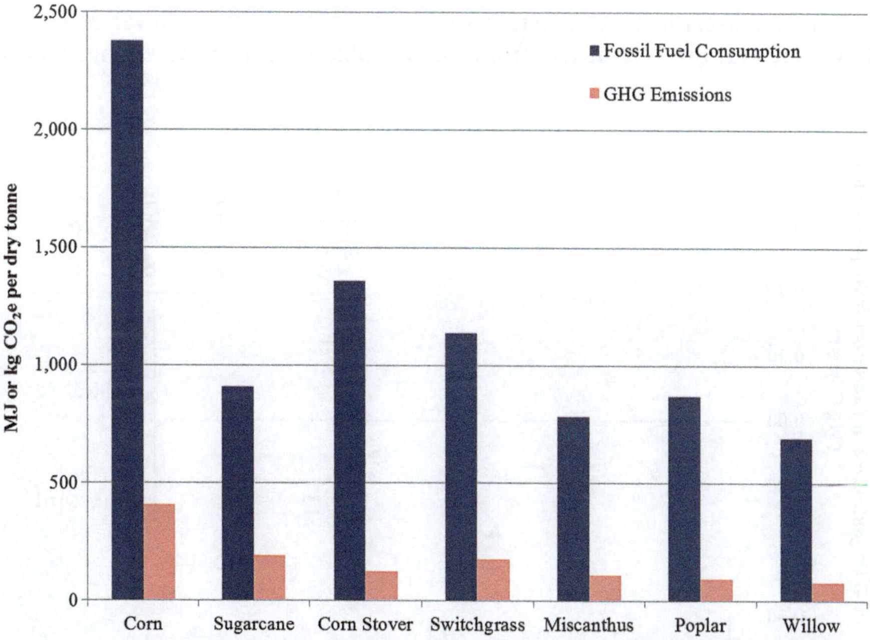

Figure 11.3 Fossil fuel consumption (MJ per dry tonne) and GHG emissions (g CO_2e per dry tonne) for different feedstocks.[52] (Moisture content of sugarcane and corn grain assumed to be 75% and 15%, respectively.)

Fertilizer consumption and energy expended during harvest drive the differences in fossil energy and GHG intensity of the feedstocks compared in Figure 11.3. Of these feedstocks, producing corn incurs the highest fossil fuel and GHG burdens. It is important to note that although corn is the most burdensome feedstock to produce, bioproduct yields in processes using this feedstock may be higher than in processes using more recalcitrant cellulosic feedstocks. Of the cellulosic feedstocks, willow and poplar are the least burdensome to produce.

The values in Figure 11.3 do not include changes in SOC that occur as a result of feedstock production. Soil dynamic models such as CENTURY can be used to estimate changes in SOC that result from land transitions to feedstock production.[36] These SOC changes depend strongly on a number of factors including climate conditions, soil type, past land use history, and land management practices under feedstock production. In previous work, we have estimated SOC changes upon cropland, cropland-pasture, grasslands, or forests converting to production of corn, corn with 30% by mass stover removal, switchgrass, and miscanthus at a county-level.[4] Figure 11.4 illustrates results at an agricultural-ecological zone (AEZ) level for these different land transitions. SOC implications of land transitions depend strongly on the feedstock that will be grown. In every case in Figure 11.4, except for conversion of forest to switchgrass, conversion of land to energy grass production sequesters carbon. On individual plots, of course, SOC changes may not always reflect this trend. For conversion of either forest or grassland to production of corn with or without stover removal, however, carbon is emitted from soils. Taking these SOC changes into

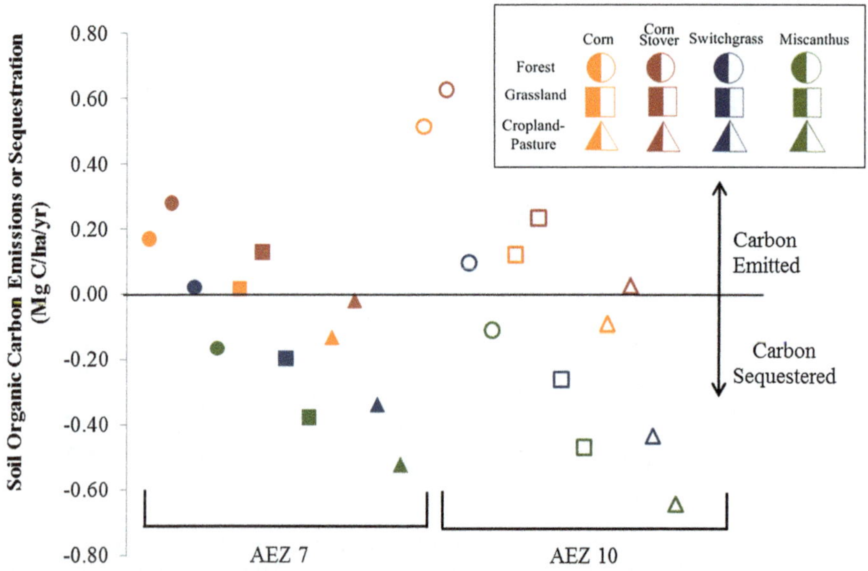

Figure 11.4 SOC changes upon land transitions in two US AEZs.

account will alter the carbon intensity of the different feedstocks shown in Figure 11.3.

The values reported in Figure 11.3 include the burden associated with transporting the feedstock to the conversion facility. Usually, the main burden in this life cycle stage is the consumption of diesel fuel in trucks for feedstock transport. The amount of fuel these trucks consume is a function of the payload and the distance traveled. Feedstock transportation parameters in GREET are based on the assumption that distance is a function of feedstock yield and plant capacity and that the payload is a function of feedstock density. Specific bioproduct producers may know the distance feedstock travels from farm-to-biorefinery gate and the feedstock density, which would enable use of pathway-specific data in an LCA of a specific product in their portfolio.

The conversion process itself is the focus of the next step in LCA. Here bioproduct producers have access to facility electricity and natural gas bills as well as records of quantities of process inputs that are consumed, such as pretreatment chemicals, enzymes, and solvents that may be used in separation processes. Moreover, the product slate, facility air and water emissions, and waste streams may be known and quantified. These real-world data can be used to construct the material and energy flows for the conversion phase in the bioproducts supply chain. A discussion of supply chains is presented in Chapter 12.

When utility and material bills are unavailable, it is possible to develop estimates of conversion stage burdens with one of two approaches. First, the literature may contain data to facilitate analysis of this stage. For example, the literature contains several studies in addition to ours with results for the conversion stage of bioproducts (see Tables 11.1 and 11.2). Second, it may be necessary to develop a process model with software such as Aspen Plus.

Given the lack of conversion stage data in the literature for the compounds in which we were interested, we used Aspen Plus to develop process models, relying on the technical literature and patents for the required process data. Patents can lend insight into which technologies show most industrial promise, but often the level of detail is limited, requiring, for example, a process engineer to select a reaction temperature based on a range of temperatures the patents provides. In addition to processing conditions, yield and selectivity data are key. Yield optimization, however, may not be the objective of a literature study. Academic papers may be published to show that a particular pathway is possible or to examine the influence on yield of key variables such as microorganism or catalyst type. Yield data extracted from the academic literature may therefore not represent achievable or optimal yields. Additional key inputs for process modeling are the level of different process inputs such as hydrogen or nutrients for fermenting microorganism growth. Finally, we sought information about separations techniques used in these types of processes, although these data are scarce. In our simulations we used conventional distillation, liquid–liquid extraction, and electrodeionization to purify product streams.

Similarly, we turned to the literature for data on the production of conventional products. Many of these compounds had been examined in the literature and whenever possible, we used existing information. For propylene glycol, however, we built a process model because we were unable to find any publicly available information on its production.

One important consideration in the analysis of the conversion step is the production of co-products. If a process produces more than one product, the process burdens (*e.g.*, energy and materials consumption) can be divided among them on the basis of mass, energy content, or economic value. This approach is termed mass, energy, or market value allocation. For chemical compounds, allocation by energy is often not relevant because the products are not energy carriers like biofuels, gasoline, or electricity. Economic allocation essentially takes the viewpoint that the most valuable product, the production of which may be the motivation for the operation of an entire facility, should bear the majority of process burdens. A drawback to this approach, however, is that the market value of the co-products will change over time. In the case of a facility producing solely chemical compounds, mass allocation may be the best choice for handling co-products. A second co-product handling option is to treat the co-products as displacing a conventional product and credit the main product with the energy, GHG, and air emissions savings that stem from this displacement. This approach could be quite relevant in the case of a facility that produces a biofuel as its main product and a chemical compound as its co-product, which could be viewed as displacing a conventional compound. One drawback of the displacement technique is that the energy consumption and emissions of the displaced product must be well-understood. In this study, we did not consider facilities that co-produced biofuels alongside chemicals and so used the mass allocation technique to allocate process burdens among co-products.

Figure 11.5 illustrates the influence of the co-product handling method on life-cycle GHG results for corn ethanol co-produced with dried distiller's grains with solubles (DDGS). This animal feed co-product can be treated as displacing a certain amount of conventional animal feed (*e.g.*, corn grain, soy meal).[54] In that case, the farm-to-wheels, or life-cycle, GHG emissions are 52 g CO_2e MJ^{-1}. In this scenario, the feedstock production stage receives a credit for avoided production of conventional animal feed and for avoided methane emissions from cattle fed with DGS *versus* conventional feeds. This credit helps reduce the GHG impact of the feedstock production stage compared to the other scenarios in Figure 11.5. When energy allocation is used as the co-product handling technique, life-cycle GHG emissions drop to 45 g CO_2e MJ^{-1}. The applicability of this technique, however, is arguable because DDGS is not really an energy product. In a dry mill, corn ethanol is about 61% of the mill's products on an energy basis, but is approximately 76% on a market value basis. As a result, when burdens are allocated between co-products based on their market value, fuel production emissions rise as compared to when the energy allocation technique is used. Life-cycle GHG emissions in the economic allocation scenario are 55 g CO_2e MJ^{-1}.

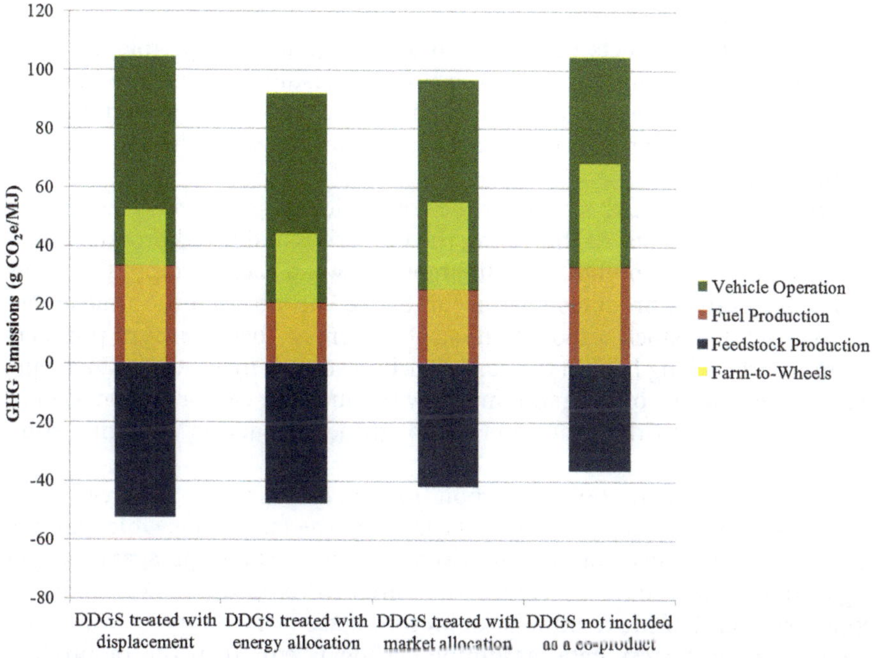

Figure 11.5 Life-cycle GHG emissions for corn ethanol with different treatments of the DDGS co-product. (Land-use change GHG emissions excluded.)

Finally, if DDGS is not considered at all as a co-product, corn ethanol life-cycle GHG emissions are nearly 70 g CO_2e MJ^{-1}. For corn ethanol, co-product treatment differences can therefore alter results by a significant 25 g CO_2e MJ^{-1}.

There are several key limitations to building material and energy flows for the conversion stage in the absence of real-world data from a biorefinery. First, selection of critical parameters such as yield is based on imperfect information. Second, the process steps and the extent of their integration may not reflect what would occur in a well-designed industrial facility. For example, heat integration can significantly reduce process energy demands by transferring high-quality rejected heat between unit operations that generate it and those that could use this rejected heat rather than heat supplied, for example, by combusting natural gas in a boiler to generate steam. For these reasons, estimates of life-cycle energy consumption, GHG emissions, and air emissions must be regarded as useful for two very specific purposes. First, if LCA results are generated for multiple bioproducts with a consistent methodology, it is possible to develop a sense for which bioproducts might be more energy- and GHG-intensive. Second, for an individual bioproduct, it is possible to develop a sense of the key factors that influence results. This deeper understanding could be used to select research and development priorities to minimize the life-cycle impacts of individual products. In this report, we also compare LCA results for bioproducts and their conventional counterparts. This comparison is important to assess whether the bioproducts may offer an advantage to their petroleum counterparts, but given the uncertainty in the LCA results for both the bio- and fossil-derived compounds, the results should be taken as general indicators rather than providing exact differences in impacts between the two routes to a final product.

If the analysis were to stop after the conversion stage, it could be called a cradle-to-gate analysis. At this point, the bio- or fossil-derived products store all the carbon they contain. The difference between these two product types is that bio-derived products have accumulated carbon from the atmosphere during the feedstock production stage whereas fossil-derived products contain carbon long held in the earth. Carbon storage in either product type, however, could be considered temporary because the carbon at some point will be released to the atmosphere through a degradation or combustion process.

Figure 11.6 sketches out the carbon flows in a bioproduct's life cycle and shows three options for treatment of carbon at the end of life, which are the same for bioproducts or fossil-derived products. Different agencies and organizations who have developed standards for bioproduct LCA differ on appropriate end-of-life treatment for bioproducts.[55] One option is to stop the analysis at the factory gate, assuming carbon is stored in the bioproduct indefinitely. This approach, however, begs the question of what the end fate of this carbon will be and is incomplete. Another option is to assume complete degradation in a landfill that results in complete release of the carbon in the product. For a bioproduct produced from a feedstock that has a short growth cycle (*e.g.*, switchgrass, corn), the uptake of carbon during the growth phase is generally seen as offsetting the carbon emissions at end of life. Fossil-derived products do not receive a credit for atmospheric carbon uptake earlier in the supply chain because the carbon they contain has been in the ground for many years. It is possible, however, that the bioproduct could have been made from a woody feedstock, with the carbon stored in the trees for a period of several years. It would be necessary to undertake somewhat detailed accounting of carbon accumulation during growth of the feedstock, storage of that carbon in the product for its life time which could be short (*e.g.*, 10 years) or long (*e.g.*, 100 years), and then its release to paint a complete picture of woody-derived bioproduct life-cycle GHG emissions. In fact, the literature is experiencing a debate over whether woody-derived bioproducts and biofuels can be considered carbon neutral.[56]

A temporal analysis would account for the delayed radiative forcing as a result of carbon storage in bioproducts. If the carbon is stored for a long period of time, such as in some plastics like polyethylene, which may take roughly a century to degrade, it could be considered that a significant delay of radiative forcing has occurred and the product could receive some credit for carbon storage. Another end-of-life option is to burn the bioproduct. In that case, carbon-containing ash that may be produced could store carbon for a long time and deserve some credit. Additionally, it could be possible to

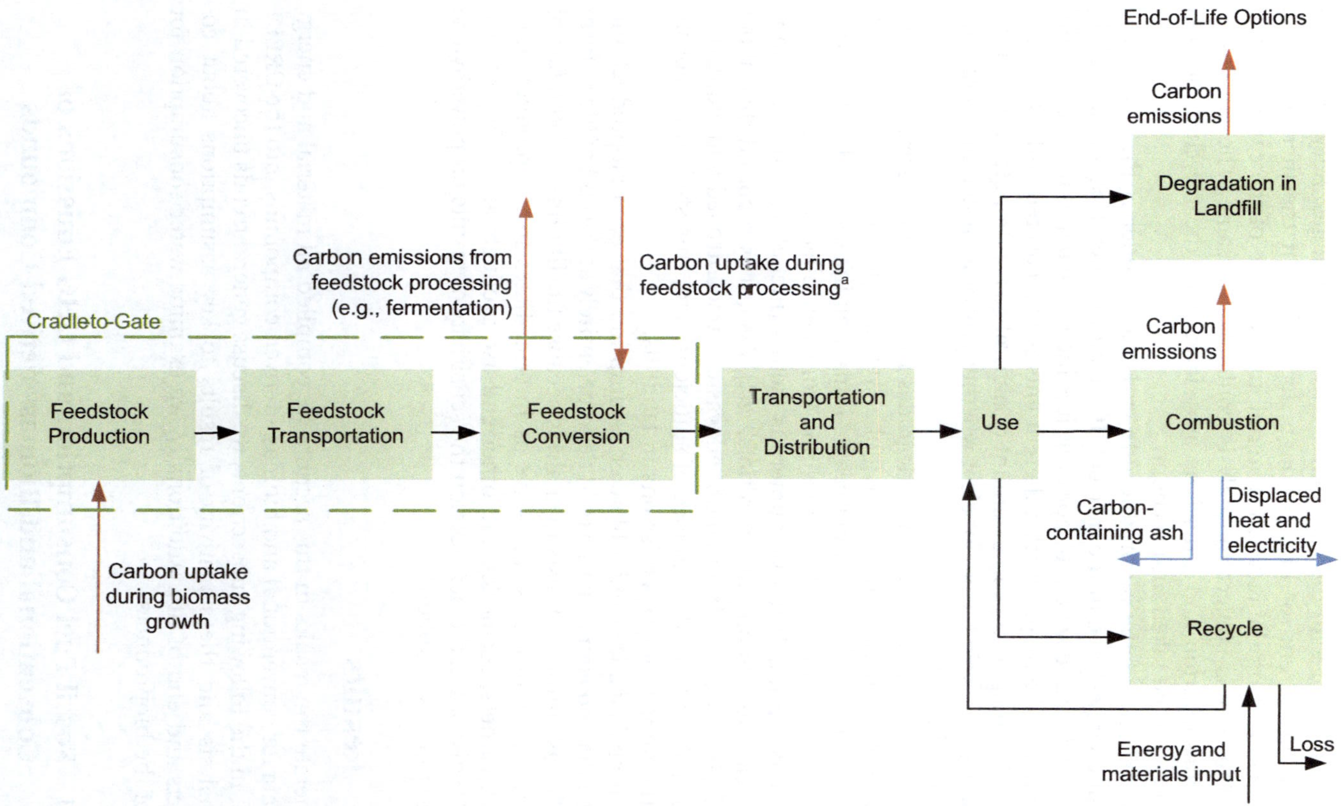

Figure 11.6 Carbon flows in the life cycle of a bioproduct with end-of-life treatment options. ([a]For example, CO_2 is consumed in the production of succinic acid by sugar fermentation.)

recover heat and use it to generate steam or electricity to supplement the energy requirements of the conversion facility. In that case, the LCA could credit the product with displaced electricity or natural gas. Recycling is another option for end of life, although depending on the lifetime of the product containing the recycled material, the end fate of the carbon in the compound that is subject to analysis could be the same. The additional energy consumption and emissions associated with the recycling process would need to be taken into account.

Another facet of the bioproduct life cycle to consider is the fate of the inherent energy stored in the product. If the bioproduct decays in a landfill, that energy is unrecovered. If, however, it is incinerated and produces useful heat, some of the bioproduct's inherent energy is recovered and, again, can be treated as displacing some amount of natural gas and/or electricity. If the bioproduct is recycled, it is not clear whether any of its inherent energy will be recovered at end of life.

When landfilled, organic waste degrades *via* anaerobic digestion mostly, generating CH_4. Because of its high global warming potential, any amount of CH_4 emissions would have a noticeable impact on life-cycle GHG emissions results. The amount of CH_4 emissions from landfilled waste, however, is highly uncertain, affected by numerous factors (including climate, acidity, microorganism community, materials, *etc.*). Also, it is reported that about 25% of CH_4 generated from landfilled waste is uncontrolled and emitted to the atmosphere.[57] In the absence of sufficient data, this study does not include the impact of CH_4 emissions in landfills.

In our analysis, we used a blanket assumption that each bioproduct and its fossil-derived counterpart would degrade entirely at the end of life to form CO_2. This assumption assists in analysis because the lifetime and end fate of some of the bioproducts that serve as intermediates in the production of plastics, carpets, cosmetics and other end-use products are difficult to assess. We discuss in the next section the sensitivity of results for polyethylene to end-of-life assumptions.

11.3 Results

To generate the results in this section, we compiled all material and energy flow data for conventional and biomass-derived compounds into the GREET model. In the following subsections, we discuss general trends uncovered in the analysis and the sensitivity of results to key assumptions about co-products and end of life. Additionally, we examine water consumption for each of the bioproducts.

11.3.1 Fossil Fuel Consumption and GHG Emissions of Conventional and Biomass-derived Compounds

Figure 11.7 and 11.8 display cradle-to-gate fossil energy consumption and cradle-to-grave GHG emissions, respectively, for all of the products we

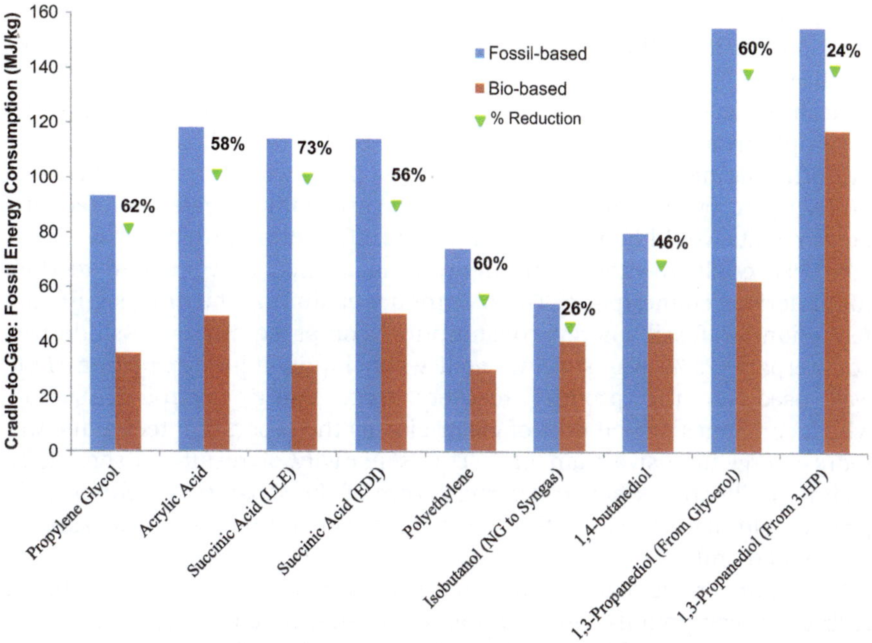

Figure 11.7 Cradle-to-gate fossil fuel consumption.

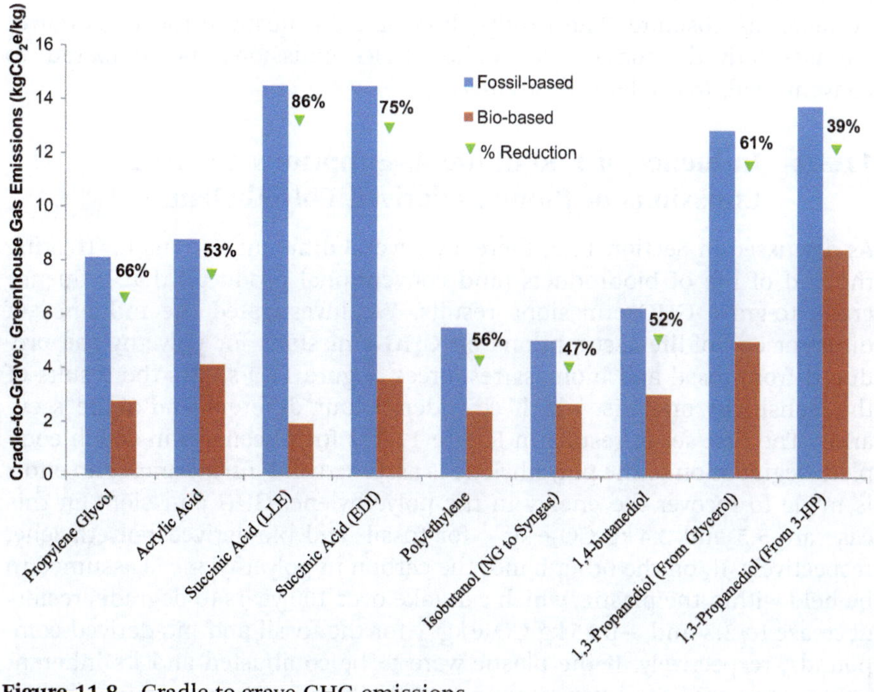

Figure 11.8 Cradle-to-grave GHG emissions.

considered. We provide an overview of these results in this section. Further information, including detailed sensitivity analyses, is available in a separate publication.[58]

Results for energy consumption are reported as cradle-to-gate because assumptions about energy recovery at end of life are difficult to make in a uniform manner for all bioproducts. One option would be to assume that no energy recovery is attempted, in which case cradle-to-gate and cradle-to-grave results would be identical. In Figure 11.7, all bioproducts have lower (26–73%) cradle-to-gate fossil energy consumption as compared to their fossil-derived counterparts. The bioproduct exhibiting the most significant reduction in fossil energy consumption compared to its fossil-derived counterpart (73%) was succinic acid when liquid–liquid extraction (LLE) was used for the product recovery step. Using electrodeionization, which consumes a good deal of electricity, as the separation technique was more energy-intensive than LLE. This sensitivity of results to separation technique illustrates how much the choice of the separation technology in process simulation can influence results. Separations are discussed in Chapters 6 and 7.

We report cradle-to-grave GHG reductions in Figure 11.8, assuming all carbon in compounds whether fossil- or bio-derived is released to the atmosphere at end of life. We investigate the effect of this assumption on results for polyethylene in Section 11.3.2. GHG reductions compared to fossil conventional compounds range from 39–86%. As discussed earlier, these results are based on conceptual process simulations and so should not be taken as absolute. The results, however, do illustrate the potential of biomass-derived products to reduce GHG emissions as compared to conventional, fossil-derived products.

11.3.2 Influence of End-of-life Assumptions on GHG Emissions of Biomass-derived Polyethylene

As discussed in Section 11.2, there are several different options for treating the end of life of bioproducts (and conventional products) that influence cradle-to-grave GHG emissions results. We investigated the influence of different end-of-life assumptions on GHG emissions for polyethylene produced from fossil and biomass resources. Figure 11.9 shows the results of this sensitivity analysis, which considered four different end-of-life scenarios. The first set of results in Figure 11.9 is for a scenario in which complete degradation of the polyethylene is assumed and, furthermore, no effort is made to recover the energy in the polyethylene. GHG emissions in this case are 5.5 and 2.4 kg CO_2e kg^{-1} for fossil- and bio-derived polyethylene, respectively. If, on the other hand, the carbon in polyethylene is assumed to be held within the plastic, which can take over 100 years to degrade, results decrease to 2.3 and −0.75 kg CO_2e kg^{-1} for the fossil and bio-derived compounds, respectively. If the plastic were to be combusted and its inherent energy recovered as either steam or electricity, all carbon contained in the

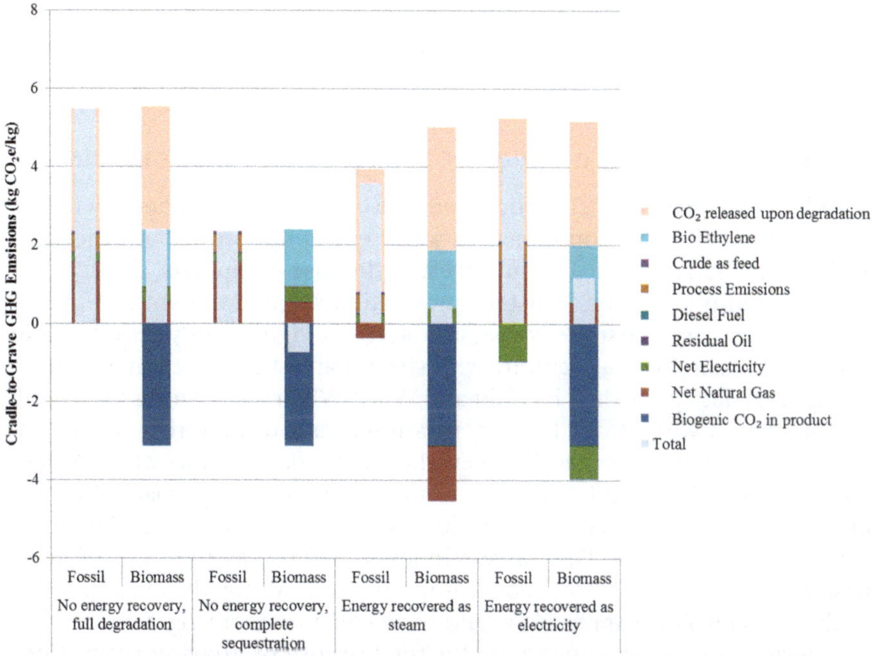

Figure 11.9 Cradle-to-grave GHG emissions for fossil- and bio-derived polyethylene under different end-of-life assumptions.

plastic could be assumed to be released to the atmosphere. Some natural gas use, however, would also be offset if the steam from the polyethylene combustion were used to supplant natural gas-derived steam. Alternatively, recovered electricity could be used in the place of grid electricity. We investigated these energy recovery options for both fossil and bio-derived polyethylene. We assumed that the boiler combusting the polyethylene has a 50% efficiency. Because the lower heating value of polyethylene is 45 MJ kg^{-1}, the total energy recovered from its combustion is 22 MJ kg^{-1}. This energy is assumed to displace an equivalent amount of energy converted to steam produced from natural gas combustion in an 80% efficient natural gas boiler. The amount of displaced natural gas is then about 28 MJ natural gas kg^{-1} polyethylene. If all energy is recovered as steam, bio- and fossil-derived polyethylene both receive a credit of 1.9 kg CO_2e kg^{-1} from natural gas displacement. Total GHG emissions for the fossil- and bio-derived compounds in this case are 3.6 and 0.47 kg CO_2e kg^{-1}, respectively.

If the steam from polyethylene is converted to electricity with a 30% efficiency, about 6 MJ electricity kg^{-1} polyethylene can be displaced. In our analysis, we assume the displaced electricity is representative of the US' average grid, which comprises 23% natural gas, 46% coal, and 20% nuclear power plants. The remaining 11% of power plants use biomass, wind, solar, and hydropower as energy sources. In this scenario, total GHG emissions for fossil- and bio-derived PE are 4.3 and 1.2 kg CO_2e kg^{-1}, respectively.

This case study demonstrates the importance of end-of-life treatment for bioproducts and fossil counterparts to which they are compared. Certainly, end-of-life assumptions must be reasonable and transparent.

11.3.3 Cradle-to-gate Water Consumption of Bioproducts

In addition to energy consumption and GHG emissions, another key metric for the production of bioproducts is water consumption in both the feedstock production and conversion stages. In the case of biofuels, the driver of water consumption is the feedstock production stage. Comparatively, water consumed in the feedstock conversion stage is significantly less.[15] Water consumed during corn agriculture with stover harvest can be estimated from county-level data published by Chiu and Wu.[11] When considering the overall system of corn with stover harvest, it is important to note that water consumption must be allocated between the main corn product and the harvested stover. Chiu and Wu take a mass allocation approach for this purpose, assigning water consumption between the grain and the stover with the harvest index. The harvest index is defined as the ratio of grain mass to the mass of the total above ground plant. Although we adopt their results, an alternative approach would be to consider that irrigation water in corn agriculture is consumed only for the purpose of growing grain. Corn stover generation could be considered to require no irrigation water. Herein, we report blue water consumption. Blue water constitutes consumptive irrigation water that is lost as a result of conveyance, crop evapotranspiration, and process water losses. Chiu and Wu also present results for green water, or rainfall lost through evapotranspiration, and grey water, a volume of water calculated to dilute pollutant levels in discharge streams to a permissible level. Chiu and Wu report blue water consumption attributable to corn stover as a biofuel feedstock is between 0.01 and 1283 L kg^{-1}. For this reason, Figure 11.10 shows a minimum and maximum value for blue water consumption in the feedstock production life-cycle stage.

In the case of algae, a different study[13] considers water demand for producing algal oil in the southern United States. This study reports water demand for production of algal oil to range between 1036 and 1666 L water L^{-1} algal oil. Some of that demand (3–91%) could be met with municipal waste water (MWW) effluent. Water demand and MWW effluent availability are location-dependent parameters. For this analysis, we calculated a range of possible water demands as between 90 and 1000 L fresh water L^{-1} algal oil based on MWW effluent availability. Figure 11.10 reports the minimum and maximum water consumption for producing algae as a glycerol feedstock, which is subsequently converted to 1,3-PDO, 3-hydroxypropionic acid, or propylene glycol.

We do not report water consumption in the conversion stage because the process simulations we developed did not include wastewater treatment or water recovery steps nor did they incorporate technologies that would optimize water recycling and recovery. In most cases, however, conversion stage

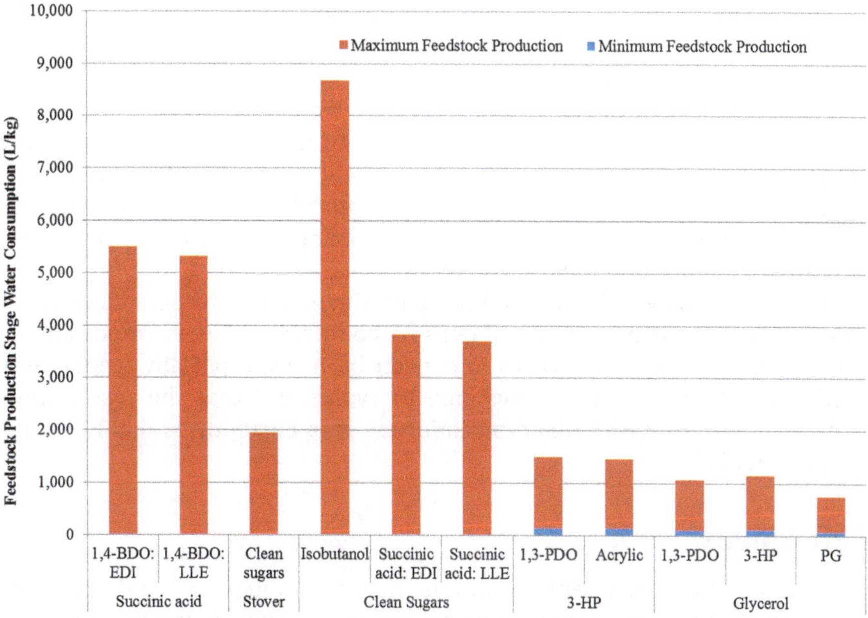

Figure 11.10 Water consumption for feedstock production (algae or corn stover) for selected bioproducts.

water consumption was vanishingly small in comparison to feedstock production stage water consumption. The process to produce clean sugars was, on the other hand, notably water-intensive because a large amount of steam is needed in the pretreatment reactor. In the production of 1,3-PDO, the fermentation step was a significant contributor to water consumption. Overall, conversion process design should be carried out with close attention to water consumption and opportunities to reduce it through process intensification (*e.g.*, increased yields) and water recycling.

Figure 11.10 displays water consumed in the feedstock production stage for each bioproduct. The results are driven by two factors: the water intensity of feedstock production and the product yield. As described earlier, corn stover is significantly more water intensive than algal glycerol when water consumption is split between corn grain and corn stover with mass allocation. If corn stover is not burdened with any water consumption, it would be less water intensive than algal glycerol. Of the compounds produced from clean sugars, isobutanol has the highest water consumption because it consumes 4.5 ton clean sugar per ton isobutanol whereas the succinic acid processes consume about 2 ton clean sugars per ton succinic acid.

11.4 Conclusions

LCA is a useful tool for evaluating the contributions of each stage of a bioproduct's life cycle to overall energy consumption, GHG emissions, and water consumption. As researchers and corporations develop and publish bioproduct LCAs, it is incumbent upon them to be transparent regarding assumptions both about material and energy flows in the product's life time and about treatment of the end-of-life stage, which can significantly influence results. As understanding of conversion processes improve, results for life-cycle GHG emissions and other impacts will evolve to reflect changes in both feedstock production and conversion processes. It is likely that conversion process yields, for example, will continue to increase, improving process economics and environmental metrics.

Overall, our analysis indicates that bioproducts offer lower cradle-to-grave GHG emissions than their fossil-derived counterparts. It is important to note, however, that technologies to convert biomass, especially second and third generation feedstocks, continue to evolve and that the results presented herein are based on information as it is currently available in the public domain.

References

1. Thomas Content. Virent bottling its new technology, *Milwaukee Journal Sentinel*, 2013, at <http://www.jsonline.com/business/virent-developing-environmentally-friendly-bottles-for-coke-b9994995z1-223770051.html>.
2. J. Moye, Coca-Cola PlantBottle Packaging, *The Coca-Cola Company*, at <http://www.coca-colacompany.com/stories/15-billion-and-counting>.
3. C. K. Wright and M. C. Wimberly, Recent land use change in the Western Corn Belt threatens grasslands and wetlands, *Proc. Natl. Acad. Sci.*, 2013, **110**, 4134–4139.
4. J. B. Dunn, S. Mueller, H. Kwon and M. Q. Wang, Land-use change and greenhouse gas emissions from corn and cellulosic ethanol, *Biotechnol. Biofuels*, 2013, **6**, 51–63.
5. J. B. Dunn, J. Han, J. Seabra and M. Wang, in *Handbook of Bioenergy Economics and Policy: Modeling Land Use and Greenhouse Gas Implications*, ed. M. Khanna and D. Zilberman, Springer Publishers, 2014.
6. S. Ahlgren and L. D. Lucia, Indirect land use changes of biofuel production – a review of modelling efforts and policy developments in the European Union, *Biotechnol. Biofuels*, 2014, 7, 35–44.
7. Z. Wang, J. B. Dunn, J. Han and M. Wang, *Material and Energy Flows in the Production of Cellulosic Feedstocks for Biofuels in the GREET Model*, Argonne National Laboratory, 2013, at <https://greet.es.anl.gov/publication-feedstocks-13>.
8. US Environmental Protection Agency, *Regulation of fuels and fuel additives: changes to renewable fuel standard program; final rule*. 14669–15320, 2010.
9. CARB, *Proposed Regulation to Implement the Low Carbon Fuel Standard, Vol. I, Staff Report: Initial Statement of Reasons*, California Air Resources Board, 2009.

10. European Parliament, *Renewable Energy Directive 2009/28/E.C*, European Parliament, Ed.
11. Y.-W. Chiu and M. Wu, Assessing County-Level Water Footprints of Different Cellulosic-Biofuel Feedstock Pathways, *Environ. Sci. Technol.*, 2012, **46**, 9155–9162.
12. Y.-W. Chiu and M. Wu, The water footprint of biofuel produced from forest wood residue *via* a mixed alcohol gasification process, *Environ. Res. Lett.*, 2013, **8**, 035015.
13. Y.-W. Chiu and M. Wu, Considering water availability and wastewater resources in the development of algal bio-oil, *Biofuels, Bioprod. Biorefin.*, 2013, **7**, 406–415.
14. M. Wu, *Investigating Grey Water Footprint for the Production of Gasoline and Diesel from Biomass via Fast Pyrolysis*, Argonne National Laboratory, 2013.
15. M. Wu and Y. Chiu, *Consumptive Water USe in the Production of Ethanol and Petroleum Gasoline – 2011 Update*, Argonne National Laboratory, 2011.
16. E4tech & LCAworks, *Environmental assessment of Braskem's biobased PE resin. Summary of the life cycle assessment, land-use change and water footprint reports*, 2013, at <http://www.braskem.com.br/Portal/Principal/Arquivos/Download/Upload/Enviro%20assessment%20summary%20report%20final_148.pdf>.
17. Myriant, *Bio-Succinic Acid A High Performing Green Chemical without a 'Green Premium' Price*, at <http://www.myriant.com/media/press-kit-files/Myriant-BioSFactSheet-0613.pdf>.
18. BioAmber BioBased Platform Chemical Life Cycle Analysis, at <http://www.bio-amber.com/ignitionweb/data/media_centre_files/804/Ontario_Canada_LCA_04.16.2013.pdf>.
19. M. J. Burk, Sustainable production of industrial chemicals from sugars, 2010, **112**, 30–35.
20. DuPont, Bio-PDO Life Cycle Assessment, at <http://www2.dupont.com/Bio-based_Propanediol/zh_CN/assets/downloads/Life%20Cycle%20Analysis%20-%20Bio-PDO(tm)%2003-06-09.pdf>.
21. Dow and OPXBIO Collaborating on Renewable Route to Acrylic Acid | Business Wire, 2011, at <http://www.businesswire.com/news/home/20110411005906/en/Dow-OPXBIO-Collaborating-Renewable-Route-Acrylic-Acid#.U3Da9fldV8E>.
22. B. G. Hermann, K. Blok and M. K. Patel, Producing Bio-Based Bulk Chemicals Using Industrial Biotechnology Saves Energy and Combats Climate Change, *Environ. Sci. Technol.*, 2007, **41**, 7915–7921.
23. R. A. Urban and B. R. Bakshi, 1,3-Propanediol from Fossils versus Biomass: A Life Cycle Evaluation of Emissions and Ecological Resources, *Ind. Eng. Chem. Res.*, 2009, **48**, 8068–8082.
24. A. D. Midland, Life cycle analysis of propylene glycol, at <http://www.adm.com/en-US/products/evolution/Propylene-Glycol/Pages/Life_Cycle_Analysis.aspx>.

25. B. Cok, I. Tsiropoulos, A. L. Roes and M. K. Patel, Succinic acid production derived from carbohydrates: An energy and greenhouse gas assessment of a platform chemical toward a bio-based economy, *Biofuels, Bioprod. Biorefin.*, 2014, **8**, 16–29.
26. T. R. Brown and R. C. Brown, A review of cellulosic biofuel commercial-scale projects in the United States, *Biofuels, Bioprod. Biorefin.*, 2013, 7, 235–245.
27. T. R. Brown and M. M. Wright, Techno-economic impacts of shale gas on cellulosic biofuel pathways, *Fuel*, 2014, **117**, **Part B**, 989–995.
28. IHS, Bio-based Chemicals Getting Unexpected Boost from Shale Gas Boom in North America - See more at: http://press.ihs.com/press-release/bio-based-chemicals/bio-based-chemicals-getting-unexpected-boost-shale-gas-boom-north-#sthash.xz3pkyVt.dpuf, at <http://press.ihs.com/press-release/bio-based-chemicals/bio-based-chemicals-getting-unexpected-boost-shale-gas-boom-north->.
29. Integrated Acid: A Commercial Reality, *Myriant* at <http://www.myriant.com/platform/operating-plants.cfm>.
30. Feedstock Development, *BioAmber* at <http://www.bio-amber.com/bioamber/en/sustainability/feedstock_development>.
31. Welcome to DuPont™ Cerenol™ Polyols, *DuPont* at <http://www2.dupont.com/Cerenol_Polyols/en_US/>.
32. Embracing the Era of Renewability, *DuPont* at <http://www2.dupont.com/Renewably_Sourced_Materials/en_US/>.
33. Gevo, *Gevo White Paper on Transportation Fuels*, 2011, at <file://wales/GREET_Bioproducts/Literature/2011Gevo%20whitepaper%20on%20isobutanol.pdf>.
34. Gevo: Biofuels Digest's 2014 5-Minute Guide: Biofuels Digest, at <http://www.biofuelsdigest.com/bdigest/2014/02/16/gevo-biofuels-digests-2014-5-minute-guide/>.
35. J. B. Dunn, F. Adom, N. Sather, J. Han and S. Snyder, *Life-cycle Analysis of Bioproducts and Their Conventional Counterparts in GREET*, Argonne National Laboratory, 2014.
36. H.-Y. Kwon, S. Mueller, J. B. Dunn and M. M. Wander, Modeling state-level soil carbon emission factors under various scenarios for direct land use change associated with United States biofuel feedstock production, *Biomass Bioenergy*, 2013, **55**, 299–310.
37. D. J. Muth Jr., K. M. Bryden and R. G. Nelson, Sustainable agricultural residue removal for bioenergy: A spatially comprehensive US national assessment, *Appl. Energy*, 2013, **102**, 403–417.
38. Z. Tan, S. Liu, N. Bliss and L. L. Tieszen, Current and potential sustainable corn stover feedstock for biofuel production in the United States, *Biomass Bioenergy*, 2012, **47**, 372–386.
39. Z. Tan and S. Liu, Soil nutrient budgets following projected corn stover harvest for biofuel production in the conterminous United States, *GCB Bioenergy*, 2015, 7(2), 175–183.

40. US Environmental Protection Agency, Regulation of Fuels and Fuel Additives: Identification of Additional Qualifying Renewable Fuel Pathways Under the Renewable Fuel Standard Program, *Fed. Regist.*, 2013, **78**, 14190–14217.
41. J. Han, A. Elgowainy, J. B. Dunn and M. Q. Wang, Life cycle analysis of fuel production from fast pyrolysis of biomass, *Bioresour. Technol.*, 2013, **133**, 421–428.
42. M. Wang, J. Han, J. B. Dunn, H. Cai and A. Elgowainy, Well-to-wheels energy use and greenhouse gas emissions of ethanol from corn, sugarcane and cellulosic biomass for US use, *Environ. Res. Lett.*, 2012, **7**, 045905.
43. PRNewsire, Global 1,3-Propanediol Market is Expected to Reach USD 683.2 million in 2019: Transparency Market Research, at <http://www.prnewswire.com/news-releases/global-1-3—propanediol-market-is-expected-to-reach-usd-6832-million-in-2019-transparency-market-research-253235491.html>.
44. H. Chim and T. Kumamoto, *CEH Product Review Propylene Glycols*, 2011.
45. DuPont Tate and Lyle BioProducts, at <http://www.duponttateandlyle.com/>.
46. J. J. Bozell and G. R. Petersen, Technology development for the production of biobased products from biorefinery carbohydrates—the US Department of Energy's 'Top 10' revisited, *Green Chem.*, 2010, **12**, 539–554.
47. *Renewable, Low-Cost Carbon Fiber for Lightweight Vehicles Summary Report from the June 4-5, 2013, Workshop Detroit, MI*, US Department of Energy, at <http://www1.eere.energy.gov/bioenergy/pdfs/carbon_fiber_summary_report.pdf>.
48. W. Bauer, in *Kirk-Othmer Encyclopedia of Chemical Technology*, 2003.
49. Gevo, Isobutanol: A Gateway to Chemicals and Fuels, at <http://www.gevo.com/our-markets/isobutanol/>.
50. *Top Value Added Chemicals from Biomass Volume I - Results of Screening for Potential Candidates from Sugars and Synthesis Gas*, 2001.
51. S. Davis, T. Kalin and T. Kumamoto. *CEH Marketing Research Report 1,4-Butanediol.*, 2010.
52. Argonne National Laboratory, *GREET1_2013*, Argonne National Laboratory, 2013, at <http://greet.es.anl.gov/>.
53. J. Han, J. B. Dunn, H. Cai, A. Elgowainy and M. Q. Wang, *Updated Sugarcane Parameters in GREET1_2012, Second Revision*, Argonne National Laboratory, at <https://greet.es.anl.gov/publication-greet-updated-sugarcane>.
54. S. Arora, M. Wu and M. Wang, Estimated displaced products and ratios of distillers' co-products from corn ethanol plants and the implications of lifecycle analysis, *Biofuels*, 2010, **1**, 911–922.
55. P. Pawelzik *et al*, Critical aspects in the life cycle assessment (LCA) of bio-based materials – Reviewing methodologies and deriving recommendations, *Resour., Conserv. Recycl.*, 2013, **73**, 211–228.

56. T. Helin, L. Sokka, S. Soimakallio, K. Pingoud and T. Pajula, Approaches for inclusion of forest carbon cycle in life cycle assessment – a review, *GCB Bioenergy*, 2013, **5**, 475–486.
57. US Environmental Protection Agency. *Compilation of Air Pollutant Emission Factors, Vol. 1: Stationary Point and Area Sources*, 1995.
58. F. Adom, J. B. Dunn, J. Han and N. Sather, Life-cycle fossil energy consumption and greenhouse gas emissions of bioderived chemicals and their conventional counterparts, *Environ. Sci. Technol.*, 2014, **48**(24), 14624–14631.

CHAPTER 12

Design and Planning of Sustainable Supply Chains for Biobased Products

JAESUK PARK, DAJUN YUE AND FENGQI YOU*

Northwestern University McCormick School of Engineering, 2145 Sheridan Rd, Evanston, IL 60208, USA
*Email: you@northwestern.edu

12.1 Introduction

Many of the chemicals that various parts of our current society utilize are petrochemicals.[1] As fossil fuel dependence continues to be an issue and as renewable resource technologies continue to rise in sophistication and efficiency, utilizing biomass as a raw material to synthesize various chemical products (biobased products) naturally becomes one of the potentially viable topics to consider.[2] A significant amount of current research focuses on converting biomass into fuel sources (biofuels). However, because of the low fossil fuel costs, biofuels can have a major challenge gaining a comparative market advantage over those nonrenewable energy sources.[3] On the other hand, commercial products derived from the same amount of biomass can be sold typically at a higher price and thus generate a higher profit. Furthermore, production of a biofuel often results in byproducts that can be further processed into commercial products thereby reducing industrial waste and generating additional profits simultaneously.[4]

The rationale for modeling and optimizing supply chain for biobased products is twofold: primarily, because biomass supply varies greatly from

RSC Green Chemistry No. 43
Commercializing Biobased Products: Opportunities, Challenges, Benefits, and Risks
Edited by Seth W. Snyder

Published by the Royal Society of Chemistry, www.rsc.org

time to time in a given year, the supply chain must take seasonality into account for adequate and feasible delivery of biobased products into market.[5–7] Also, because the total profit from any supply chain can vary depending on its operational optimality, careful selection and a well-developed insight on various aspects of a supply chain—including which technology is more appropriate, onto which capacity each refinery should be built, and, above all, how much of which biobased product to distribute into each demand region—are essential to maximize operational efficiency.[8–11]

However, an economically optimized supply chain for biobased products would be incomplete and could defeat its purpose if environmental issues—in particular, greenhouse gas (GHG) emissions—are not simultaneously taken into account during the modeling and optimization process. Therefore, the model should have two simultaneous main objectives—economic and environmental—rather than one; such a multi-objective model would utilize the properties of the Pareto distribution curve as its underlying principle.[12–14]

This chapter introduces how to optimize supply chains in the case of biobased products. More specifically, under a few assumptions, we devise a mathematical model that simulates an industrial-scale conversion of biomass into various biobased products through a supply chain superstructure, composed of suppliers, refineries, and demand regions.

12.2 Biomass-to-chemical Pathways

Based on the report by the Argonne National Laboratory (also see Chapter 11),[15] we have selected a number of promising biomass-to-chemical pathways to be considered in this work. As shown in Figure 12.1, we consider one type of biomass feedstock (corn stover) and five types of bioproducts (clean sugar, ethylene, succinic acid, 1,4-butanediol, and isobutanol). Corn stover can be first converted into sugar through pretreatment and saccharification processes. Then sugar can be used as a building block to produce a variety of chemicals. We consider production of ethylene from sugar *via* fermentation and dehydration. We consider production of isobutanol from sugar *via* fermentation. We also consider production of succinic acid from sugar *via* fermentation, product separation and recovery. Furthermore, 1,4-butanediol can be produced from succinic acid *via* aqueous-phase hydrogenation.

12.3 Problem Statement

The problem addressed in this chapter on optimized biobased product supply chain design and planning is formally stated in Figure 12.2.

We are provided with a biobased product supply chain superstructure (Figure 12.2), including a set of suppliers ($s \in S$), a set of refineries ($r \in R$), and a set of demand regions ($d \in D$) at a given time period ($t \in T$), which takes seasonality of biomass availability into account. A set of biomass ($b \in B$) can be purchased from the suppliers. For the refineries in the superstructure,

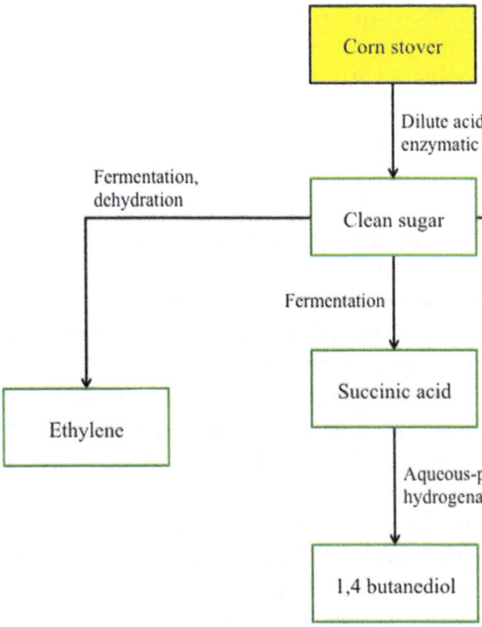

Figure 12.1 Process models for biobased products.

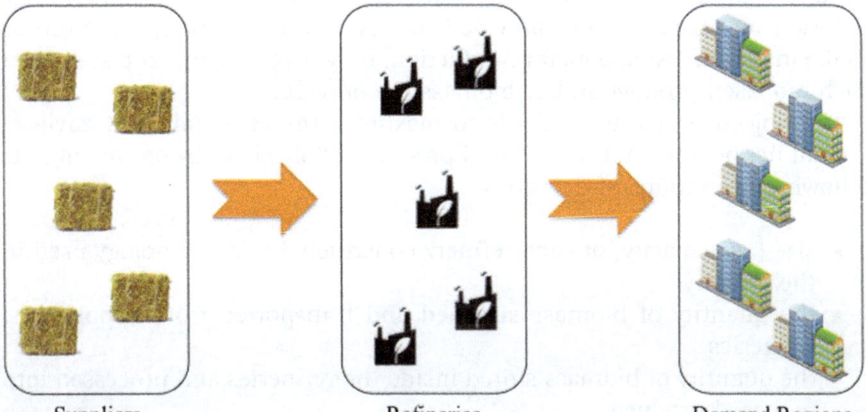

Figure 12.2 Superstructure diagram for the problem.

each refinery can choose one technology from the set of technologies ($\tau \in \tau$), through which a set of biobased products ($p \in P$) can be produced. Finally, to transport biomass and biobased product from one location to another, elements from a set of transportation modes ($m \in M$) are considered.

For each supplier, we are provided with the amount of available biomass for purchase at a given time period. Regarding transportation of the purchased biomass from the supplier site to refineries, we are also provided with the maximum capacity of each mode of transportation in terms of biomass weight.

For all refineries, we are provided with the proportion of biomass degrading during each time period, as well as conversion efficiencies (*i.e.* efficiencies of converting biomass into biobased products) corresponding to all technologies utilized in the refineries. Also, for each refinery, we are provided with lower and upper bounds of capacity for the technology used in the refinery. Finally, for all refineries, we are provided with the minimum proportion of refinery that needs to run whenever the refinery is on operation, as well as the number of days of operation in a year.

For every demand region, we are provided with the lower and upper bounds of quantity demanded for each biobased product. Also, concerning transportation of the finished biobased products from each refinery to demand regions, we are provided with the maximum capacity of each mode of transportation—just as in the case of biomass, but this time in terms of biobased product (not biomass) weight.

Finally, to calculate costs associated with this superstructure, financial data values, such as annuity factor, number of project years, and the annual market interest rate, are provided. Also, reference capacities for each technology in terms of biobased product weight and reference capital costs are also provided. The cost data associated with biomass purchase, transportation (which includes distance data and fixed/variable cost data for unit distance traveled), storage, and production, in addition to the market price for each biobased product provided for revenue calculation, are given as well. Finally, emission data for production, harvest, transport, and storage of each biobased product and/or biomass are provided.

The objective of this model is to maximize the economic and environmental net benefits of the biobased product supply chain by optimizing the following five groups of decisions:

- size (*i.e.* capacity) of each refinery constructed and technology used in the refinery
- the quantity of biomass supplied and transported from suppliers to refineries
- the quantity of biomass stored inside the refineries and processed into biobased products
- the quantity of biobased products produced and stored inside the refineries
- the quantity of biobased products transported into demand regions and distributed within

12.4 Model Formulation

A multi-objective, multi-period mixed-integer nonlinear programming (MINLP) model has been constructed to represent the optimization of biobased product supply chain. Constraints (12.1)–(12.3) model the purchase of biomass from the suppliers, taking the seasonality of biomass into account, while constraints (12.4)–(12.7) model the processing of biomass into biobased products using the available technologies, and constraints (12.8)–(12.10) model the distribution of biobased products into demand regions. The set of eqn (12.11)–(12.22) calculates the total cost/benefit and the various sub-costs/benefits incurred throughout the project, including initial capital costs, production cost, transportation cost, regional revenue, *etc.*, and eqn (12.23) represents the profit, *i.e.* an economic objective to be maximized. Finally, the set of eqn (12.24)–(12.28) calculates the total GHG emission and its components incurred across the superstructure, and eqn (12.29) represents the GHG emission, *i.e.* an environmental objective to be minimized. A complete list of indices and sets, parameters, and decision variables is given in the Nomenclature section. In general, a parameter is denoted with two initial uppercase letters, and a decision variable is denoted with one initial uppercase letter.

12.4.1 Constraints

12.4.1.1 Supply System of Biomass

The total amount of biomass b purchased from supplier s at time period t (denoted $Qsu_{b,s,t}$) should not exceed the biomass's availability in terms of dry weight.

$$Qsu_{b,s,t} \leq ABms_{b,s,t} \quad \forall\ b \in B,\ s \in S,\ t \in T \tag{12.1}$$

All of the quantity of biomass b collected from supplier s at time period t should be transported to refineries (*i.e.* there should be no biomass lost during transportation). This mass-balance is modeled as follows:

$$Qsu_{b,s,t} = \sum_{m}\sum_{r} Qtr_{m,b,s,r,t} \quad \forall\ b \in B,\ s \in S,\ t \in T \tag{12.2}$$

where $Qtr_{m,b,s,r,t}$ represents the quantity of biomass b transported *via* mode m from supplier s to refinery r at time period t.

In addition, the quantity of any biomass (in wet tons) transported *via* mode m from supplier s to refinery r at any time period should not exceed the maximum capacity of transportation mode m from supplier s to refinery r (denoted $KMTr_{m,s,r}$, measured in *dry* tons), modeled as follows:

$$\frac{Qtr_{m,b,s,r,t}}{(1-\mu_b)} \leq KMTr_{m,s,r} \quad \forall\ m \in M,\ b \in B,\ s \in S,\ r \in R,\ t \in T \tag{12.3}$$

where μ_b represents the moisture content of biomass transported.

12.4.1.2 Processing of Biomass into Biobased Products

In each refinery, the quantity of biomass b transported into the refinery r from all suppliers *via* all modes of transportation at time period t, when added to the biomass b remaining viable in storage at the same refinery for the previous time period, should equal in amount to the quantity of biomass processed at the same time period (denoted) plus the quantity of the biomass to be stored in the refinery at time period t (denoted $Qst_{b,r,t}$). This is modeled as:

$$\begin{aligned} &\sum_{s}\sum_{m} Qtr_{m,b,s,r,t} + (1-\delta_{b,t--1})Qst_{b,r,t--1} \\ &= \sum_{\tau} Qpc_{b,r,\tau,t} + Qst_{b,r,t} \quad \forall\ b\in B,\ r\in R,\ t\in T \end{aligned} \tag{12.4}$$

where $Qpc_{b,r,\tau,t}$ represents the quantity of biomass b processed at refinery r using technology τ at time period t, and $\delta_{b,t--1}$ denotes the proportion of biomass b degraded in the storage during time period $t-1$. The notation "$t--1$" takes the year cycle into account, *i.e.* the last time period is followed by the first time period of the following year.

For each unit of biobased product p produced *via* conversion of biomass b, there exists a number between 0 and 1 that represents the yield using technology τ (*i.e.* conversion efficiency, denoted $\varepsilon_{b,p,\tau}$), which can be multiplied to the quantity of biomass processed to calculate the quantity of biobased product p produced at the refinery r using technology τ at time period t (denoted $Qpd_{p,r,\tau,t}$), according to the following equation:

$$Qpd_{p,r,\tau,t} = \sum_{b} Qpc_{b,r,\tau,t} \cdot \varepsilon_{b,p,\tau} \quad \forall\ p\in P,\ r\in R,\ \tau\in\tau,\ t\in T \tag{12.5}$$

If a certain process utilization rate is required, there is a lower bound to the proportion of active refinery capacity r using technology τ (denoted $\theta_{r,\tau}$). In addition, if refineries operate only for portion of a year (denoted ΔY_t), those days in a year when refineries are inactive also certainly affect the total quantity produced. Eqn (12.6) takes all these into account:

$$\theta_{r,\tau}\frac{\Delta Y_t}{\Delta Y}Cap_{r,\tau} \leq \sum_{b} \omega_b \cdot Qpc_{b,r,\tau,t} \leq \frac{\Delta Y_t}{\Delta Y}Cap_{r,\tau} \quad \forall\ r\in R,\ t\in T,\ \tau\in\tau \tag{12.6}$$

where ΔY is the total number of days in a year, ω_b is the weight factor for different types of biomass, and $Cap_{r,\tau}$ denotes the capacity of refinery r using technology τ.

Once biobased products are produced at time period t, its quantity plus the quantity of biobased product stored in the refinery for the previous time period should add up to a quantity equal to the quantity transported to all demand regions *via* all available modes of transportation at the same time period plus the quantity stored in the refinery r at time period t for the next round of shipment ($Qstp_{p,r,t}$). This mass-balance is modeled as follows:

$$\begin{aligned} &\sum_{\tau} Qpd_{p,r,\tau,t} + (1-\delta_{p,t--1})Qstp_{p,r,t--1} \\ &= \sum_{m}\sum_{d} Qtrp_{m,p,r,d,t} + Qstp_{p,r,t} \quad \forall\ p\in P,\ r\in R,\ t\in T \end{aligned} \tag{12.7}$$

where $Qtrp_{m,p,r,d,t}$ represents quantity of biobased product p transported *via* mode m from refinery r to demand region d at time period t, and $\delta_{p,t--1}$ denotes the proportion of biobased product p degraded in the storage during time period $t-1$, again with the subscript notation implying the year cycle.

12.4.1.3 Distribution of Biobased Products in Demand Regions

For every demand region d, quantity of biobased product p transported from all refineries *via* all available modes of transportation at time period t defines the quantity of the biobased product distributed in the demand region at that time period (denoted $Qdist_{p,d,t}$):

$$Qdist_{p,d,t} = \sum_{m}\sum_{r} Qtrp_{m,p,r,d,t} \quad \forall\ p \in P,\ d \in D,\ t \in T \tag{12.8}$$

Also, the quantity of biobased product p distributed in a given demand region at a given time period should fall within the predetermined demand range:

$$DBnd^{L}{}_{p,d,t} \leq Qdist_{p,d,t} \leq DBnd^{U}{}_{p,d,t} \quad \forall\ p \in P,\ d \in D,\ t \in T \tag{12.9}$$

where $DBnd^{L}{}_{p,d,t}$ and $DBnd^{U}{}_{p,d,t}$ represent the lower and upper bounds, respectively, of quantity of biobased product p demanded in demand region d at time period t.

Finally, each mode of transportation, just as in biomass transportation, has a capacity limit ($KMTrp_{m,r,d}$):

$$Qtrp_{m,p,r,d,t} \leq KMTrp_{m,r,d} \quad \forall\ m \in M,\ p \in P,\ r \in R,\ d \in D,\ t \in T \tag{12.10}$$

12.4.2 Costs, Revenues, and Economic Objective

12.4.2.1 Costs

The following sub-costs comprise total cost (C_{Total}): capital cost (C_{capital}), biomass acquisition cost ($C_{\text{acquisition}}$), production cost ($C_{\text{production}}$), storage cost (C_{storage}), transportation cost ($C_{\text{transportation}}$), and distribution cost ($C_{\text{distribution}}$). Each is modeled below:

$$C_{\text{Total}} = C_{\text{capital}} + C_{\text{acquisition}} + C_{\text{distribution}} + C_{\text{production}} + C_{\text{transportation}} + C_{\text{storage}} \tag{12.11}$$

$$C_{\text{capital}} = AN \sum_{r}\sum_{\tau} Tci_{r,\tau} \tag{12.12}$$

$$C_{\text{acquisition}} = \sum_{b}\sum_{s}\sum_{t} CBms_{b,s,t} \cdot Qsu_{b,s,t} \tag{12.13}$$

$$C_{\text{distribution}} = \sum_{p}\sum_{d}\sum_{t} CDist_{p,d,t} \cdot Qdist_{p,d,t} \tag{12.14}$$

$$C_{\text{production}} = \sum_p \sum_r \sum_t \sum_\tau CPd_{r,\tau} \cdot Qpd_{p,r,\tau,t} \tag{12.15}$$

$$C_{\text{transportation}} = \left(\sum_m \sum_s \sum_r \sum_t CTr_{m,s,r,t} \sum_b \frac{Qtr_{m,b,s,r,t}}{(1-\mu_b)} \right) + \left(\sum_m \sum_s \sum_r \sum_t CTrp_{m,r,d,t} \sum_p Qtrp_{m,p,r,d,t} \right) \tag{12.16}$$

$$C_{\text{storage}} = \sum_b \sum_r \sum_t CSt_{b,r,t} \cdot Qst_{b,r,t} + \sum_p \sum_r \sum_t CStp_{p,r,t} \cdot Qstp_{p,r,t} \tag{12.17}$$

where AN is the annuity factor, $Tci_{r,\tau}$ is the total capital investment for refinery r that uses technology τ, $CBms_{b,s,t}$ is the cost of purchasing biomass b from supplier s at time period t, $CDist_{p,d,t}$ is the cost of distributing biobased product p in demand region d at time period t, $CPd_{r,\tau}$ is the cost of production at refinery r with technology τ, $CTr_{m,s,r,t}$ and $CTrp_{m,r,d,t}$ are cost of transporting biomass from supplier s to refinery r and biobased product from refinery r to demand region d, respectively, *via* transportation mode m at time period t, and $CSt_{b,r,t}$ and $CStp_{p,r,t}$ denote the cost of storing biomass b and biobased product p, respectively, in refinery r at time period t.

To set values for the parameter AN and decision variables $Tci_{r,\tau}$ and $Cap_{r,\tau}$, some more parameters and variables are used, as below:

$$\sum_\tau Bin_{r,\tau} \leq 1 \quad \forall\, r \in R \tag{12.18}$$

$$\frac{Tci_{r,\tau}}{REFcapit_\tau} = \left(\frac{Cap_{r,\tau}}{REFcapa_\tau} \right)^{0.6} \quad \forall\, r \in R,\ \tau \in \tau \tag{12.19}$$

$$Bin_{r,\tau} \cdot M_\tau^L \leq Cap_{r,\tau} \leq Bin_{r,\tau} \cdot M_\tau^U \quad \forall\, r \in R,\ \tau \in \tau \tag{12.20}$$

where $Bin_{r,\tau}$ is a binary variable that equals 1 if refinery r using technology τ is constructed (0 otherwise), $REFcapit_\tau$ and $REFcapa_\tau$ correspond to reference capital cost and reference capacity, respectively, for technology τ, and M_τ^L and M_τ^U represent lower and upper bounds, respectively, of capacity for a biorefinery utilizing technology τ.

12.4.2.2 Revenues

The total revenue (Rv_{Total}) is calculated by a summation of all regional revenue at a given time for a given biobased product ($Rv_{p,d}$):

$$Rv_{\text{Total}} = \sum_p \sum_d Rv_{p,d} \tag{12.21}$$

$$Rv_{p,d} = \sum_t MP_{p,d,t} Qdist_{p,d,t} \quad \forall\, p \in P,\ d \in D \tag{12.22}$$

where $MP_{p,d,t}$ is the market price of biobased product p in demand region d at time period t.

12.4.2.3 Economic Objective: Maximizing Profit

Intuitively, given revenue and cost calculations, the economic objective is to maximize the total revenue minus total cost, or profit, from this supply chain superstructure:

$$\max \text{PROFIT} = Rv_{\text{Total}} - C_{\text{Total}} \tag{12.23}$$

12.4.3 GHG Emissions and Environmental Objective

12.4.3.1 GHG Emissions

The total GHG emission (E_{Total}) consists of GHG emissions from production ($E_{\text{production}}$), transportation ($E_{\text{transportation}}$), storage ($E_{\text{storage}}$), and harvest ($E_{\text{harvest}}$). Each is modeled below:

$$E_{\text{Total}} = E_{\text{Production}} + E_{\text{transportation}} + E_{\text{storage}} + E_{\text{harvest}} \tag{12.24}$$

$$E_{\text{production}} = \sum_p \sum_r \sum_\tau \sum_t EPd_{r,\tau} \cdot Qpd_{p,r,\tau,t} \tag{12.25}$$

$$E_{\text{transportation}} = \sum_m \sum_r \sum_t \left(\sum_s \sum_b \frac{ETr_{b,m} \cdot DTr_{s,r} \cdot Qtr_{m,b,s,r,t}}{(1 - \mu_b)} + \sum_d \sum_p ETrp_{p,m} \cdot DTrp_{r,d} \cdot Qtrp_{m,p,r,d,t} \right) \tag{12.26}$$

$$E_{\text{storage}} = \sum_b \sum_r \sum_t ESt_{b,r,t} \cdot Qst_{b,r,t} + \sum_p \sum_r \sum_t EStp_{p,r,t} \cdot Qstp_{p,r,t} \tag{12.27}$$

$$E_{\text{harvest}} = \sum_b \sum_s \sum_t EHv_{b,s,t} \cdot Qsu_{b,s,t} \tag{12.28}$$

where $EPd_{r,\tau}$ is the emission during production at refinery r using technology τ, $ETr_{b,m}$ and $ETrp_{p,m}$ are the emission during transportation of biomass b and biobased product p, respectively, *via* transportation mode m, $DTr_{s,r}$ and $DTrp_{r,d}$ are the distances from supplier s to refinery r and from refinery r to demand region d, respectively, $ESt_{b,r,t}$ and $EStp_{p,r,t}$ are the emission during storage of biomass b and biobased product p, respectively, at refinery r at time period t, and $EHv_{b,s,t}$ is the emission during harvest of biomass b at the site of supplier s at time period t. All emission data are in terms of an equivalent amount of CO_2.

12.4.3.2 Environmental Objective: Minimizing Emissions

The environmental objective, therefore, is to minimize GHG emissions, which are, as previously mentioned, quantified as the equivalent amount of carbon dioxide:

$$\min \text{EMISSION} = E_{\text{Total}} \tag{12.29}$$

12.5 Solution Algorithm

The proposed MINLP model includes two objective functions, thus multi-objective optimization techniques are required to solve the problem. We employ the ε-constraint method and transform the environmental objective function (12.29) into an ε-constraint given below, thus transforming the problem into a single-objective MINLP.[16]

$$\text{EMISSION} \leq \varepsilon \tag{12.30}$$

By solving the profit maximization problem at different ε-values, we can obtain a series of solutions that constitute an approximated Pareto curve.

We note that the single-objective MINLP problem can be solved using the branch-and-refine algorithm. The branch-and-refine algorithm constructs piecewise linear approximations for the concave terms in eqn (12.19) and the resulting mixed-integer linear programming (MILP) problem provides an upper bound to the original MINLP. The lower bound can be obtained by directly evaluating the nonlinear economic objective function. The advantage of the branch-and-refine algorithm is that it requires only MILP solvers to solve a nonconvex MINLP problem.[17]

12.6 Case Study

The following hypothetical supply chain design scenario illustrates how the framework described can be implemented. All computations were performed on a computer with Intel Core i5 2.4 GHz CPU and 4GB RAM. The MILP described in the previous section was coded in GAMS 24.3.3[18] and solved with the solver CPLEX 12. The optimality tolerances were all set to be zero.

12.6.1 Input Data

In this generic scenario, five biomass suppliers provide corn stover (an example of crop residue) to be processed into any of five different biobased products—clean sugar, ethylene, succinic acid, isobutanol, and

1,4-butanediol—in up to five refineries. These biobased products are then transported and distributed into five nearby demand regions *via* the two transportation modes available—rail and truck—which are the same transportation modes available from suppliers to refineries.

The biomass is harvested only in October and November.[19] The quantity of biomass available from each supplier at either of the two months is set between 50 and 300 kton per year in dry weight. We assume that there is sufficient transportation capacity for both biomass and biobased products. Also, the moisture proportion of corn stover was set to be 30%.[20]

It was assumed that 1% of corn stover stored in refineries at the beginning of each month degrades,[21] whereas it was also assumed that degradation of biobased products is negligible. Also, each of the five biobased products produced corresponds to a different biomass processing technology. Each of these technologies converts biomass (corn stover) into a biobased product with a certain yield. As mentioned earlier, these conversion efficiency data were calculated based on data from the report by the Argonne National Laboratory (see Chapter 11).[15] Additionally, to simplify calculations, all refineries are assumed to operate 30 days for each of 12 months, and it was assumed that refineries are able to shut down completely as necessary (*i.e.* there is no minimum capacity required to be operational at all times).

Each of the five demand regions is assumed to have demand for a different biobased product. The demand data for each demand region were pulled from various sources and are listed in Table 12.1, along with the conversion efficiency data. All data for the economic objective we were able to access—market price of biobased products, cost of producing each biobased product, costs of transporting biomass and biobased products *via* each mode of transportation (both fixed and variable), and cost of storing biomass in refineries at a given time—are pulled from various sources and summarized in Tables 12.2 and 12.3, while data that were inaccessible or are difficult to standardize—cost of distributing a biobased product in each demand region, cost of storing biobased products in refineries, cost of purchasing one dry kiloton of biomass (corn stover) from each supplier at both months in a given year—are set to be zero or a random number under uniform distribution between reasonable lower and upper bounds. Finally, the discount rate was set to be 10%, and the number of project years was set to be 10 years. Data for the environmental objective—emissions from

Table 12.1 Biobased product conversion efficiency and demand data.

Biobased product	Conversion efficiency (ton/ton corn stover)	Demand region	Demand (kton)
Ethylene	0.156	1	50
1,4-Butanediol	0.238	2	2.5
Succinic acid	0.333	3	0.3
Clean sugar	0.667	4	10
Isobutanol	0.148	5	10

Table 12.2 Market price and production cost data for each biobased product.

Biobased product	Market price ($/ton)	Production cost ($/ton)
Ethylene	1700	580
1,4-Butanediol	2000	560
Succinic acid	2500	280
Clean sugar	500	150
Isobutanol	2100	700

Table 12.3 Transportation cost for each mode of transportation.

Mode of transportation	Load type	Fixed cost ($/ton)	Variable cost ($/ton · km)
Rail	Biomass	14.15	0.023
Truck	Biomass	4.39	0.12
Rail	Biobased product	5.48	0.017
Truck	Biobased product	3.01	0.07

Table 12.4 Biomass harvest and biobased product production emission data.

Biomass/biobased product	Harvest/production emission (kg CO_2-eq./ton)
Corn stover (biomass)	79
Ethylene	5259
1,4-Butanediol	1126
Succinic acid	626
Clean sugar	302
Isobutanol	2069

Table 12.5 Transportation emission data.

Mode of transportation	Load type	Emission (kg CO_2-eq./ton · km)
Rail	Biomass	0.06016
Truck	Biomass	0.19464
Rail	Biobased product	0.06016
Truck	Biobased product	0.19464

production, transportation, and harvest—are listed in Tables 12.4 and 12.5. Emission from storage of biomass is equal to zero, assuming open atmosphere storage.

12.6.2 Results and Discussion

The graph of approximate economic objective value as a function of maximum environmental objective value (as in the ε-constraint method) is presented in Figure 12.3, with the horizontal axis corresponding to the maximum GHG emission permitted (the epsilon) in tons CO_2 and the vertical axis corresponding to the economic profit in thousand dollars.

The point in the lower left-hand corner of Figure 12.3 represents the objective value of 14.8 million dollars for the environmentally optimal solution at the epsilon value of approximately 438.9 million tons of CO_2, whereas the upper right-hand corner point represents the objective value of 73.2 million dollars for the economically optimal solution at the epsilon value of 781.3 million tons of CO_2. The curve has a positive slope, *i.e.* the objective value increases as the emission limit increases; intuitively, as more emission is permitted, the supply chain can produce more economic profit.

Tables 12.6 and 12.7 summarize the supply chain structure for an economically optimal solution and environmentally optimal solution, respectively; for example, in the economically optimal solution, at location 2, a refinery that receives corn stover from supplier 3 to produce up to 60 kilotons of ethylene for demand region 1 is built.

Figure 12.4 shows the cost breakdown of the two extreme solutions (economically optimal solution and environmentally optimal solution).

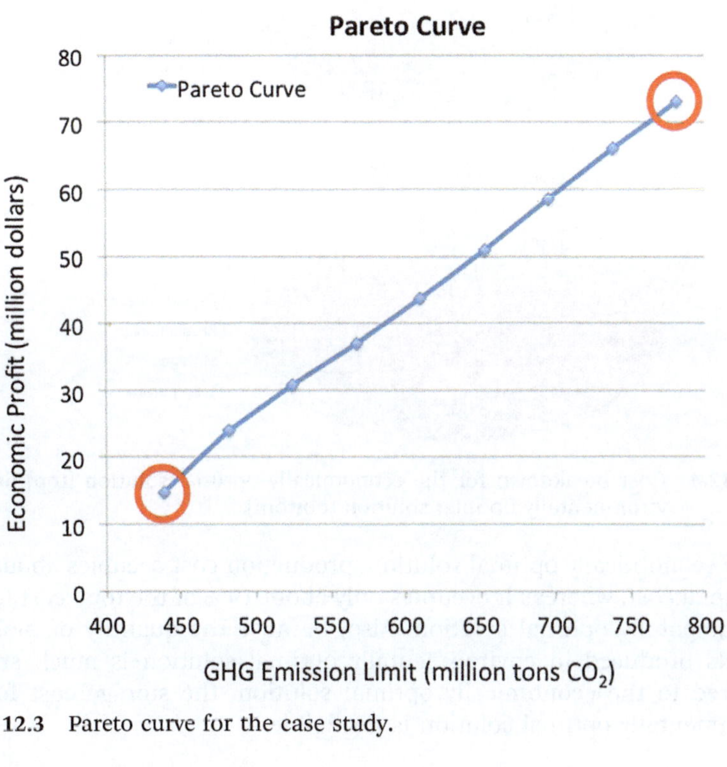

Figure 12.3 Pareto curve for the case study.

Table 12.6 Supply chain structure for economically optimal solution.

Refinery location	Supplier	Annual capacity (kton per year)	Biobased product	Demand region
1	5	120	Clean sugar	4
2	3	60	Ethylene	1
3	1, 2, 4, 5	112.655	Isobutanol	5
4	5	3.6	Succinic acid	3
5	3, 5	30	1,4-Butanediol	2

Table 12.7 Supply chain structure for environmentally optimal solution.

Refinery location	Supplier	Annual capacity (kton per year)	Biobased product	Demand region
1	4	72	Clean sugar	4
2	1	18	1,4-Butanediol	2
3	2, 4	72	Isobutanol	5
4	2	462.679	Ethylene	1
5	4	2.16	Succinic acid	3

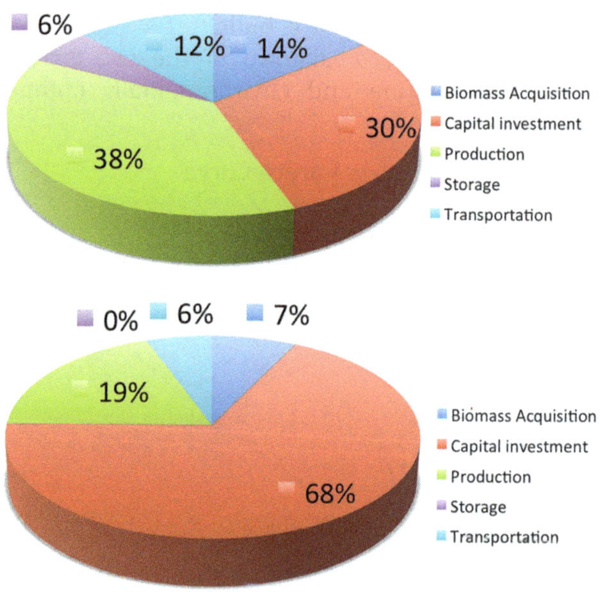

Figure 12.4 Cost breakdown for the economically optimal solution (top) and environmentally optimal solution (bottom).

For the economically optimal solution, production cost occupies about 38% of the total cost, whereas it occupies only about 19% of the total cost for the environmentally optimal solution. Also, because the quantity of biobased products produced in environmentally optimal solution is much smaller compared to the economically optimal solution, the storage cost for the environmentally optimal solution is negligible.

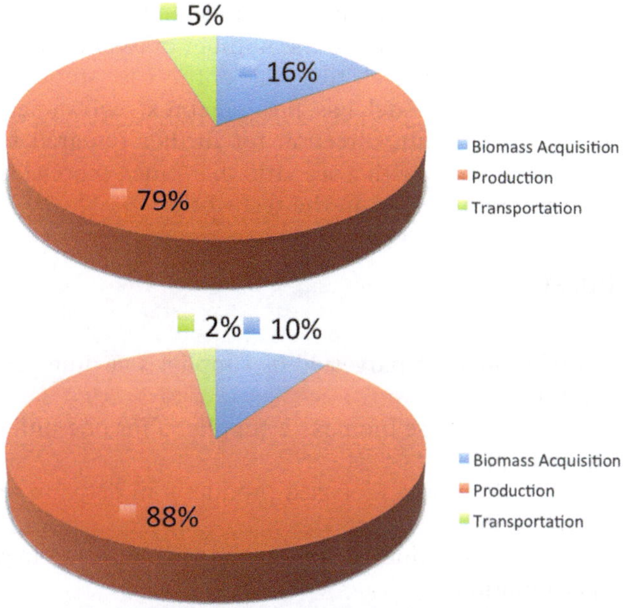

Figure 12.5 Emission breakdown for the economically optimal solution (top) and environmentally optimal solution (bottom).

Likewise, Figure 12.5 shows the emission breakdown of the two extreme solutions. It should be noted that the percentage difference in emission during the production stage between the two extreme solutions is much smaller than the percentage difference in cost for production between the two solutions; the biomass acquisition and transportation emissions only slightly differ as well.

12.7 Conclusion

In this chapter, we presented a mathematical framework to model and optimize a supply chain for biobased products. A multi-objective, multi-period MILP model that takes various aspects of the supply chain into account was designed and implemented. The MILP model used the branch-and-refine solution algorithm along with the ε-constraint method to simultaneously optimize economic objective (maximizing profit) and environmental objective (minimizing greenhouse gas emissions). The proposed model then underwent a small case study scenario to demonstrate its functionality. The result confirmed that as more greenhouse gas emission is allowed, higher profit is generated.

This approach, however, needs more testing for validation, both at a greater scale to ensure feasibility in terms of computational difficulty and with more parameters as more data become accessible. We acknowledge that the case study presented here has a great limitation in terms of reflecting a real-world scenario, due to the number of assumptions needed to be made, considering the relatively tiny scale of this case study. Nonetheless, it demonstrated that the model has not generated contradictory results; therefore, the most important direction for further research should be a further testing of this model with a scenario that much more closely reflects a biobased product supply chain in the real world.

Nomenclature

I. Parameters

$ABms_{b,s,t}$	Total biomass *b* harvested by supplier *s* at time period *t*
AN	Annuity factor
$CBms_{b,s,t}$	Cost to purchase biomass *b* supplier *s* for one unit of biomass at time period *t*
$CDist_{p,d,t}$	Cost to distribute biobased product *p* in demand region *d* at time period *t*
$CPd_{r,\tau}$	Cost of production at biorefinery *r* using technology permutation τ
$CSt_{b,r,t}$	Cost to store biomass *b* in biorefinery *r* at the end of time period *t*
$CStp_{p,r,t}$	Cost to store biobased product *p* in biorefinery *r* at the end of time period *t*
$CTr_{m,s,r,t}$	Cost to transport biomass *via* mode *m* from supplier *s* to biorefinery *r* at time period *t*
$CTrp_{m,r,d,t}$	Cost to transport biobased product *via* mode *m* from biorefinery *r* to demand region *d* at time period *t*
$DBnd^{L}{}_{p,d,t}$	Lower bound of quantity of biobased product *p* demanded in demand region *d* at time period *t*
$DBnd^{U}{}_{p,d,t}$	Upper bound of quantity of biobased product *p* demanded in demand region *d* at time period *t*
$KMTr_{m,s,r}$	Capacity of transportation mode *m* from supplier *s* to biorefinery *r*
$KMTrp_{m,r,d}$	Capacity of transportation mode *m* from biorefinery *r* to demand region *d*
M^{L}_{τ}	Lower bound of capacity for any biorefinery using any technology
M^{U}_{τ}	Upper bound of capacity for any biorefinery using any technology
$MP_{p,d,t}$	Price of biobased product *p* in demand region *d* at time period *t*
$REFcapa_{\tau}$	Reference capacity for given technology τ
$REFcapit_{\tau}$	Reference capital cost for given technology τ, based on corresponding $REFcapa_{\tau}$
ΔY	Number of operational days per year
ΔY_{t}	Number of operational days in time period *t*

$\delta_{b,t}$	Proportion of biomass *b* degraded inside storage within time period *t*
$\delta_{p,t}$	Proportion of biobased product *p* degraded inside storage within time period *t*
$\varepsilon_{b,p,\tau}$	Yield (*i.e.* efficiency) of conversion of biomass *b* into biobased product *p* using technology τ
$\theta_{r,\tau}$	Minimum proportion of capacity required to be operational in biorefinery *r* utilizing technology τ
ω_b	Weight factor for biomass *b*
μ_b	Moisture content for biomass *b*

II. Decision Variables

$Bin_{r,\tau}$	1 if biorefinery *r* using technology τ was constructed; 0 otherwise
$C_{\text{acquisition}}$	Total cost of acquiring biomasses incurred in the project
C_{capital}	Total capital investment cost incurred in the project
$C_{\text{distribution}}$	Total distribution cost incurred in the project
$C_{\text{production}}$	Total production cost of biobased products incurred in the project
C_{storage}	Total storage cost, both biomass and biobased product, incurred in the project
C_{Total}	Total cost incurred in the project
$C_{\text{transportation}}$	Total transportation cost, both biomass and biobased product, incurred in the project
$Cap_{r,\tau}$	Capacity of biorefinery *r* using technology τ
$Qdist_{p,d,t}$	Quantity of biobased product *p* distributed into demand region *d* at time period *t*
$Qpc_{b,r,\tau,t}$	Quantity of biomass *b* processed at biorefinery *r* using technology τ at time period *t*
$Qpd_{p,r,\tau,t}$	Quantity of biobased product *p* produced at biorefinery *r* using technology τ at time period *t*
$Qst_{b,r,t}$	Quantity of biomass *b* stored in biorefinery *r* at the end of time period *t*
$Qstp_{p,r,t}$	Quantity of biobased product *p* stored in biorefinery *r* at the end of time period *t*
$Qsu_{b,s,t}$	Quantity of biomass *b* purchased from supplier *s* at time period *t*
$Qtr_{m,b,s,r,t}$	Quantity of biomass *b* transported *via* mode *m* from supplier *s* to biorefinery *r* at time period *t*
$Qtrp_{m,p,r,d,t}$	Quantity of biobased product *p* transported *via* mode *m* from biorefinery *r* to demand region *d* at time period *t*
$Rv_{p,d}$	Total revenue from biobased product *p* sold in demand region *d*
Rv_{Total}	Total revenue generated from the project
$Tci_{r,\tau}$	Capital investment cost incurred for biorefinery *r* that uses technology τ

References

1. I. E. Grossmann, Challenges in the new millennium: product discovery and design, enterprise and supply chain optimization, global life cycle assessment, *Comput. Chem. Eng.*, 2004, **29**(1), 29–39.
2. T. Werpy and G. Petersen, *Top Value Added Chemicals from Biomass Volume I—Results of Screening for Potential Candidates from Sugars and Synthesis Gas*, U.S. Department of Energy, 2004.
3. D. Yue, F. You and S. W. Snyder, Biomass-to-bioenergy and biofuel supply chain optimization: Overview, key issues and challenges, *Comput. Chem. Eng.*, 2014, **66**(0), 36–56.
4. J. J. Bozell and G. R. Petersen, Technology development for the production of biobased products from biorefinery carbohydrates-the US Department of Energy's "Top 10" revisited, *Green Chem.*, 2010, **12**(4), 539–554.
5. F. You and B. Wang, Life Cycle Optimization of Biomass-to-Liquid Supply Chains with Distributed–Centralized Processing Networks, *Ind. Eng. Chem. Res.*, 2011, **50**(17), 10102–10127.
6. J. A. Elia, R. C. Baliban and C. A. Floudas, Nationwide energy supply chain analysis for hybrid feedstock processes with significant CO2 emissions reduction, *AIChE J.*, 2012, **58**(7), 2142–2154.
7. J. Kim, M. J. Realff, J. H. Lee, C. Whittaker and L. Furtner, Design of biomass processing network for biofuel production using an MILP model, *Biomass Bioenergy*, 2011, **35**(2), 853–871.
8. I. Grossmann, Enterprise-wide optimization: A new frontier in process systems engineering, *AIChE J.*, 2005, **51**(7), 1846–1857.
9. J. Schmidt, S. Leduc, E. Dotzauer, G. Kindermann and E. Schmid, Cost-effective CO2 emission reduction through heat, power and biofuel production from woody biomass: A spatially explicit comparison of conversion technologies, *Appl. Energy*, 2010, **87**(7), 2128–2141.
10. C. Papapostolou, E. Kondili and J. K. Kaldellis, Development and implementation of an optimisation model for biofuels supply chain, *Energy*, 2011, **36**(10), 6019–6026.
11. N. Parker, P. Tittmann, Q. Hart *et al.*, Development of a biorefinery optimized biofuel supply curve for the Western United States, *Biomass Bioenergy*, 2010, **34**(11), 1597–1607.
12. I. E. Grossmann and G. Guillén-Gosálbez, Scope for the application of mathematical programming techniques in the synthesis and planning of sustainable processes, *Comput. Chem. Eng.*, 2010, **34**(9), 1365–1376.
13. A. Azapagic, Life cycle assessment and its application to process selection, design and optimisation, *Chem. Eng. J.*, 1999, **73**(1), 1–21.
14. A. Azapagic and R. Clift, The application of life cycle assessment to process optimisation, *Comput. Chem. Eng.*, 1999, **23**(10), 1509–1526.
15. J. B. Dunn, F. Adom, N. Sather, J. Han and S. Snyder, *Life-cycle Analysis of Bioproducts and Their Conventional Counterparts in GREET*, Argonne National Laboratory: ANL/ESD-14/9;2014.

16. F. You, L. Tao, D. J. Graziano and S. W. Snyder, Optimal design of sustainable cellulosic biofuel supply chains: Multiobjective optimization coupled with life cycle assessment and input–output analysis, *AIChE J.*, 2012, **58**(4), 1157–1180.
17. F. You and I. E. Grossmann, Stochastic inventory management for tactical process planning under uncertainties: MINLP models and algorithms, *AIChE J.*, 2011, **57**(5), 1250–1277.
18. R. E. Rosenthal, *GAMS – A User's Guide*, Washington, DC, USA, GAMS Development Corporation, 2014.
19. USDA, Field Crops Usual Planting and Harvesting Dates, United States Department of Agriculture, National Agricultural Statistics Service, 2010, Agricultural Handbook Number 628.
20. Oregon Deparment of Energy, *Biomass Resource Assessment and Utilization Options for Three Counties in Eastern Oregon*, 2003.
21. A. A. Rentizelas, A. J. Tolis and I. P. Tatsiopoulos, Logistics issues of biomass: The storage problem and the multi-biomass supply chain, *Renewable Sustainable Energy Rev.*, 2009, **13**(4), 887–894.

CHAPTER 13

US Government Bioproducts Policy "Watch What We Do, Not What We Say"

ROBERT E. KOZAK[a,b]

[a] Advanced Biofuels USA, 507 North Bentz Street, Frederick, MD 21701, USA; [b] Atlantic Biomass Conversions, Inc., 507 North Bentz Street, Frederick, MD 21701, USA
Email: AtlanticBiomass@aol.com

"Watch what we do, not what we say"
US Attorney General John Mitchell, 1969.

Shortly after taking office, Richard Nixon's Attorney General made this statement to the press explaining how they should measure the policies and actions of the incoming Administration. While this statement became infamous during the Watergate investigation, it remains a very straightforward way to measure government policy—look past the speeches and press releases and instead examine the legislation, regulations, and spending policies that were or were not enacted to implement the pronouncements.

13.1 US Bioproducts Policy: Words, but No Deeds

In recent years, much has been written in the trade press, and even in mainstream media outlets, about the potential for a US bioproducts industry to become a substantial economic force that would also make a significant contribution to climate change mitigation. While much of this reporting

RSC Green Chemistry No. 43
Commercializing Biobased Products: Opportunities, Challenges, Benefits, and Risks
Edited by Seth W. Snyder

Published by the Royal Society of Chemistry, www.rsc.org

depended on press releases, major sources, reports commissioned by the US Department of Energy (DOE), were often written by US National Laboratories.[1] The imprimatur given to this topic by the involvement of the National Laboratories and the DOE resulted in a strong impression that bioproducts were actually an important policy priority of the United States government.

If such a policy existed, then a strong legislative and regulatory framework should exist as well. Following the pattern of other important US policy priorities—the reduction of ozone causing pollutants, the production of "tight oil," or the destruction of terrorist organizations—a policy framework would include such elements as enforceable standards, mandatory government funding, and favorable tax policies.

However, a search for the Congressional legislation or Executive Branch regulations that would have established such a US bioproducts policy found that neither Congress nor a Presidential Administration ever created the necessary framework.

Instead, the only enabling legislation passed by Congress to include anything on bioproducts is the series of US agriculture enabling legislation referred to as "Farm Bills". Beginning with the *Farm Security and Rural Investment Act of 2002*, Public Law 107–171 (and continuing up to the *Agricultural Act of 2014*, Public Law 113–79), something called *Section 9002-Federal Procurement of Biobased Products* was included.

While the title of this section sounds as if Congress is directing the US government to use its substantial buying power to create an "early adopters" market for bioproducts, that is not the case. Section 9002 continues to this day to lack any legally enforceable requirements for government purchases of biobased products. Nor did it establish legally enforceable requirements for the inclusion of biobased products in consumer goods. Instead, the language of section 9002 includes numerous methods for government agencies to avoid purchasing bioproducts.

> (b) PROCUREMENT SUBJECT TO OTHER LAW—Any procurement, by any Federal agency, which is subject to regulations of the Administrator [of the Environmental Protection Agency] under section 6002 of the Solid Waste Disposal Act (42 U.S.C. 6962), **shall not be subject to the requirements of this section to the extent that such requirements are inconsistent with such regulations.**
>
> (2) AGENCY FLEXIBILITY—Notwithstanding paragraph (1), an agency **may decide not to procure** such items if the agency determines that the items—
> (A) **are not reasonably available** within a reasonable period of time;
> (B) **fail to meet the performance standards set forth in the applicable specifications** or **fail to meet the reasonable performance standards of the procuring agencies**; or
> (C) **are available only at an unreasonable price.**

The clause "*fail to meet the reasonable performance standards of the procuring agencies*" is especially telling in its generality. Rather than requiring an agency to use product specifications to make purchasing decisions, it allows agencies to adopt their own performance standard that could rule out bioproducts in general.

Furthermore, while the argument can be made that flexibility is needed to implement government programs, compare the directive clause in (2) Agency Flexibility, "*may decide not to procure,*" with the directive clause "*shall make*" used in other sections of the Farm Bill when Congress wanted to make clear that an action was to happen.

> **SEC. 1116. PRICE LOSS COVERAGE.**
> Agricultural Act of 2014, Public Law 113–79
>
> (a) PRICE LOSS COVERAGE PAYMENTS.—If all of the producers on a farm make the election under subsection (a) of section 1115 to obtain price loss coverage or, subject to subsection (c)(1) of such section, are deemed to have made such election under subsection (c)(2) of such section, the **Secretary shall make price loss coverage payments** to producers on the farm on a covered commodity-by-covered-commodity basis if the Secretary determines that, for any of the 2014 through 2018 crop years—(1) the effective price for the covered commodity for the crop year; is less than (2) the reference price for the covered commodity for the crop year.

A further indication of the lack of priority given to bioproducts by Congress and Presidential Administrations is that the quantity or value of government bioproducts purchased under Section 9002 is not known by the US Government. The legislation did not include any mandatory reporting and the first effort by a President to require Section 9002 purchasing information was through a Presidential directive by Barack Obama in 2014. As of October 2014, according to Agriculture Secretary Tom Vilsack,[2] only the Department of Agriculture is preparing a report.

A similar approach of appearing to establish a bioproducts policy while also providing provisions to limit implementation is present in a change enacted to Section 9003, *Biorefinery Assistance,* of the *Agricultural Act of 2014.* The revised Section 9003, now called *Biorefinery, Renewable Chemical, Biobased Product Manufacturing,* extends the US biorefinery federal loan guarantee program to bioproducts production. However, additional new wording in Section 9003 capped the bioproducts portion of the loan program to 15 percent and allowed a lower percentage at the prerogative of the Secretary of Agriculture.

> (B) BIOBASED PRODUCT MANUFACTURING.—Of the total amount of funds made available for fiscal years 2014 and 2015 under subparagraph (A), the Secretary **may use for the cost of loan guarantees under this**

section not more than 15 percent of such funds to promote biobased product manufacturing.

The effect of this new provision has not yet been seen. However, given that the hardware and chemical/enzyme systems needed to convert biomass to bioproducts are more complex than those needed to produce ethanol or biodiesel, the maximum of 15 percent of funds available for all biorefinery-related loan guarantees will most likely not cover the costs of bringing innovative bioproducts systems on line.

In addition, given the lack of industrial funding by financial institutions since the 2008–2009 bank bailout, it is hard to see the 15 percent investment cap bringing much private funding to innovative bioproduct production (see Chapter 15 for discussion of financing).

13.2 Why Is There a Lack of US Bioproduct Policy?

In the US, putting a new legislative and regulatory framework that "has teeth" in place has become increasingly difficult because of the entrenched nature of those interest groups and industries benefitting from existing programs. Two approaches are generally used to overcome these roadblocks:

- Assemble coalitions to exert simultaneous pressure on legislative and executive branches. The Iron Triangle approach.
- Attach a new product or policy to existing legislation or regulations, often in a non-related area. The Accidental Policy approach.

In the case of bioproducts, the emerging bioproducts industry is not large enough to fund the type of coalition needed to make an Iron Triangle approach effective. In addition, this industry has not yet identified a high priority (and funded) US program where bioproducts could make a critical contribution.

13.2.1 In Washington, Size Matters

A widely used term to describe how the federal government works in Washington is Iron Triangle (Figure 13.1).

Iron Triangles exist in many different policy areas: national defense, banking, petroleum, and health care, to name a few. As shown in Figure 13.1, startup and maintenance costs of Iron Triangles run high. For example, PACS and "527" fundraising apparatuses were substantial contributors to the multimillion dollar campaign budgets of most 2014 Congressional races.

At present, the US bioproducts industry, and for that matter the US biochemical industry, simply are not large enough to assemble an Iron Triangle on their own behalf.

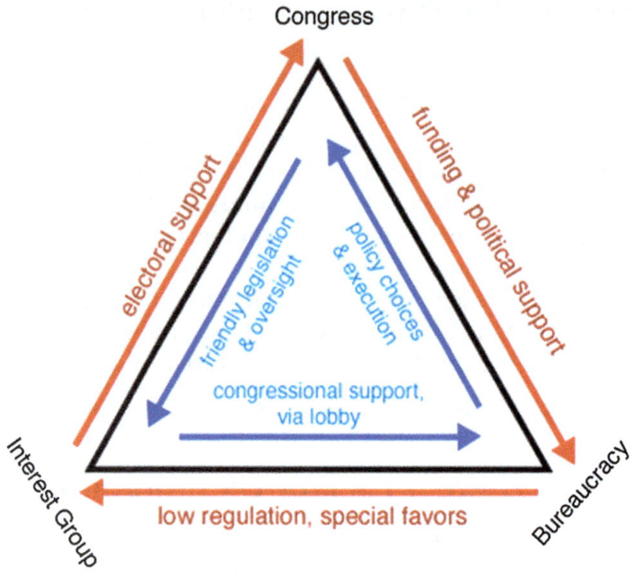

Figure 13.1 The closed, mutually supportive relationships that often prevail in the United States between the government agencies, the special interest lobbying organizations, and the legislative committees or subcommittees with jurisdiction over a particular functional area of government policy.[3]
Diagram from Wikipedia entry on Iron Triangle.

13.2.2 US Biochemical Production: A New, Small Industry

In recent years, the petroleum industry in the US has divided into two largely separate sectors.

- Upstream: large petroleum exploration and crude production companies
- Downstream: small fuel and biochemical refining companies

This separation came about through the sales of many refining and biochemical assets that were seen as being low-margin operations by their large international parent companies, ExxonMobil, BP, Shell, *et al.* As stated in a 2012 Congressional Research Service (CRS) report:

> Although the five major oil companies are integrated firms, the majority of their earnings come from exploration and production activities. For example, in 2011, ExxonMobil earned about 84% of its corporate profits from upstream activities. Chevron earned 92%, and ConocoPhillips earned 66% from upstream activities in 2011.
>
> Some evidence of transition in the U.S. market has been observed. ConocoPhillips announced in 2010 a decision to split into two independent companies, ConocoPhillips, an upstream company, and

Table 13.1 US employment by 2011 US NAICS employment code.

	Total US employment	113 425 965	
NAICS code	Industry name	Number of employees	% US Employment
11	Agriculture	156 520	0.14
3361	Motor vehicle production	136 676	0.12
211	Oil and gas extraction	118 959	0.10
32 511	Petrochemicals	10 398	0.01
325 193	Ethyl alcohol	10 299	0.01

Phillips 66, a downstream company. The company also plans to either sell or close its refinery in Trainer, PA. Sunoco, an independent refining and marketing corporation, has left the refining sector, to concentrate on logistics and marketing, closing and attempting to sell its two refineries in the Philadelphia area.[4]

The resulting US biochemical industry, which is part of the refining industry, is quite small.

As shown in Table 13.1, which uses 2011 (the latest available) US Census North American Industry Classification System (NAICS) employment codes as a data source, the petrochemical industry employs approximately 10 000 people. By comparison, the farming, motor vehicle, and oil/gas extraction industries each employ over 100 000 people or more than 10 times those in the petrochemical industry. In addition, ethanol production industry employment equals that of petrochemical production.

In addition, while the assumption may be made that the interests of the downstream petroleum industry will be represented by the upstream "majors", that is often not the case because their interests are not necessarily mutual.

The 2014 Congressional debate over expanding US oil exports provides an example. Upstream petroleum companies are pushing strongly to increase US exports and their income, while four downstream refiners have formed a group asking Congress to retain current restrictions so that US oil supplies will remain relatively plentiful and lower priced than in the world market.[5] (US West Texas crude averaged about 7 percent less than Brent crude in 2014.[6])

As of 2014, the bioproducts portion of the US biochemical industry is miniscule. Therefore, even if they were able to ally with the petroleum-based portion of the industry, it is hard to see them being able to close the loopholes in Sections 2002 and 2003 of the *Agricultural Act of 2014* or to enact other enforceable legislation through the creation and maintenance of a viable Iron Triangle.

13.2.3 Accidental Policy in the United States

More often than most people in the US policy establishment would admit, significant policy decisions have been a product of an accidental confluence of actions not directly related to the final policy outcome. US ethanol policy is a prime example of this type of policy making.

13.2.3.1 US Ethanol Policy Resulted from 1990 Clean Air Act Ozone Compliance

The US has the world's largest market for a biofuel (10 percent of all gasoline fuel mixtures are biobased ethanol) and is generally seen as a leader in biomass utilization.

However, the nationwide use of 10 percent ethanol did not come about because of an energy or climate change policy enacted by either Congressional legislation or executive branch regulation. In fact, the inclusion of biobased ethanol in gasoline nationwide was a case of a struggling emerging industry being given the market because: (1) a cancer-causing element of a petroleum competitor was found in water supplies; and (2) ethanol had a key characteristic that would save petroleum refiners money. It was much more accidental than planned.

In the late 1990s, the US Environmental Protection Agency (EPA) discovered that the MTBE (methyl tertiary butyl ether), used as an oxygenate to comply with 1990 Clean Air Act ozone reduction requirements, was a persistent carcinogen that was leaking from underground gasoline storage tanks into natural water systems. Citizen groups nationwide pushed for action and New York and California passed legislation to ban MTBE beginning in 2004.

Midwestern cornstarch ethanol producers that already supplied ethanol as an oxygenate in their home states were more than happy to expand production to replace MTBE in ozone non-attainment regions nationwide. Ethanol plants appeared almost overnight throughout Iowa, Minnesota and other states. By 2007 ethanol (at 10 percent total volume) had not only replaced MTBE in ozone control areas but had become a national *de facto* standard because of its high octane properties.

The 100+ octane anti-knock rating of ethanol, much higher than that of MTBE or similar compounds, allowed petroleum refiners to lower the octane of the gasoline base blend from 87 (US regular grade octane) to 84 and use the 10 percent ethanol to increase the octane. This resulted in a considerable production cost saving for petroleum refiners since the change from 84 to 87 was at the start of the logarithmic portion of the production cost curve.

It should be noted that this refining cost decision was made when the major oil companies owned most of the refining assets in the US. As the previously cited 2012 Congressional Research Service (CRS) report indicated, the major oil companies now focus on total oil sales rather than on maximizing the value of refinery output; thus, their position has changed to oppose increased ethanol content in gasoline or any substitution of biobased fuels for petroleum-based ones as evidenced by their promoting the repeal of the Renewable Fuel Standard (RFS) of the Energy Independence and Security Act of 2007.[7]

13.3 Could a Similar Approach Be Used for Bioproducts?

Could a similar set of circumstances as ethanol experienced come together for bioproducts? That is doubtful given the relatively small GHG and other environmental impacts that replacement of petroleum sources for bioproducts with biobased sources would have.

13.3.1 Cost Effectiveness of Using Bioproducts to Reduce GHG Impacts

As shown in Figure 13.2, Tables 13.2 and 13.3, the portion of US oil and natural gas used by the US petrochemical industry is relatively low. In 2010, the latest year for complete US EIA data:

- Approximately 7.4% of US petroleum production was used for petrochemicals (2.8% as feedstock and 4.9% as products).
- About 4.2% of natural gas was consumed by the chemical industry as non-heating feedstock.

By comparison:

- Transportation fuels accounted for over 74% of petroleum use (10 times the amount of biochemical feedstocks).
- Gasoline alone consumes about 46% of US petroleum use.

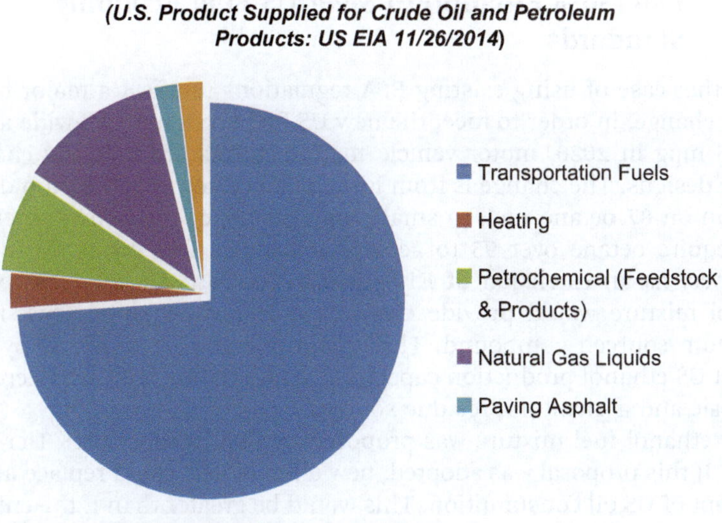

Figure 13.2 US Petroleum Use by Sector: 2010.

Table 13.2 Sources for Figure 13.2. US product supplied for crude oil and petroleum products: US EIA 11/26/2014.

1. Transportation fuels	74.2%
1.a. Gasoline	(46%)
2. Heating	2.8%
3. Total petrochemical	7.7%
3.a. Petrochemical feedstocks	(2.8%)
3.b. Petrochemical products	(4.9%)
4. Natural gas liquids	11.8%
5. Paving asphalt	1.9%
6. Coking	2.0%
Total	100.3%

Table 13.3 2010 US natural gas consumption (US DOE/EIA, March 2013 MECS report, Table 2.1).

Total US NG 2010 consumption (million ft^3)	24 477 425
2010 Chemical consumption (million ft^3)	1 026 000
Chemical % total	4.2%

The cost-effectiveness argument for converting the entire biochemical industry to biomass sources (maximum 7.7 percent of petroleum use), which would require billions in research and construction costs, also has weaknesses when compared to other available biofuel strategies—especially ones that could increase the use of fuels with excess installed production capacity.

13.3.2 A Competing GHG Reduction Approach: Using E30 (30% Ethanol) to Meet US Fuel Economy Standards

In another case of using existing EPA regulations to affect a major technological change, in order to meet the new US fuel economy fleet-wide average of 54.5 mpg in 2020, motor vehicle manufacturers are radically changing engine designs. The change is from low-efficiency, low-compression designs that run on 87 octane fuel, to small, high efficiency turbocharged engines that require octane over 95 to achieve maximum fuel economy and performance. As an extension of E10's use to increase octane, a 30 percent ethanol mixture would provide the octane at a much lower cost than a petroleum sourced compound. This quantity could be provided by using current US ethanol production capacity in combination with near-zero GHG cellulosic and agricultural residue sources.

This ethanol fuel mixture was proposed by EPA in their 2013 Tier 3 proposal.[8] If this proposal was adopted, new ethanol use would replace another 9 percent of US oil consumption. This would be greater than if the entire US chemical industry, 7.7 percent, were to switch to biobased sources.

13.4 Possible Strategies to Develop a US Bioproducts Policy Framework

Since the US bioproducts industry has a difficult case to make when it comes to creating new jobs or providing a cost-effective GHG mitigation strategy as a reason to invest resources in its development, a possible alternative approach would be to use national security or national strategic resources as a way to develop bioproducts production systems.

An example of this is the rubber industry. Portions of the industry and some European governments are investing in natural rubber crops and production systems that could function outside of the narrow band of East Asian geography where all natural rubber currently comes from.[9] While the geopolitical reasons for this research are self-evident, the technical reason is that despite over seventy years of research, only natural rubber is able to provide the performance characteristics needed in aircraft and high-performance motor vehicle tires.

Another example is the decision by the US Navy to use biofuels for Navy jets and non-nuclear surface vessels (The Great Green Fleet). The goals established by US Navy Secretary Ray Mabus, Table 13.4, and the solicitations for fuel purchase beginning in 2014[10] have almost singlehandedly pushed bio-jetfuel into commercial production. This US Navy project is a classic example of using the substantial buying power of the US Government to create an "early adopters" market for bioproducts.

13.5 Conclusion: "Watching What They Do, Not What They Say"

By looking past the speeches, commissioned reports, and press releases and instead examining the applicable legislation, it is clear that the US government has not created the legislative, regulatory, or spending frameworks necessary to implement the bioproducts policy pronouncements made in those documents.

Furthermore, the divestment of downstream refining and biochemical assets by the petroleum majors has left the US biochemical sector without the political and financial assets to create an effective bioproducts policy on their own.

Unless the supporters of bioproducts in and out of the industry pursue creative approaches, including identifying bioproducts production

Table 13.4 Biofuel goals of the US Navy (http://greenfleet.dodlive.mil/energy/great-green-fleet).

Year	Biofuel consumption (million gallons)
2016	3 360 000
2020	336 000 000

technologies or specific products that could fulfill high priority national security needs, it is doubtful that any significant change in US bioproducts policy or significant growth in the US bioproducts industry will occur in the near future.

References

1. For example: (a) *Top Value Added Chemicals From Biomass*, Produced by Staff at the Pacific Northwest National Laboratory (PNNL) and the National Renewable Energy Laboratory (NREL), August 2004; (b) *Low Cost Carbon Fiber Overview*, 9 May 2011, C. David (Dave) Warren, Oak Ridge National Laboratory; (c) *Biological Conversion of Sugars to Hydrocarbons Technology Pathway*, Ryan Davis, Mary Biddy, Eric Tan, and Ling Tao National Renewable Energy Laboratory, Susanne Jones Pacific Northwest National Laboratory, Technical Report NREL/TP-5100-58054, PNNL-22318 March 2013, Prepared for the U.S. Department of Energy Bioenergy Technologies Office; (d) *Renewable, Low-Cost Carbon Fiber for Lightweight Vehicles*, Summary Report from the June 4–5, 2013, Workshop, Detroit, Workshop and Summary Report sponsored by the U.S. Department of Energy, Office of Energy Efficiency and Renewable Energy Bioenergy Technologies Office.
2. Speech by US Agriculture Secretary Tom Vilsack at Bioproducts World Conference, October 6, 2014, Columbus, Ohio.
3. A Glossary of Political Economy Terms, Paul M. Johnson, Auburn University, copyright © 1994–2005.
4. *Financial Performance of the Major Oil Companies, 2007-2011*, Robert Pirog, February 17, 2012, US Congressional Research Service, 7-5700, R42364.
5. Poltico's Morning Energy, Alex Guillén, December 9, 2014.
6. Average of November 2013 to November 2014. 52-Week high and low prices for Brent and West Texas crude were $88.94 (7.4% higher) and $82.39, respectively.
7. *Congress Should Repeal Unworkable Renewable Fuels Standard*, American Petroleum Institute, November 27, 2012 http://www.api.org/news-and-media/news/newsitems/2012/nov-2012/congress-should-repeal-unworkable-renewable-fuels-standard.aspx.
8. ENVIRONMENTAL PROTECTION AGENCY, 40 CFR Parts 79, 80, 85, 86, 600, 1036, 1037, 1065, and 1066, [EPA–HQ–OAR–2011–0135; FRL_XXXX-X],RIN 2060-AQ86, Control of Air Pollution from Motor Vehicles: Tier 3 Motor Vehicle Emission and Fuel Standards.
9. *Tire Makers Race to Turn Dandelions into Rubber*, Ludwig Burger, Aug 20, 2014, Reuters.
10. Solicitation Number: SP060014R0061, Agency: Defense Logistics Agency, Office: DLA Acquisition Locations.

CHAPTER 14

Study on Investment Climate in Bio-based Industries in the Netherlands†

LARA DAMMER* AND MICHAEL CARUS

nova-Institute for Ecology and Innovation, Chemiepark Knapsack, Industriestraße 300, 50354 Hürth, Germany
*Email: lara.dammer@nova institut.de

14.1 Introduction and Executive Summary

This chapter presents the results of a short study that investigated the barriers faced by small companies active in bio-based economy when they want to acquire investment for their businesses. The study was conducted from January to May 2014 in the form of interviews with start-ups or other SMEs in the Netherlands, Germany, France and Belgium and was complemented by selected literature. The focus was exclusively on bio-based chemicals and materials, not on food, feed or energy produced from biomass.

The objective of this study was to assess the investment climate for bio-based industries in the Netherlands in comparison to other countries. The main research questions were: which reasons move investors to provide money for bio-based entrepreneurs? What are the advantages of the Netherlands as a location for bio-based industries? Which hurdles might

†Adapted from a report written for: Ministry of Economic Affairs, Directorate-General for Enterprise and Innovation, Biobased Economy Department, P.O. Box 20401, 2500 EK Den Haag, The Netherlands.

RSC Green Chemistry No. 43
Commercializing Biobased Products: Opportunities, Challenges, Benefits, and Risks
Edited by Seth W. Snyder

Published by the Royal Society of Chemistry, www.rsc.org

prevent investors from placing their funds in the Netherlands and which conditions make other locations potentially more attractive for investing?

The interviews highlighted several issues that make investors insecure about the new bio-based industries and therefore constitute barriers for companies that need to acquire funding for their businesses. Barriers were found in the structure of public funding programmes, the lack of will of banks to take over risk financing, lack of tax incentives for innovation, missing standard of bio-based products, labelling, GMO (genetically modified organisms) regulations and infrastructure.

The last parts address the strengths and weaknesses of the Netherlands as a location for bio-based industries, the level playing field and then derive recommendations, mainly focussing on political will and clear measures that can support the infrastructure of bio-based industries, create market demand and give security to investors.

14.2 Objective and Methodology

Lack of innovation investment is currently a widely debated topic in Europe. Many stakeholders observe that research and development is often carried out and supported with large public funding in Europe, but that the next step of pilot facilities or commercialization is realized more and more in the Americas or Asia. The graph in Figure 14.1 underlines this finding for the bio-based polymers sector, shifting more and more to South America and Asia.

This study was carried out to assess the investment climate for bio-based industries in the Netherlands in comparison to other countries. It investigated the reasons that move investors to provide money for bio-based entrepreneurs, the advantages of the Netherlands as a location for biobased industries, the hurdles that might prevent investors from placing their funds in the Netherlands and the conditions that make other locations potentially more attractive for investing.

In order to get a more comprehensive picture and due to the political realities of the European Union, it was not only industries in the Netherlands that were examined, but also companies from surrounding countries: Germany, Belgium and France. The perspective during the interviews was global, taking into account also developments in Asia or the Americas.

In order to achieve the objective of the study, we conducted interviews with companies active in the bio-based economy. The main focus was on start-up companies and other small or medium enterprises with a view on expansion, since they are in regular exchange with potential investors and know the reasons pro and contra investment in the focus regions very well.

Thirteen interviews were conducted in total, five with companies in the Netherlands, five in Germany, two in France and one in Belgium. The interviews were held in personal meetings, *via* phone or communication software such as Skype or go-to-meeting. The guideline underlying the talks was sent out to most interview partners before the talk in order to enable optimum preparation. The questions covered topics from the general

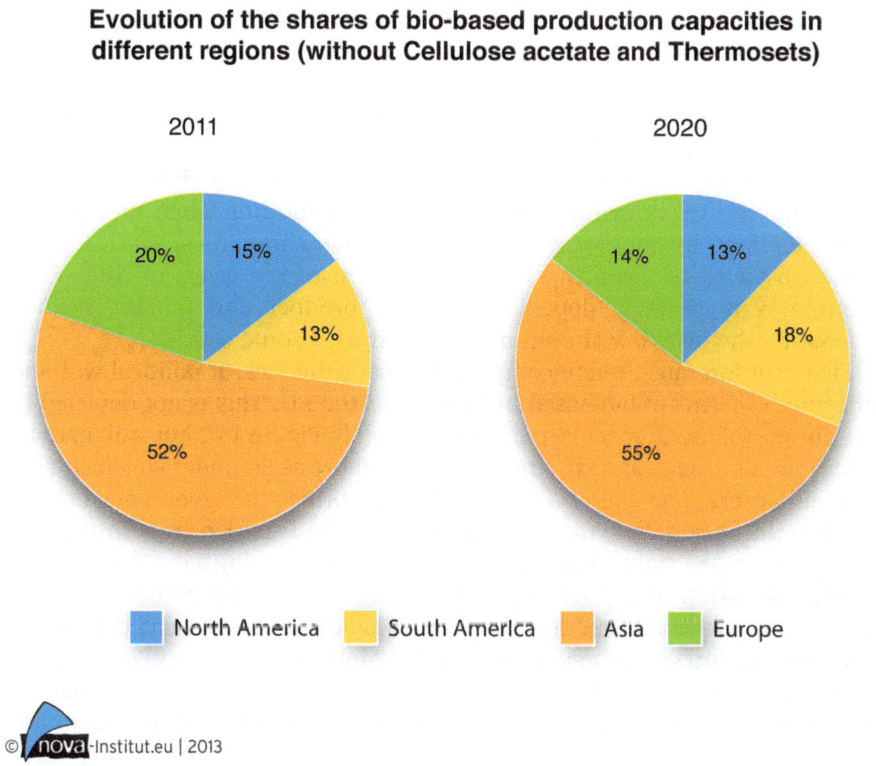

Figure 14.1 Evolution of the shares of bio-based polymer production capacities in different regions.

background of the company, their current size and employees, their product portfolio and manufacturing infrastructure to their experiences with investors and what hinders or enables investment in their bio-based business activities. The questions were posed with a focus on those circumstances that can be addressed by policy-makers.

In addition to the results of the interviews, some insights into the structures of the Dutch bio-based economy from previous projects were used to shed light on the advantages and disadvantages of the Netherlands as a location for bio-based production. This made it possible to narrow down the broad issue of investment climate and to derive concrete recommendations at the end of the report.

14.3 Investment Climate and Barriers for Investment

The investment climate experienced by bio-based start-ups varied to some degrees, mostly depending on the product, but several impressions were shared by a large majority of interview partners. Interestingly, there were no significant differences in general statements between the four selected countries. The market and investment in bio-based chemicals and materials is at least a European issue, and often even global.

According to the statements of the interview partners, small companies active in bio-based economy in Europe face several issues when acquiring funding. The graph in Figure 14.2 depicts the frequency with which the different topics were mentioned during the interviews. It should be noted that some of the companies already had partial funding when they started their bio-based business (*e.g.* from selling former companies), so they faced fewer obstacles in acquiring their funding. Also, the levels of difficulties seem to vary strongly depending on the product and market sectors. Therefore, the overview should be perceived with some caution.

First and foremost, many companies lament the lack of political will and direction in terms of bio-based industries for the EU. This is not depicted in the number of mentioned barriers in the graph in Figure 14.2 but still matters, since it was not seen as a specific hurdle, but more as an underlying factor for several barriers. There are no clear goals or priorities that guarantee a stable market, thus not giving any security for investors. Experiences with the biofuels policy make stakeholders even more cautious, fearing a quick change in political direction once public opinion changes or new scientific findings are made. This makes Europe a less attractive place for investment than more pragmatic countries such as the US, Brazil, Malaysia, *etc.*

Another general tenor found in most interviews was the fact that an abundance of public funding is available for research and development, but

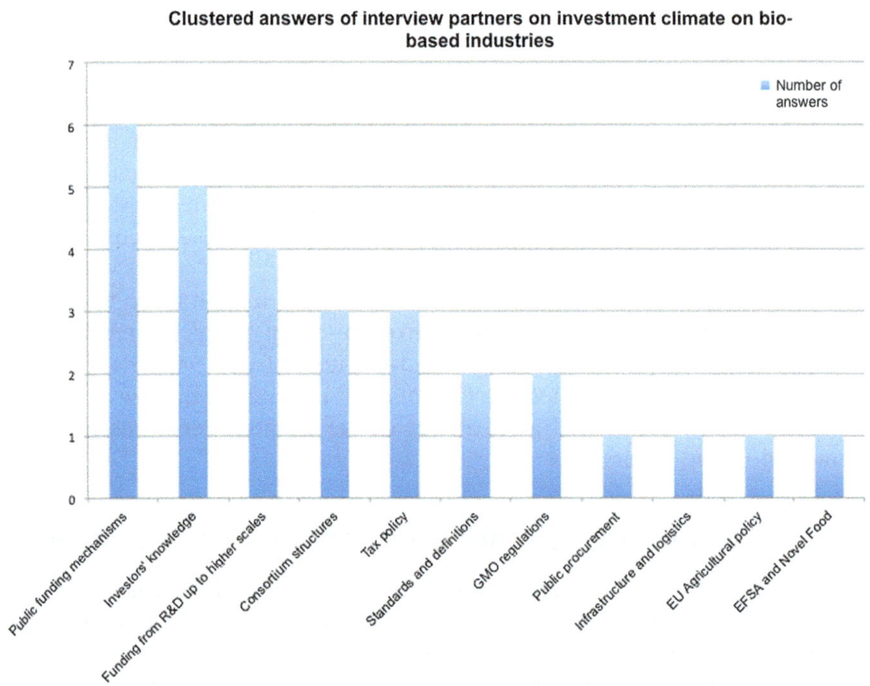

Figure 14.2 Clustered answers of interview partners—relative weight of barriers.

almost none for the later stages. Some funds exist for pilot and demonstration scale, but the closer a product gets to commercialization, the less support is available, even though it would be especially necessary to cross the "valley of death". Some more concrete insights into the problems with public funding structures can be found later.

This presents a clustered summary of the specific hurdles faced by the companies interviewed (for all see Figure 14.2).

14.3.1 Definitions and Standards

Some interview partners stated it was a barrier for the development of a bio-based market in Europe that there are not enough definitions and standards in place for the new business sector. It was stated that there is no definition of what constitutes a bio-based product, that the term "bio-based" is not protected, that many logos and claims confuse the consumers and that agencies take a long time to update definitions and classify new products.

These claims hold up only partially to a "reality check". In fact, a dedicated Technical Committee within the European standardization body CEN is working on bio-based products (CEN/TC411).[1] They have defined a set of relevant terms in standard (EN 16575 Bio-based products[1]) that was adopted in September 2014:

Bio-based: derived from biomass

Biomass: material of biological origin excluding material embedded in geological formations and/or fossilized

- Biomass can have undergone physical, chemical or biological treatment(s).
- The correct spelling of 'bio-based' is with a hyphen ('-'). It is, however, in common usage sometimes spelt without a hyphen.
- The methods to determine and communicate "bio-based" as a characteristic are detailed in specific standards of CEN/TC 411.
- The commonly used biomass, also called bio-based resources, is starch, sugar, vegetable oils, (hemi)cellulose (timber, natural fibres, straw and other by-products) and special biomolecules such as lignin or natural rubber.

Bio-based product: product wholly or partly derived from biomass

- The bio-based product is normally characterised by the bio-based carbon content or the bio-based content. For the determination and declaration of the bio-based content and the bio-based carbon content, see the relevant standards of CEN/TC 411.
- Product can be an intermediate, material, semifinished or final product. "Bio-based product" is often used to refer to a product which is partly bio-based. In those cases the claim should be accompanied by a quantification of the bio-based content.

Figure 14.3 DinCertco and Vinçotte bio-based labels, indicating a range of shares of bio-based carbon.

The work on standardizing bio-based content measurement in Europe is on-going, so in that regard the claim of missing standards is factually correct. However, in terms of practical application, the ASTM standard D6866 Test Methods to Determine the Bio-based Carbon Content of Materials Using Radiocarbon and Isotope Ratio Mass Spectrometry is already in usage in Europe. Two certifiers, DinCertco and Vinçotte have introduced "bio-based" labels, indicating a range of shares of bio-based carbon. These labels are already widely used and globally accepted (Figure 14.3).

This means that there are already a few helpful definitions and labels in place, but they are not established and well-known enough by players in the sector.

14.3.2 Knowledge and Education

Several interview partners quoted a lack of knowledge about the new business sector of bio-based chemicals and materials within the investor scene as an important barrier to acquiring funding. Since many banks and investment funds do not know enough about the new products and market dynamics, they cannot make an informed decision and instead often "go with their guts", which is mostly a decision against too much unknown risk. One interviewee compared the circumstances of industrial biotechnology to the pharmaceutical sector and explained that in the latter, a whole range of specialized and knowledgeable potential investors can be addressed with ideas of new projects, while this is not the case for bio-based chemicals or materials. More and more emerging family offices are positive exceptions and seem to counter this situation to some degree, but the general problem still exists. France was mentioned to possess a slightly livelier investor scene for industrial biotechnology.

Also, consumers' perceptions were mentioned repeatedly and interview partners stressed the importance of continued awareness about environmental issues, such as waste reduction, resource efficiency, saving energy and protecting the climate, many goals that can be supported by increasingly using bio-based products. This matches the necessity for a clear prioritization of bio-based materials by the policy makers mentioned earlier, in order to be clear about the reasons for deciding in favour of bio-based materials and to improve their visibility and image on the market.

14.3.3 Infrastructure

For many small companies and start-ups it would be a tremendous help if they could start their business in an existing infrastructure so they could focus their attention exclusively on developing their product and bringing it to the market instead of setting up facilities, complying with waste regulations, *etc.* One interview partner who set up their pilot facilities in Europe, but then established their first commercial plant in Malaysia told us that among others, the support by the Malaysian government in terms of infrastructure was one of the main reasons for this decision. They had the possibility to open their facilities at a dedicated bio-hub, thus relieving them of many arduous tasks and decisions that are related to these steps.

Also for Europe, it seems to be a good opportunity to bring new life to industrial sites such as chemical parks by equipping them with specialized features for bio-based industries. Supply routes for biomass as raw materials, special pipelines for bio-based intermediates or waste removal systems are factors that can make one location very attractive for the settling of new companies. Modern energy systems would ensure that the companies' energy consumption was efficient, thus saving money and contributing to a sustainable image. It would be a win-win situation to support small, innovative companies while at the same time reviving the industrial infrastructure of Europe and keeping jobs, which is also in line with the proclaimed goal of the Commission to strengthen Europe's producing industry again so that it will contribute 20% to GDP by 2020 (European Commission 2014).[2]

14.3.4 Public Procurement

European public authorities spend almost 2000 billion Euros on goods and services every year, which means that public procurement can be a powerful tool for creating market pull, also for innovative bio-based products. Interview partners noted, however, that at the moment it is not an effective tool and that even though sustainability and environmental factors are to be considered under the Green Public Procurement of the EU, often buyers in the end decide just for the most economical option.

It is true that at the moment, bio-based products are barely covered by public procurement, but there is movement to change this. It should be noted that the property of being bio-based is not an advantage *per se* and does not automatically mean that the product is "green". Furthermore, most people in charge of public procurement do not know about the state-of-the-art of bio-based products, therefore they do not consider these options in their buying decisions.

The BioPreferred® program of the USDA is a very pragmatic example of how public authorities can promote bio-based products. Depending on the product group, products need to have a minimum share of bio-based carbon content, certified according to the already mentioned ASTM standard D6866. Then they are placed on an online product list where they can be found by public authorities in charge of procurement. Also, there is a directive that these products should be treated preferentially, thus giving legal certainty to decision makers. The US policy is discussed in Chapter 13.

The situation is a bit more complicated in Europe. Currently, there are several initiatives, mostly aiming at improving knowledge about bio-based products and how they are compatible with procurement law. On the European level, public procurement is covered in a working group of the Expert Group on Bio-based Products and also in a Horizon2020 call that aims at building procurement networks for innovative bio-based products (European Commission 2013).[2] There are already a multitude of national and regional platforms that support sustainable procurement, and some of them have dedicated information on bio-based products, too. The FP7 project Open-Bio has collected these product information platforms and found 25, mostly in Europe, but also elsewhere in the world (publication forthcoming on www.open-bio.eu).

If these initiatives manage to make public procurement officials really aware of the possibilities of bio-based products and to give them legal certainty for their buying decisions, then public procurement might become a truly effective tool to create market demand for bio-based materials.

14.3.5 Public Funding Structures

Several public funding structures were mentioned very positively by the interview partners, for example the GO and GROWTH programmes in the Netherlands, KfW or FNR programmes in Germany and support by the Flemish Agency for Entrepreneurs in Belgium. However, complaints were also made about the general structures and some inefficiencies. As stated earlier, one structural issue was the fact that a lot of funding is available for R&D, but almost none for up-scaling to pilot and demonstration, or even to commercial levels.

Furthermore, many funding programmes make it (almost) obligatory to have universities or big companies as partners, or to form big consortiums in general. This makes the acquisition process very lengthy, research becomes more inflexible and small companies are often worried about their intellectual property, too.

Even without the necessity to form big consortiums, many application processes are very complicated and bureaucratic and the approval takes a long time. One company stated that by the time their applications were processed, the direction of their research had changed significantly and their previous proposal was not applicable anymore. Flexibility is one of the most crucial advantages of small companies, and that is taken away when public funding is slow. Also, in many cases, small companies do not have the resources to bridge a longer amount of time when expected funding is delayed due to administrative complexities.

Another problem mentioned were the "trends" of public funding which force many companies to do research on subjects where funding is available instead of what they originally wanted to do research on or what makes sense on the market (a current example would be the hype on second-generation lignocellulosic feedstocks).

14.3.6 Tax Policy

There was wide agreement that tax policy in Europe does not give sufficient support to innovation in small companies. First of all, tax incentives are only advantageous for larger profitable companies, because company taxes only apply to profits. So a tax reduction is not useful for start-ups with losses only due to investment in their technology, but interesting for bigger companies only. Someone stated that tax legislation and loss carryforward are unnecessary administrative burdens.

Also, the tax structure does not make it attractive to take any risks with investments, since one has to pay as much taxes on return on investment (ROI) from innovation as one has to do on ROI from established technologies. The recent report on biotechnology in Germany by Ernst & Young has found that only very limited risk capital is available for this sector, despite Germany's excellent competitive situation in the world market (World Economic Forum 2014).[3] A press release by Ernst & Young[4] therefore demands a reform of tax policy for innovation and supports a concept by two entrepreneurs who suggest that private investors should be allowed to invest one per cent of their assets into biotechnology without paying any capital gains tax.

Another concept would be to directly give tax cuts to companies that establish a new innovative bio-based business. Malaysia, for example, offers several incentives for investment in various industries for the biotechnology sector, among others. Companies active in biotechnology that have been accredited with the BioNexus status by the Malaysian government are eligible for measures such as 100% exemption from statutory income tax for a duration of 10 years, a concessionary tax rate of 20% on statutory income from qualifying activities for 10 years upon the expiry of the tax exemption period, exemptions from taxes on dividends distributed by the company, double deduction on expenditure incurred for R&D *etc.* (MIDA 2012).[5]

14.3.7 Other

Several smaller aspects were mentioned during the interviews, which shall be summarized here. Since they were mostly quoted only by one or two participants, the urgency of these factors seems to be less than for the ones in the previous sub-chapters, which were all brought up by several interviewees.

GMO regulations, which are important for numerous biotechnological processes, are seen to be too strict and out-dated by a few market players. The regulations are written with a view on modified crops, but do not take into consideration the use of GMO in processes. The use of GM bacteria is even further complicated by the fact that the regulations are different in the Member States. For example, the use of plasmid bacteria is allowed in Belgium, but not in the Netherlands.

European legal procedures in terms of permits *etc.* are seen as too complicated, similar to the public funding structure. One company told us that even though they had acquired a pledge for funding, the legal procedures that were necessary to approve their production process before really obtaining the money took so long that they almost lost the funds and were only able to survive because of short-term (and quite expensive) bridge financing. They are now moving more to the US where permits and similar documents take much less time.

Feedstock security was not stressed as a strong barrier in Europe, but one interview partner mentioned it and especially lamented the protectionist agricultural policy of the EU, especially the sugar quota that impedes feedstock availability. Since the latest CAP reform abolished the quotas, the situation will probably change in 2017; however, it is not yet clear how that will affect prices and supply structures.

Some policies concerning valorisation of biomass are not harmonized among Member States, or are changed quite quickly (*e.g.* the EU biofuels policy). Another example is that in some countries, it is allowed to process glycerol from Cat. I fats into biogas and then wastes into fertilizers, but in others it is not.

Lastly, Novel Food and EFSA regulations are a huge problem for companies that make new food products. However, this does not concern the materials sector very strongly (*e.g.* cosmetics made from oil extracts), but more the production of food supplements.

14.4 Strengths and Weaknesses of the Netherlands as a Location for Bio-based Economy

In a previous study, we investigated the position of the Dutch bio-based economy on the global market, coming to the conclusion that the

Netherlands is the location of many top players in the sectors of bio-based polymers and plastics, building blocks and composites (see Dammer *et al.* 2013,[6] report for NL Agency, reference number 52 202). For that purpose, we interviewed several experts who have been active in the Dutch bio-based economy for many years.[7] They assessed the strengths and weaknesses of the Netherlands with the following conclusions that will—in combination with the results of the interviews—be the basis for the recommendations in the next chapter:

Strengths of Dutch economy for bio-based materials and chemicals

Several of the interviewees in this previous project named the presence of a large and state-of-the-art chemical industry infrastructure in the Netherlands as one of the key advantages. In the vision paper of the VNCI, the search for alternative feedstocks is mentioned as an element of further developing the chemical industry in the Netherlands. However, some of the experts also pointed at the circumstance that often, existing industries have vested interests in keeping up the status quo and are not very willing for new developments, which also mean new competitors. One expert correspondingly criticized the VNCI report for not expressing real interest in true innovation, but focusing more on making existing things better. So, an established chemical industry can present advantages, but the motivation for changing to bio-based feedstocks needs to be strong in order to create change.

On the other hand, there are already a multitude of bio-based industries in the Netherlands (agriculture, food industry, horticulture, paper industry) that are looking for new outlets in order to diversify and develop. Here, new bio-based materials and chemicals are very attractive options, since the value added created from biomass is much higher for these products than for energy products. In the Netherlands, there is great potential for a stronger cooperation between agricultural, chemicals and plastic industries, which are all already existing and for which also the know-how is there. Competencies that were not associated before (like chemistry and agriculture) now need to become strongly associated to enhance innovative power and subsequent economic activity.

The knowledge base is further improved by excellent university programmes and top-notch research, which was stressed by all experts.

Furthermore, the Netherlands is a trading hub and thus possesses an excellent logistics network via sea, air, road, rail and rivers, which provides access to all European markets. Especially for biorefineries, this could be a decisive advantage, since they need to be built where biomass is easily available, which is often next to ports. Especially the port of Rotterdam as the most important trading port of Europe therefore gives a crucial asset to the Dutch bioeconomy.

Access to aquaculture might be an important advantage for the development of 3rd generation feedstocks, too.

But also "soft" factors have come up during discussions. It was mentioned that in the Netherlands, there is a culture of cooperation between government, industry, research institutions, NGOs and the public. This also leads to a very strong awareness of the necessity of improving the public's knowledge and perception of the bio-based economy. The qualitative study "My 2030s", which was carried out at four different locations in the Netherlands, for example, found that consumers do not have a clear definition of the term "bio-based" in mind, but mostly associate good things with it. Even though the study also found that many consumers do not really know much about what is going on in bio-based production and development, the existence of such a study shows that the Netherlands is one of the pioneers in this field. Also the willingness of government and public authorities to consult with expert knowledge from different sources signals openness for development, change and innovation.

Weaknesses of Dutch economy for bio-based materials and chemicals

Although the experts in this previous study identified many strong points, they also saw room for improvement in some aspects of the Dutch bioeconomy. One more general remark was that still, the political awareness of opportunities for prosperous future developments is inadequate, or that politicians are not willing to make tough choices, which results in far too low support levels for several areas, including the bioeconomy.

Strong criticism was expressed about the unlevelled playing field between bio-based materials and bio-energy as well as between bio-based and fossil materials. More concretely, bio-based energy receives a lot of subsidies, whereas materials do not receive funding on the commercial level.

As a potential weak point, the lack of available feedstock was mentioned. However, this seems to be somewhat controversial, since other experts expressed the view that biomass is available in sufficient amounts in the Netherlands, especially sugar and starch plants. According to them, the true problems lie with the investment conditions.

Concerning incentives for bio-based businesses, all experts agreed that the conditions need to be improved in order to make the Netherlands more attractive for setting up new bio-based productions. The example of Corbion[8] *illustrates very well that innovative and successful bio-based companies often do not choose the Netherlands as a production location, due to much better conditions offered in other countries, e.g. Thailand. Also, other Asian or North and South American countries offer stronger incentives for these companies. Experts explained this lack of incentives with the strong mind-set that is convinced of a free market, which is predominant in the Netherlands. Therefore, subsidies for industries are frowned upon (and would—at current legislation—often lead to legal disputes over distortion of competition). All experts, however, stressed how important it would be to support especially small and medium enterprises in their innovative ventures—by providing more financial incentives and capital, space for experimentation and by reducing bureaucracy. This was seen as the crucial weakness of the Netherlands, which fits well into the overall structures in Europe and which results in more and more investments taking place elsewhere in the world.*[6]

14.5 Conclusions and Recommendations

The evaluation of the interviews has shown that small companies active in bio-based economy in Europe face several issues when acquiring funding.

Clearly, the criticism of public funding schemes dominates the answers (see Figure 14.2). Many companies made positive mention of public funding mechanisms, mainly for the research that they had done. However, when it came to subsequent steps in the direction of pilot plants or even commercialization, they felt that support from the public side was lacking. For small companies that have no resources to fall back on, the first years of commercialization are crucial and in need of support. Several interview partners also complained about bureaucracy and suggested solutions similar to the ones in Canada or the US. In Canada, a one-stop shop—the SDTC—helps innovative entrepreneurs with all their questions about setting up a business and acquiring funding for their activities. In the US, Delaware in particular was mentioned as a positive example of public administrative procedures—one company reported that it took them only 24 hours to obtain a business permit, instead of the eight weeks that it had taken in Germany. In cases where funding is dependent on timely implementation of business activities, such aspects might be decisive for the survival of a small company.

The other two aspects receiving quite a lot of attention are lack of investors' knowledge about the new industries and tax policy. They seem to be part of a bigger issue in Europe: lack of political commitment to the bio-based industries. It was also mentioned in the introduction to the barrier analysis that many companies lamented the lack of clear goals or priorities that guarantee a stable market for bio-based products. Several countries have introduced bio-based economy roadmaps, and also the EU is focusing more and more on the topic. However, in terms of concrete measures, there is not much to be seen in Europe. Experiences with the biofuels policy make stakeholders even more cautious, fearing a quick change in political direction once public opinion changes or new scientific findings are made. The debate about the use of food crops, or first generation feedstocks, is also detrimental for the development of these new industries. Many companies fear that they will get negative press for using plastics made from starch or sugar, for example.

All of this creates insecurity for investors. These impressions were also supported by statements by the German biotechnology cluster CLIB2021. During discussions it was mentioned that in general, there is more than enough risk capital available in Europe, but that it does not go to new, bio-based industries. For instance, there are pensions funds containing hundreds of millions of Euros. Often, these funds are simply too big for investments in small, innovative firms. But it is also a very important factor that politics in Europe have not made it a clear priority to improve the framework conditions for bio-based chemicals and materials.

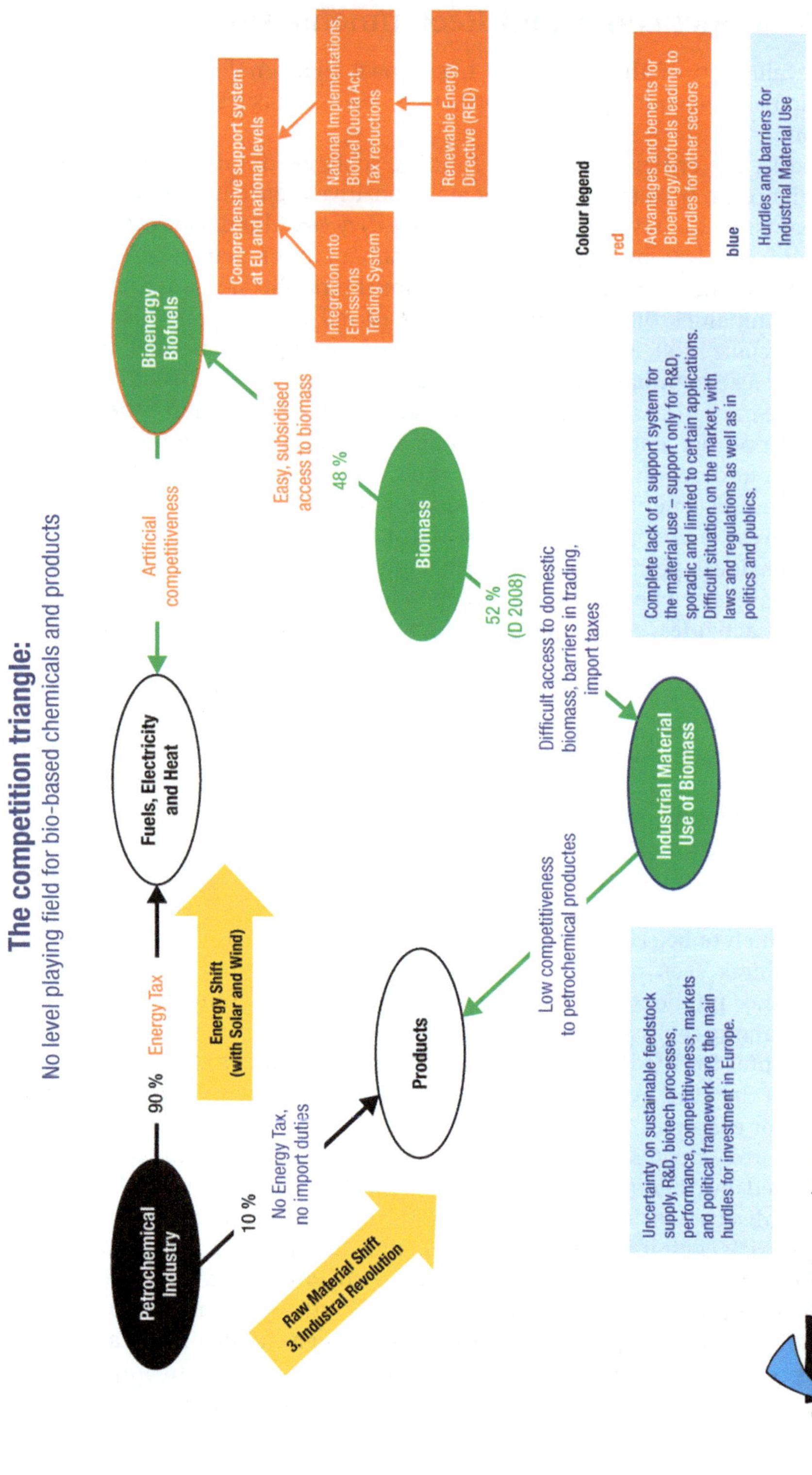
The competition triangle:
No level playing field for bio-based chemicals and products
Petrochemical Industry
90 %
Energy Tax
Fuels, Electricity and Heat
Artificial competitiveness
Bioenergy Biofuels
Comprehensive support system at EU and national levels
Integration into Emissions Trading System
National Implementations, Biofuel Quota Act, Tax reductions
Renewable Energy Directive (RED)
Energy Shift (with Solar and Wind)
10 %
No Energy Tax, no import duties
Products
Easy, subsidised access to biomass
48 %
Biomass
52 % (D 2008)
Difficult access to domestic biomass, barriers in trading, import taxes
Industrial Material Use of Biomass
Low competitiveness to petrochemical productes
Raw Material Shift
3. Industrial Revolution
Uncertainty on sustainable feedstock supply, R&D, biotech processes, performance, competitiveness, markets and political framework are the main hurdles for investment in Europe.
Complete lack of a support system for the material use – support only for R&D, sporadic and limited to certain applications. Difficult situation on the market, with laws and regulations as well as in politics and publics.
Colour legend
red
Advantages and benefits for Bioenergy/Biofuels leading to hurdles for other sectors
blue
Hurdles and barriers for Industrial Material Use
© nova-Institut.eu | 2013

14.5.1 Level Playing Field—the Competition Triangle

Still, there is no level playing field for these materials, since they need to compete for resources with a heavily subsidized energy sector, and for market shares with a stable established fossil-based chemical industry that does not have to pay any taxes on the material use of fossil carbon. This state of affairs can be summed up in a "competition triangle" (Figure 14.4).

14.5.2 Sustainability and Incentives—Two Sides of One Coin

Another interesting aspect that is rarely talked about in the discussion about a level playing field is the aspect of feedstock sustainability. Recently, a few organisations and initiatives have been working on defining standards for sustainability certification of bio-based feedstocks for industrial material use (CEN/TC411 WG4 on a European level, INRO in Germany) (Table 14.1). ISCC+ and RSB already offer sustainability certification for bio-based feedstocks for industrial material use, which is used by some industrial players.

From a general environmental perspective, it should be welcomed that any use of resources should be determined by sustainability factors. And the

Figure 14.4 The competition triangle: industrial material use of biomass—biofuels/bioenergy—petrochemical industry.
Right side: bioenergy/biofuels and material use competing for biomass
Material use is competing with bioenergy for biomass that is not used for food or feed. As a result of the comprehensive support system for bioenergy and biofuels, which was ultimately created by the EU RED, the prices for biomass and land have greatly increased.[9,10] This makes access to biomass for material use much harder and more expensive, but this is not compensated for by support measures. This market distortion hinders the competitiveness of producers of materials from biomass.
Left side: petrochemical products competing with bio-based products
The bio-based chemistry and plastics industries are exposed to full competition from chemical industry products. Without any accompanying measures, new, bio-based industries must be developed that can prove their viability in the face of the well-established and long-optimized mass production of the chemical industry. Then there are high biomass prices resulting from the promotion of energy use, which are not counteracted by taxes on fossil carbon sources as a raw material for the chemical industry. All of this creates an extremely tough competitive environment.
Upper side: fossil energy competing with bio-energy/biofuels
Due to the comprehensive support system for the energetic use of biomass, originating from the RED and its national implementations, an artificial competitive situation compared to fossil energy sources has been created over the years. Furthermore, the latter are subject to a substantial energy tax—this makes for extremely favourable, artificially created competitive conditions for bio-energy and biofuels.

Table 14.1 Sustainability criteria for different sectors and applications (energy and material) (nova 2014).

	Industrial material use, CEN TC 411 on bio-based products (draft version)	Biofuels	Biofuels	Bioenergy	Bioenergy, proposal for solid and gaseous biomass (draft version)	Fossil-based energy, fuels & products (petro-chemicals)
Incentives	No	No	With RED incentives (accounted for in the) renewable fuel quota)	No	With RED incentives (accounted for in the) renewable fuel quota)	Different subsidies in many countries
Number of environmental criteria	9	0	5	0	7	0
Number of social criteria	6	0	0	0	0	0
Number of economic criteria	1	0	0	0	0	0
Total number of sustainability criteria	16	0	5	0	7	0

discussion mostly follows the logic that biofuels have to fulfil sustainability criteria, so material use of biomass should do the same.

However, that is not a fair depiction of the situation. Biofuels and bioenergy only have to prove any kind of sustainability certification if they want to be eligible for the substantial support mechanisms available in the EU. Incentives and sustainability are inextricably linked for biofuels. This fact is not acknowledged at all when suddenly multiple stakeholders demand sustainability certification for bio-based materials as well. So far, companies voluntarily obtain this certification in order to fetch a Green Premium price on the market. However, when looking at the proposed sustainability schemes for bio-based materials, they even suggest more obligatory sustainability criteria to be fulfilled than is asked from the heavily subsidized biofuels.

Furthermore, the petro-based sector is completely left out of any sustainability discussion. Except for the specific agricultural aspects, many criteria could also be applied to the use of fossil resources, but this is never seriously taken into consideration. In terms of a level playing field, there is still much to be done in this regard.[11]

14.5.3 Recommended Measures

14.5.3.1 Prioritization

The first step towards a strengthening of the bio-based economy would be a clear definition of why this is a priority of European policy. "Bio-based" in itself is not accepted as an advantage in many regards, because of the food competition, negative environmental impacts such as eutrophication or other aspects. The US has a very pragmatic approach and makes it clear that bio-based materials create jobs and value added, especially in the faltering agricultural sector. Secondly, an increased use of renewable raw materials grants more independence from fossil resources, which is accepted as an advantage, too. Environmental aspects are appreciated as well, but are not the only reason why bio-based materials are supported. The new Farm Bill 2014 grants funding for biorefineries for chemicals and materials under the same mechanism that supports biofuels plants.[12]

The Bioeconomy Strategy published by the EU in 2013 is a good step in the right direction. As one result, the Bio-based Industries Initiative (BBI) was founded as a Public Private Partnership (PPP) by the Commission and the Bio-based Industries Consortium (BIC). This initiative has already raised 3.7 billion Euros of investment in bio-based innovation and is therefore one important measure to support this sector. However, less ambiguity about the motives and priorities for bio-based chemicals and materials would be helpful also in national policies. One example is also the issues of resource efficiency, cascading use and circular economy which are stressed more and more often by the EU as well as by national governments, while at the same time the subsidies support direct energy use of raw materials, preventing them from entering cascades or circles. This is very contradictory and makes investors more insecure. Just very recently, the European Parliament has commented on the Bioeconomy Strategy by emphasizing "that bioeconomy policies must be better designed to ensure a cascading use of biomass [and called] in this respect, for the development of a legal instrument that will pave the way for a more efficient and sustainable use of this precious resource".[13] Apparently, also the Parliament sees a need for more concrete actions.

14.5.3.2 Public Procurement of Bio-based Products

Green public procurement could also be designed to strengthen the public procurement of bio-based products, thus creating market demand and providing more security for manufacturers and investors. However, the doubts about the general environmental advantageousness of bio-based products and an unclear legal situation make it currently difficult for public procurement officials to prefer bio-based products. As mentioned before, the European Commission has included activities on this issue in a Horizon2020 Call in order to educate procurement officials about bio-based products and to make sure that such action will comply with European procurement law. The Dutch Public Procurement Expertise Centre PIANOo is furthermore starting a pilot project on public procurement of bio-based products very soon.

14.5.3.3 Infrastructure: Dedicated Bio-hubs and Chemical Parks

One of the strengths of the Netherlands mentioned in the expert interviews was the existing infrastructure in the chemical industry as well as in agriculture that can be combined to create new value added with innovative bio-based outlets. In combination with the excellent logistics mentioned also and the knowledge present in the Netherlands, it could be a very valuable step to create more "bio-hubs" at strategically interesting points in order to facilitate the set-up of small innovative companies. Combining state-of-the-art production facilities with dedicated supply routes of bio-based resources, specialized waste removal systems and efficient energy systems would make it very attractive for companies not only to develop the ideas and technologies in the Netherlands, but also to set up production there, thus creating more jobs and value added.

In terms of infrastructure, it is also interesting to note that bio-based chemicals can be profitable on a much smaller scale according to some recent studies, thus relieving the pressure on feedstock transports and making local, small-scale production of high-value specialty chemicals possible. This is a very distinct property of bio-based materials compared to both energy and the petrochemical sector. The following quote highlights the advantages this could mean for the Dutch economy:

> *The conclusion I is that the chemical industry is the only sector that allows a higher cost price for its raw materials because these can be compensated by lower capital costs.*

A second important conclusion II has been made in the recent study, that because of the lower capital costs per ton of product, also the scale of operation is less an important factor in the competition between companies of the future. It is anticipated that factories of 10 000 tonnes of product per year can become as competitive as the large petrochemical factories that have annual capacities of 200 000 and even 500 000 tonnes. The dominant competitive factor will become how to obtain the right raw material/conversion process combination. Especially in times that the biomass raw materials are not available yet as commodities, the sourcing is an additional risk factor when financing has to be done. Also for the introduction of totally new products, small factories will be a competitive factor since this gives time to develop the market. . . .

The third conclusion III is that the chemical industry 50% based on biomass will create a lot of new jobs. Employability can grow with 40 000 jobs to supply the Dutch chemical industry with 50% of biomass raw materials (now being ca. 80 000 fte). This is revolutionary in a way because in agriculture but also in the process industry we have seen only reduction of labour because of the ever-growing productivity reached by automation, taking over of human and animal labour by machines. . . .

The conclusion IV is that when the chemical industry can afford to pay higher prices for components with suitable molecular structures that lower the need for capital, the other biomass components that result from a biorefinery, can become available at a much lower price, enabling the electricity and transportation fuel sectors to obtain their raw materials at competitive prices.

Therefore, conclusion V, the chemical sector is key and could pave the way towards our biobased economy.[14]

14.5.3.4 Tax Incentives

Tax incentives could be very effective instruments in order to support small innovative companies and their bio-based products; the Malaysian example shows that exemption from taxes makes it a very attractive location for biotechnology companies. Similar measures could be put in place for other, non-biotechnological bio-based businesses. However, tax policy is always a very controversial topic and is therefore expected not to be easily implementable. Note that tax policy in the US is described in Chapter 15.

14.5.3.5 "Small" Measures for Market Pull—Directives and Bans

The following measures, in contrast, could be relatively simple to implement and are not expected to have large, undesired side effects. They would create more market pull and signalize a dedication of the Dutch government to bio-based materials.

Several directives and bans are conceivable that would promote bio-based and/or biodegradable products in a context where they have significant advantages over other products. The example of biodegradable and compostable plastic bags has been widely discussed lately and most recently been backed by the EP vote on 16 April 2014, supporting the draft rules stating that plastic bags used to wrap foods such as fruit, vegetables and confectionery should be replaced by 2019 by carrier bags made of recycled paper or biodegradable and compostable bags.[15] Other very reasonable rules could be to only use bio-based and biodegradable lubricants in environmentally sensitive surroundings or starch-based, biodegradable mulch films in agriculture.

References

1. CEN/TC411 2014: EN 16575. Bio-based products – Vocabulary. Brussels 2014.
2. European Commission 2013: H2020 Call ISIB-07-2014. Public procurement networks on innovative bio-based products. Download at: http://ec.europa.eu/research/participants/portal/desktop/en/opportunities/h2020/topics/2285-isib-07-2014.html (last accessed 2014-05-15).
3. World Economic Forum 2014: World Competitiveness Report 2013/2014. Geneva 2014. Download at: http://www3.weforum.org/docs/WEF_GlobalCompetitivenessReport_2013-14.pdf (last accessed 2014-05-15).
4. Ernst & Young 2014: Biotech Branche in Deutschland stagniert – Steuerbefreiung soll Investments in Biotech fördern. Press release 2014-04-14. Download at: http://www.ey.com/DE/de/Newsroom/News-releases/20140414-EY-News-Biotech-Branche-in-Deutschland-stagniert (last accessed 2014-05-15).
5. MIDA – Malaysian Investment Development Agency 2012. Invest in Malaysia. Incentives for Investment. Incentives for the Biotechnology Industry. Download at: http://www.mida.gov.my/env3/index.php?page=biotechnology-industry (last accessed 2014-05-15).
6. L. Dammer, M. Carus, A. Raschka, L. Scholz (2013). Market Developments of and Opportunities for bio-based products and chemicals. Report written for Agentschap NL, reference number 52202. Sittard 2013.
7. The experts who gave input to this quoted study were: Jan Ravenstijn, one of the leading international experts on bio-based polymers, Elsbeth Roelofs (and her team), consultants from TNO, as well as Christiaan Bolck and Hariette Bos, both researchers from Wageningen University.
8. Corbion (former Purac, a Dutch company) established their big LA production facility in Thailand in 2012.
9. Several concerned companies have confirmed this increase in prices (see forthcoming publication of nova-Institute: ref. 10).
10. M. Carus, L. Dammer, A. Hermann, R. Essel: Proposals for a Reform of the Renewable Energy Directive (RED) to a Renewable Energy and Materials Directive (REMD). nova-paper #4 on bio-based economy. Hürth 2014-05.

11. More details on the unlevelled sustainability discussion can be found in the publication of nova-Institute: ref. 10.
12. R. Buckhalt and J. Goodman 2014, Personal communication with Ron Buckhalt and Jeff Goodman (USDA) during a USDA webinar on the new Farm Bill, 2014-03-14.
13. European Parliament 2013, Report on innovating for sustainable growth: a bioeconomy for Europe, Committee on the Environment, Public Health and Food Safety, Rapporteur: Paolo Bartolozzi, A7-0201/2013, 2013-06-14.
14. J. Sanders 2014, Biorefinery, the bridge between Agriculture and Chemistry. Farewell speech upon retiring as Professor of Biobased Commodity Chemicals at Wageningen University, Wageningen 2014-01-23.
15. European Parliament 2014, MEPs clamp down on wasteful use of plastic carrier bags, Press release, Brussels 2014-04-16. Download at http://www.europarl.europa.eu/news/en/news-room/content/20140411IPR43461/html/MEPs-clamp-down-on-wasteful-use-of-plastic-carrier-bags (last accessed 2014-05-15).

CHAPTER 15

A Monte Carlo-based Methodology for Valuing Refineries Producing Aviation Biofuel

DAMIAN BLAZY,*[a,b] MATTHEW N. PEARLSON,[b] BRUNO MILLER[b,c] AND REBEKAH E. BARTLETT[a]

[a] Oliver Wyman, 55 23rd Street, Washington DC, 20037, USA; [b] Massachusetts Institute of Technology, 77 Massachusetts Avenue, Cambridge, MA 02139-4307, USA; [c] Metron Aviation, 45300 Catalina Court, Suite 101 Dulles, VA 20166, USA
*Email: Damian.Blazy@oliverwyman.com

15.1 Introduction

The aviation industry has stated that renewable biofuels are essential for the industry's long-term viability. By diversifying its sources of fuel, the industry intends to reduce jet fuel prices and volatility and mitigate concerns surrounding carbon emissions and the industry's environmental footprint. The objective of this chapter is to outline a methodology for valuing hydroprocessing refineries producing aviation-grade biofuel (and renewable diesel) under uncertain market and policy conditions. As such, this methodology can be used as a decision making tool to aid prospective investors in determining under what market conditions a profitable hydroprocessing refinery could be constructed.

RSC Green Chemistry No. 43
Commercializing Biobased Products: Opportunities, Challenges, Benefits, and Risks
Edited by Seth W. Snyder

Published by the Royal Society of Chemistry, www.rsc.org

Our analysis differs from previous studies in three ways: (1) it includes price uncertainty and correlation between refinery inputs and products; (2) it includes the uncertainty surrounding government mandates and price supports (*e.g.*, tax credits and Renewable Identification Number (RIN) credits); and (3) it utilizes Monte Carlo simulation to determine net present value (NPV) ranges.

15.2 Background and Review of Past Analyses

Hydroprocessing refineries are capable of producing a portfolio of sustainable fuels and can produce up to 95% middle distillate fuels (diesel, jet) by weight of sustainable feedstock. Middle distillates produced by hydroprocessing refineries include jet fuel (C8–C14) which is also known as hydrotreated esters and fatty acids synthetic paraffinic kerosene (HEFA-SPK), and renewable diesel (C10–C24). Both of these products qualify for RIN credits under the US Renewable Fuels Standard (RFS2).

At the time of writing this, HEFA-SPK had been approved by ASTM International under specification ASTM D7566 for use in aircraft in up to a 50/50 blend with fossil-based petroleum. The Boeing Company is also leading an effort to approve the blending of renewable diesel with fossil-based jet fuel. As renewable diesel contains longer hydrocarbon molecules, which can cause fuel gelling at altitude, the blend ratio of this product with fossil-based jet fuel is expected to be much smaller—likely less than 5% by volume.[1]

Previous work by one of us (Matthew Pearlson, MIT) estimated the gate price of HEFA-SPK and renewable diesel for several plant sizes and operating conditions.[2] A techno-economic model was created with material and energy requirements to estimate capital and operating expenses and the gate price of distillate fuel from the refinery. A discounted cash-flow rate-of-return (DCFROR) economic model was used to assess plant size, production profiles, project finance structure, and production ramp-up schedules.

The installed capital cost based on this model for a hydroprocessing refinery was estimated for 116, 232, and 377 MML year^{-1} (corresponding to 2000, 4000, and 6500 barrels per day, respectively).

Fixed operating expenses were estimated from capital expenses, based on heuristics in the literature, and personal interviews with industry experts familiar with the process. These expenses included insurance, taxes, maintenance, and plant staff salaries.

Variable operating expenses included utility costs and feedstocks such as natural gas, water, and soybean oil. The model was based on soybean oil because it is the primary oil produced in the United States. Gross income from refinery sales was calculated from the price that each product would be sold for at the refinery gate and included the cost of production and refiner profit. It did not include costs for distribution, transportation, retail markup, or taxes. Price supports for renewable fuel production and revenues from trading RINs were not included in gate costs.

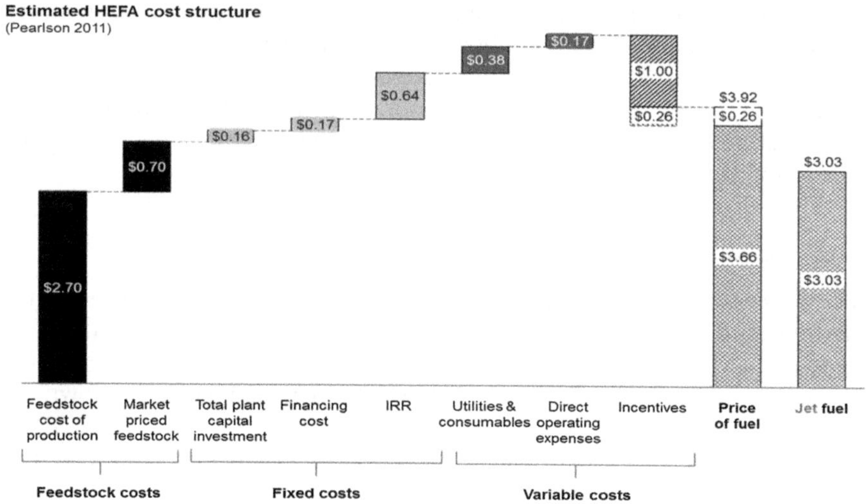

Figure 15.1 Estimated HEFA cost structure.

A DCFROR analysis was used to evaluate the gate cost of distillate fuel for each of the three plant sizes considered. The distillate fuel gate price was found to be such that the NPV of the project was zero while keeping all other product prices constant. The baseline scenario used the maximum distillate product profile. A discount rate of 15%, with a loan interest rate of 5.5% and 10-year term on 80% debt financing was used. Capital expenses during construction were spread over three years and distributed as 8%, 60%, and 32% of the total project investment, respectively. Depreciation was scheduled over 10 years, and used the variable declining balance method. The plant operated at 100% of its nameplate capacity in the base case.

The DCFROR analysis found that the baseline cost for HEFA fuel production ranged from $4 to $5 per gallon depending on the size of the facility (Figure 15.1). The current chemical profile of natural oil feedstocks is similar to the profile of diesel fuel, and longer than jet fuel molecules. When a facility maximizes the production of jet fuel, it cracks down diesel length molecules to jet length. This results in a larger number of low value, very small hydrocarbon molecules, and an overall reduction of yield per weight of feedstock to high value distillate fuel. As a result, jet fuel-focused refineries require approximately $0.25 per gallon more to cover increased operating expenses and reduced middle distillate product revenues.

15.3 Limitations of DCFROR Models in Incorporating Uncertainty

The DCFROR model developed by Pearlson contains 11 critical inputs (Table 15.1) whose future prices cannot be identified deterministically; rather they are "random" or stochastic in price behaviour and must be modelled probabilistically. These include the prices of feedstocks used to create fuel, prices that finished products are sold for, and capital costs associated with construction.

Due to stochasticity in these critical inputs, the precision of the DCFROR model for valuation is deceptive. Conventional deterministic DCF methods fail to represent this stochasticity in valuation because the calculations use averages or "most likely outcomes" for inputs into the financial model. Each input entering into the calculation of NPV is subject to a high level of uncertainty, however, especially when considered over the time scale of a prospective refinery. For example, the future price of diesel is not known with any degree of certainty and behaves stochastically. The price of diesel may be affected by a host of external factors, including global geopolitical conflicts, changing domestic policy, and growing global demand. As each of the hydrocarbon products produced in a hydroprocessing refinery will be sold at its market price, small changes in these assumptions can lead to large changes in the NPV of a hydroprocessing refinery.

For example, illustrated in Figure 15.2 is the historical monthly (July 1986 to November 2013) real price of one-month No. 2 Heating Oil futures, normalized to the price in 2013.[3]

Across discrete time periods, the price of diesel is partially governed by random or stochastic processes, where some prices are more likely to occur than others. Figure 15.3 illustrates how these historical prices for heating oil can be plotted in a frequency diagram and then fitted to a density curve. This curve provides a quantification of the likelihood of occurrence for a particular commodity price.

In addition to the underlying stochasticity in the price of commodities used as inputs to the financial model, these prices are not independent but demonstrate correlation.[4] For example, in Figure 15.4, the price of jet, crude, diesel, and typical biofuels feedstocks such as soy and palm oils

Figure 15.2 Real historical heating oil future price (HO1) (adjusted and indexed to November 2013).

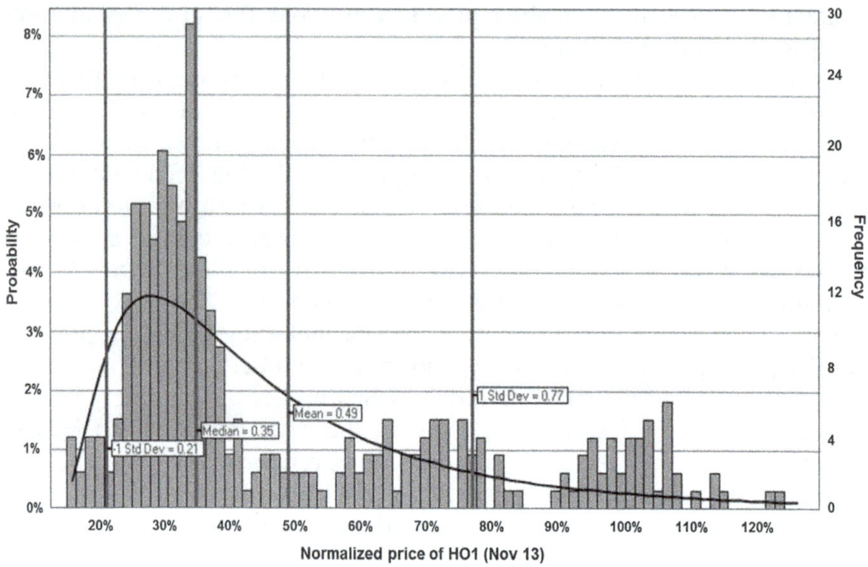

Figure 15.3 Normalized heating oil future price (HO1) (density, frequency adjusted and indexed to November 2013).

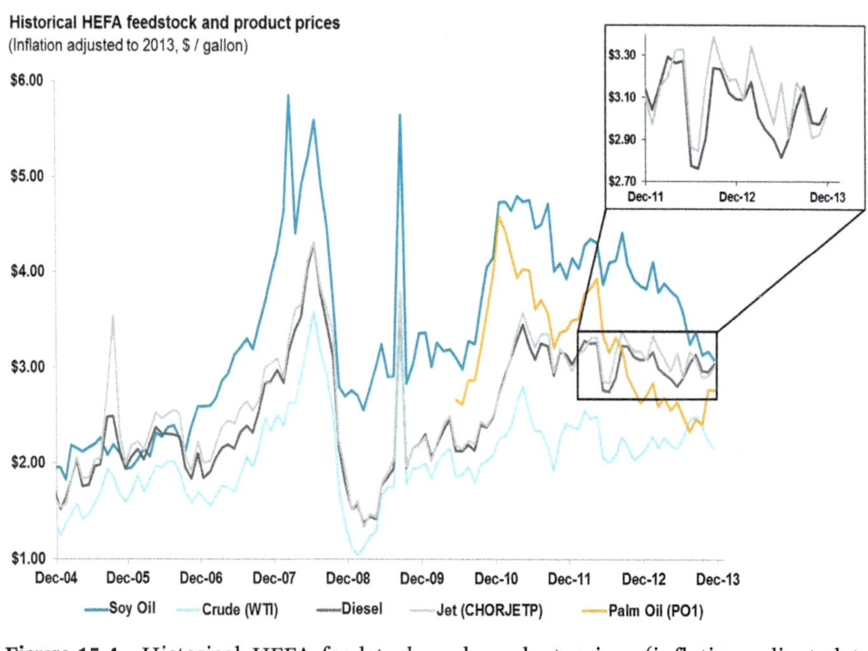

Figure 15.4 Historical HEFA feedstock and product prices (inflation adjusted to 2013, $ per gallon).

demonstrate similar price behaviour over time. Although these prices do not perfectly track, correlation must be considered when modelling to avoid price divergence.

When aggregated in the financial model, the price stochasticity of all 11 inputs results in a project's net present value (NPV) and rate of return (ROR) being described by a probabilistic distribution rather than a discrete value. When compared to the conventional approach previously developed by Pearlson, which utilizes averages or expected values, this approach provides significantly greater insight into a project or capital decision's risk and profitability.

Figure 15.5 graphically compares the difference between the conventional and stochastic approaches.

15.4 Key Methodology, Data and Assumptions Used in the Monte Carlo Model

15.4.1 Monte Carlo Methodology Overview

Following the approach outlined by David Hertz,[5] we included the underlying stochasticity in the financial inputs mentioned earlier by applying computer-based Monte Carlo simulations to the discounted cash flow (DCF) model developed by Pearlson to calculate ranges of project net present value (NPV). This approach comprised seven steps, as follows:

1. Construct uncertainty profiles for each key input into the financial model
2. Calculate time series correlation between key inputs into the financial model
3. Set assumptions such as facility size, and cost of capital
4. Use a computer to simulate the project's NPV by randomly generating input values from each input distribution
5. Deterministically calculate the simulated project NPV using the randomly generated input values from step 4
6. Use a computer to repeat steps 4 and 5 tens of thousands of times
7. Use a computer to list and aggregate the NPVs calculated in each deterministic simulation to arrive at a range or distribution of potential project NPV

15.4.2 Step 1: Constructing Uncertainty Profiles

To describe the variability in the key financial inputs (Table 15.1) to the DCF model, monthly historical price data was used to calculate both initial conditions and input variability. Initial conditions were calculated by taking the average monthly price from January 2009 to November 2013.

$$P_{0_i} = \frac{\Sigma \text{ Monthly prices}}{\text{Count of months}} \ \forall\, i \in \{1:11\} \tag{15.1}$$

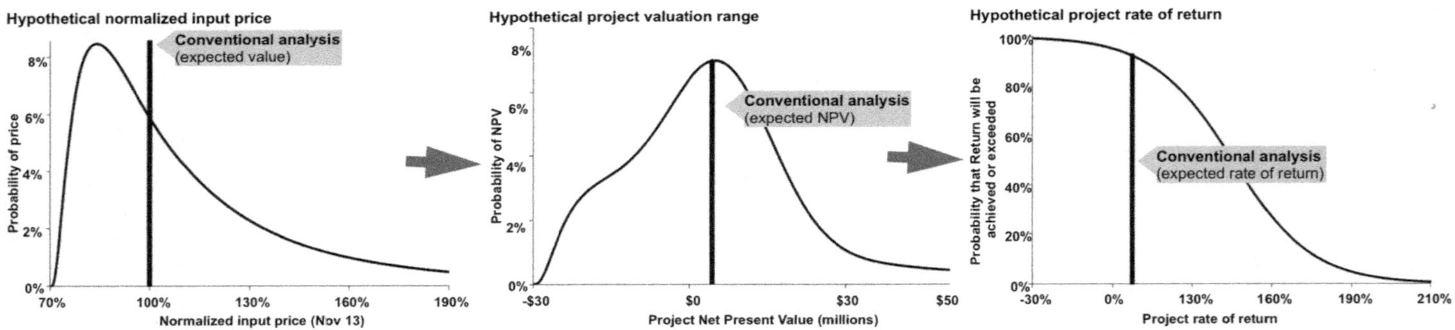

Figure 15.5 Comparison of hypothetical normalized input price, project valuation range, and project rate of return for conventional and stochastic analysis methods.

For each input, we quantified variability by first constructing a mass or probability function $f(P)$ by constructing a histogram of historical price. This was used to construct a density function $\Psi_i(P)$ which describes the probability of a discrete random price occurring.

As the number of data points for monthly price between January 2009 and November 2013 is relatively small, and the underlying stochastic nature which governs price is complex, we elected to use triangular and uniform distributions to describe behaviour. We believe that simple distributions are easier to understand by policy makers, and provide general guidance on system behaviour.

For capital expenditures, the costs calculated by Pearlson in his 2011 paper were used as a baseline, while the quarterly inflation adjusted CEP index (2000–2012) was used to construct the density curve.

In the model execution, each input's distribution of values was normalized to the mean and then multiplied against each initial baseline price P_{0_i}. Table 15.1 illustrates input data sources, time periods and distributions used.

15.4.3 Step 2: Constructing the Correlation Matrix

To determine the interdependency between key model inputs, the correlation between each time series is calculated and rounded to the nearest 20% to arrive at the "base case" (Table 15.2). To examine the sensitivity of NPV to the tightness of correlation, four scenarios were constructed and used in simulations: zero correlation, low correlation (20% less), base case, high correlation (20% more).

15.4.4 Step 3: Developing Assumptions

15.4.4.1 Product Slate

We assume that a refiner will optimize its product portfolio to maximize economic returns to its shareholders and therefore maximize the production of distillate fuels and products, which are the highest value product by weight. This will result in refiners producing higher percentages of renewable diesel rather than jet fuel to arrive at an overall higher yield of distillate by weight of feedstock (90% *vs.* 95%) as illustrated in Figure 15.6. If a refiner were to configure its assets to produce higher volumes of jet fuel, the overall return on equity and project valuation would be lower.

15.4.4.2 Cost of Capital

Investors were interviewed to determine current capital structures (debt to equity ratio), as well as the cost of debt and equity. Interviewees classified biofuel refineries as high risk, similar to a CCC bond, which is defined by credit agencies as extremely speculative, or with little prospect for financial

Table 15.1 Final inputs to discounted cash flow model.

Input	Source	P_{0_i}	Type of distribution	Min.	Max.	Mode
Power	Average retail price of industrial electricity EIA	P_{0_1} = \$0.07 per kilowatt hour	Triangular	\$0.06	\$0.08	\$0.72
Water	Gary and Handwerk	P_{0_2} = \$0.04 USD per gallon	Non variable	NA	NA	NA
Natural gas	NAGANGPL BGAP (Mid-Continent Natural Gas Spot Price/Chicago City Gate)	P_{0_3} = \$3.77 per MMBtu	Triangular	\$1.92	\$6.33	\$3.58
Naphtha	Generic Gasoline Future (XB1) with a slight discount	P_{0_4} = \$1.20 USD per gallon	Triangular	\$0.41	\$1.70	\$1.43
Light naphtha	Generic Gasoline Future (XB1) with a slight discount	P_{0_5} = \$1.20 USD per gallon	Uniform	\$1.15	\$1.25	NA
Propane	EIA Mont Belvieu, TX Propane Spot Price FOB (USD per Gallon)	P_{0_6} = \$1.20 USD per gallon	Uniform	\$0.69	\$1.70	NA
Diesel	G3ORUTLS Index (CENTS Per Gallon)	P_{0_7} = \$3.02 USD per gallon	Triangular	\$1.22	\$3.93	\$3.90
Jet	G3ORJETP Index (CENTS Per Gallon)	P_{0_8} = \$3.04 USD per gallon	Triangular	\$1.22	\$4.41	\$3.83
RINs	RINSEA12 (USD per RIN)	P_{0_9} = \$0.64 USD per gallon	Triangular	\$0.35	\$1.31	\$0.37
CAPEX	CEP Index	$P_{0_{10}}$ 2k bbl/ = \$65 million 4k bbl per day = \$ 98 million 6.5k bbl per day = \$120 million	Triangular	73%	121%	120%
Soybean oil	Soybean oil futures (BO1)	P_{011} = \$ 0.53 US cents per lb	Triangular	\$0.39	\$0.71	\$0.48

Table 15.2 Correlation matrix.

Correlation matrix	Power	Natural gas	Hydrogen	Bio oil	Propane	C4–C5 (Lnaphtha)	Naphtha	Jet	Diesel	RIN
Power	100%									
Natural gas	40%	100%								
Hydrogen	40%	100%	100%							
Bio oil	−20%	−20%	−20%	100%						
Propane	0%	20%	20%	60%	100%					
C4–C5 (Lnaphtha)	60%	80%	80%	0%	60%	100%				
Naphtha	−60%	−40%	−40%	80%	60%	−40%	100%			
Jet	−60%	−60%	−60%	60%	60%	−60%	100%	100%		
Diesel	−60%	−60%	−60%	60%	40%	−60%	100%	100%	100%	
RIN	40%	40%	40%	60%	80%	80%	0%	−20%	−20%	100%

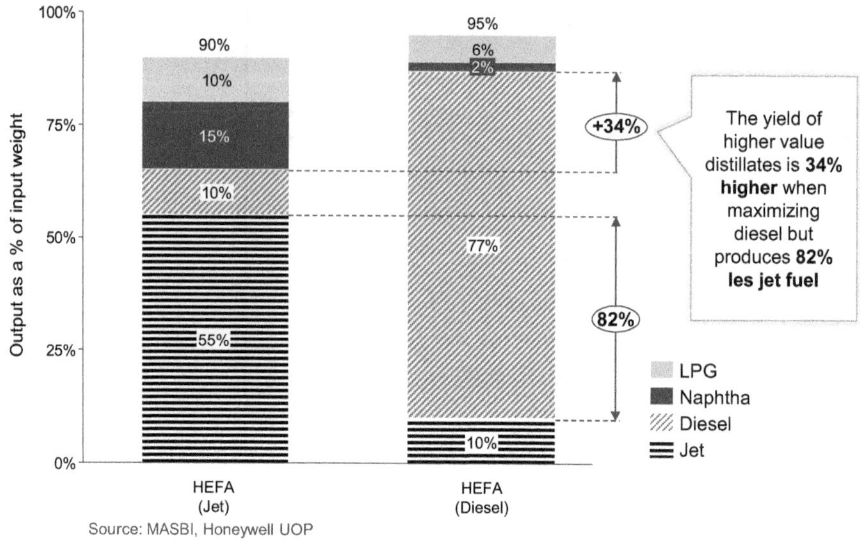

Figure 15.6 HEFA product efficiency and product output (output as a % of input weight).

recovery. As a result, a return on equity in excess of 20% and a return on debt greater than 15% are required. In addition, due to the higher risk, the majority of financing (greater than 60%) would have to come from equity rather than less costly debt. Thus, the current weighted average cost of capital for biofuel refineries can be expected to be between 20% and 25%. Given the high likelihood of large negative cash flows in the future, however, the higher costs of capital will actually increase project NPV. (In this analysis, a cost of capital of 16% is used, unless otherwise specified.)

15.4.4.3 Scale Back and Refinery Operations

A refiner will not produce blindly when economic conditions are not favourable. We assume when the marginal costs of producing products is greater than revenue, a refiner will scale back production to 50% of capacity. We do not believe a refiner will be able reduce production to less than 50%, due to off-take contracts with customers, physical limitations of the refinery and equipment, and the high cost of shutting down and restarting.

15.5 Steps 4–7: Monte Carlo Simulation—Results

To test model sensitivity to critical inputs, several scenarios were constructed. The scenarios were built by assuming different values for the following parameters:

- Facility size: 2000, 4000, and 6000 bbl. per day

- Price correlation: uncorrelated (0%), low (20% less), medium, high (20% more)
- Cost of capital: mature (6%), very optimistic (11%), optimistic (12%), base (15%), current (22%)

Although larger facilities offer a greater likelihood of producing a larger NPV (Figure 15.7—left), given the uncertainty surrounding the price of inputs and products, the standard deviation of the NPV is much larger.

The magnitude of price correlation and changing Weighted Average Cost of Capital (WACC) was not observed to have a large effect on moving the mode of a project's NPV density function. Rather, it reduced the standard deviation (Figure 15.7—middle and right).

15.6 Implications for Government Policy

To quantify the potential impact of changing government policy and subsidies on the long-term economics of a prospective hydroprocessing refinery, we analysed a hypothetical 4000 bbl. per day facility with a WACC of 16%, and medium price correlation. In this analysis, the value of RINs is a stochastic input and allowed to vary across a density function that describes historical RIN prices. Although the lifetime average RIN value (January 2014) is $0.64 per gallon, in this analysis the mean of the density curve is set to $1.00. Additionally, the value of the Biodiesel Blender Tax Credit is also assumed to be $1.00 per gallon. The resulting expected average policy benefit to a hydroprocessing refinery is $2.00 per gallon. This represents what we believe is an optimistic, best case scenario.

To determine the effects of changing policy on refinery NPV, simulations were run in which the length of time that both the Biodiesel Blender Tax Credit and RINs would be available and contribute value to a refinery was varied from zero to six operating years. (A three-year construction period was not included.) After expiration of the tax credit/RINs, the hypothetical facility would continue to produce fuels for the remainder of its operating life, assumed to be 20 years total. US government policy is also discussed in Chapter 13.

As illustrated in Figure 15.8, if RINs and the tax credit do not exist at all, there is essentially no likelihood that our hypothetical hydroprocessing refinery would achieve an economic return over its lifespan. As RINs and the tax credit are introduced on a successive annual basis, the mode of the density curves increase in value. For investors to provide capital at low enough rates of return for a refinery, a project's NPV distribution would need to have the probability of loss or loss of capital across its lifetime to be less than 5–10%. For this to occur, RINs and the tax credit would need to be available for a minimum of six years of refinery operation. If the duration of RINs and the tax credit is less than six years, the risk of the refinery losing money over its lifespan would be large enough to make financing prohibitively expensive. For example, if incentives were available for only five

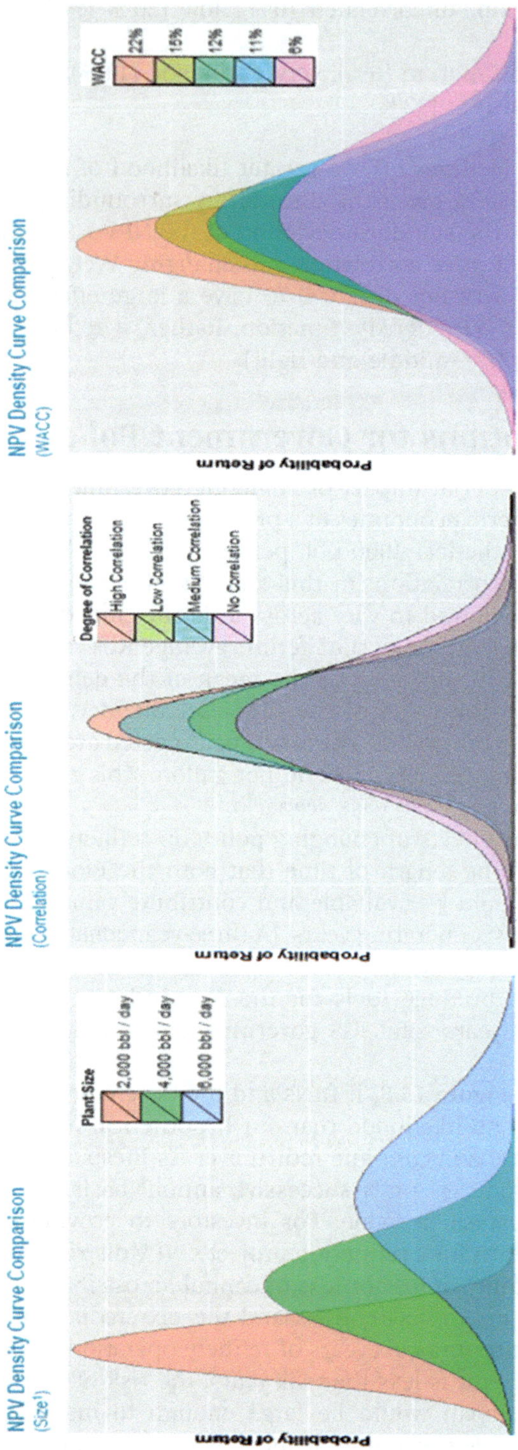

Figure 15.7 NPV density curve comparisons.

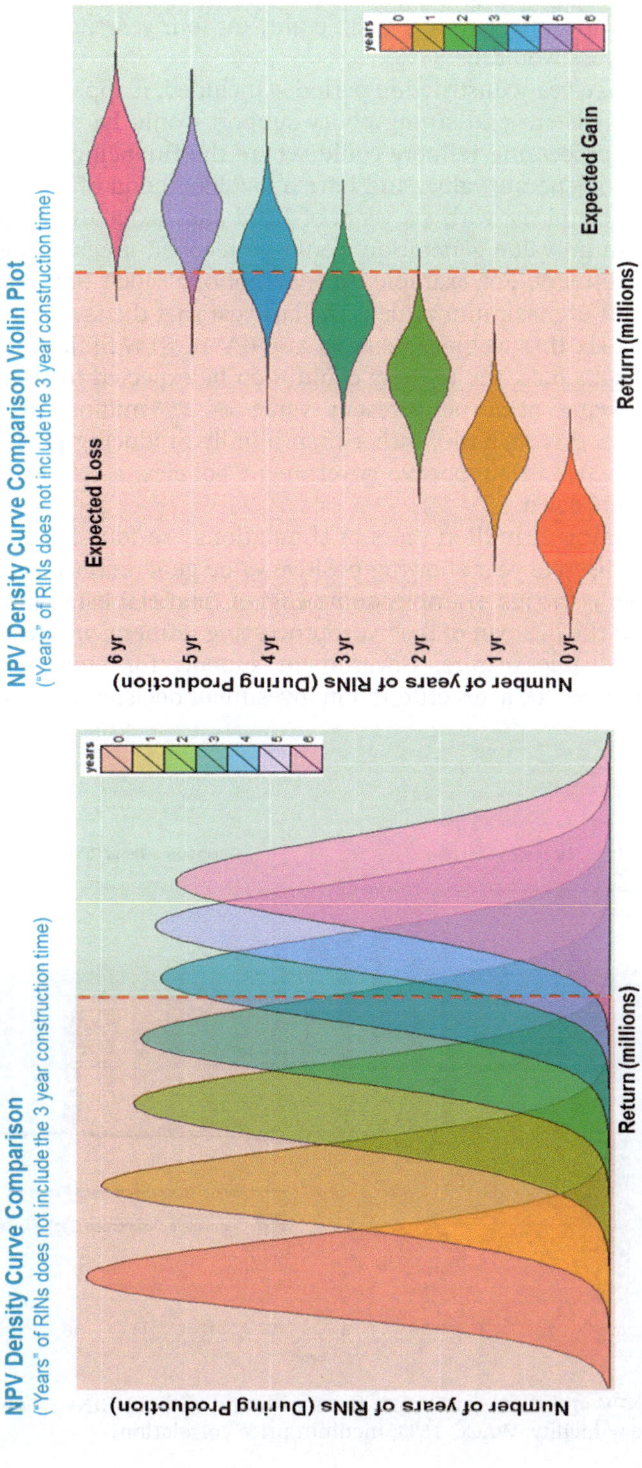

Figure 15.8 "Years" of RINs *versus* NPV.

years, the probability of expecting a loss for a 4000 bbl. per day facility would be 4%, while if incentives were available only for four years, the probability of expecting a loss would be 34%.

When the three-year construction period is included, it appears that nearly ten years of consistent and strong policy support would be required to ensure the hydroprocessing refinery could secure the financing necessary for construction, to generate value, and have a low likelihood of shortfall.

On a cash flow basis, with six years of RINs and tax credit support, the point of maximum value generation would be reached much earlier in the lifespan of a refinery. For example, for our notional 4000 bbl. per day refinery, the point of maximum value is the last year that the favourable policy exists. At year six this would equate to an NPV of $147 million. After the expiration of subsidies, the refinery could then be expected to slowly bleed off value, arriving at a net present value of $96 million in year 20 (Figure 15.9). In this scenario, rather than blindly producing after year six, and in the absence of supportive government policies, a refiner could be expected to shut down.

After performing 20 million years of simulations, we found that the likelihood of any discrete year showing positive value generation was less than 15%. As a result, in this example, some sort of financial externality would be required for the lifespan of the hydroprocessing refinery, or it would close as soon as subsidies expire. Government support for biofuels refining therefore clearly will be a key criterion in investment decisions for some time to come.

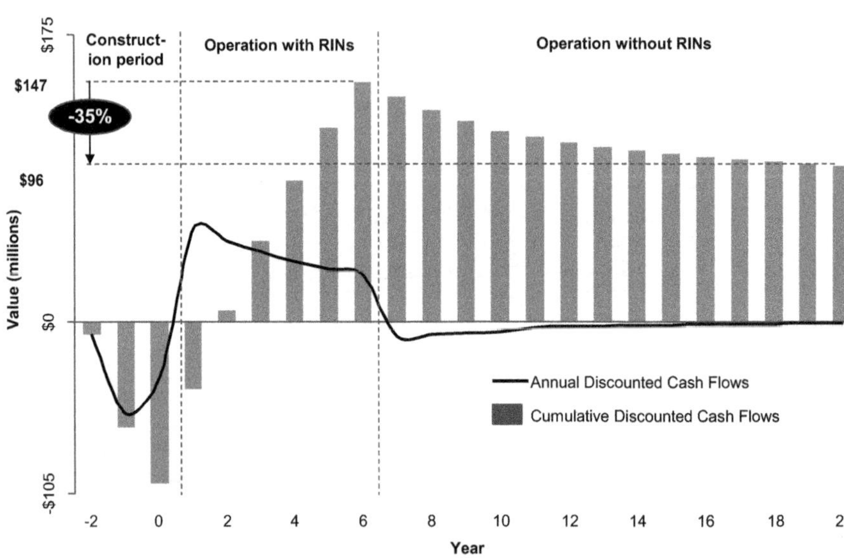

Figure 15.9 NPV and discounted cash flows with and without RINs (4000 bbl. per day facility, WACC 16%, medium price correlation).

References

1. Interview with UOP.
2. M. Pearlson, A techno-economic and environmental assessment of hydroprocessed renewable distillate fuels, Thesis, Massachusetts Institute of Technology, 2011, Available at: http://dspace.mit.edu/handle/1721.1/65508.
3. No. 2 heating oil is the commodity most widely used by the financial markets to benchmark the price of diesel, and is used to hedge both diesel and jet fuel.
4. R. S. Pindyck and J. J. Rotemberg, July 1988. *The Excess Co-Movement of Commodity Prices*, National Bureau of Economic Research, Working Paper No. 2671.
5. D. Hertz, January/February 1964. Investment Policies That Pay Off, Harvard Business Review.

CHAPTER 16

A Path Forward: Investment Cooperation between the United States and China in a Bioeconomy

SETH W. SNYDER

Energy Systems Division, 9700 S. Cass Ave., Argonne IL 60439, USA
Email: seth@anl.gov

16.1 Introduction

Is a partnership with China a route to commercializing biobased products? In this essay, the author will use the findings of the previous chapters to define the challenges to deployment and suggest a path forward by collaboration with China.

Commercializing biobased products and biofuels provides significant economic and environmental benefits. Biobased products can be designed to target specific customer markets. The bioeconomy in general, and biobased products in specific, enable consumer products to be labeled as "green" and generate a positive social reputation. With social acceptance from end use consumers, biobased products could achieve commercial success if products are manufactured near cost parity with fossil products. To achieve this goal, there has been substantial public and private research investment, which has significantly improved production efficiency. Some

RSC Green Chemistry No. 43
Commercializing Biobased Products: Opportunities, Challenges, Benefits, and Risks
Edited by Seth W. Snyder

Published by the Royal Society of Chemistry, www.rsc.org

biobased products have manufacturing costs near fossil product parity. In this essay we will explore scenarios to achieve commercial success. We will examine commercial deployment from the standpoint of the supply chain, biomass feedstocks, bioprocessing, manufacturing, market acceptance of products, and ultimately project finance. We report that project finance is the largest barrier to commercialization. We propose a path forward in which the US deploys foreign capital, notably from China. Uniting foreign capital with domestic technical experience could drive commercialization in the US and internationally.

16.2 Why Invest in the Bioeconomy?

There are strong economic, environmental and security drivers to investing in the bioeconomy. The bioeconomy reduces dependence on fossil energy sources, especially oil and gas. For energy importing nations, the bioeconomy enables domestic production of energy and large-volume commodity products, reducing both balance of payments and energy security risk. Biobased products reduce greenhouse gas emissions. From life cycle analysis, producing biobased products from biomass can reduce GHG emissions by over 60%. With landscape design, biomass feedstocks can capture nitrogen and other nutrients, decreasing agricultural impact on downstream water bodies and increasing overall agricultural productivity. The bioeconomy supply chain drives local economic development, fostering investment, creating jobs, and decreasing trade imbalances. Taken together, there are strong public, private, and environmental drivers to investing in the bioeconomy.

16.3 A Local Supply Chain

Biobased feedstocks contain about 50% oxygen in comparison to fossil sources that are close to zero % oxygen. Oxygen doesn't add energy to the feedstock; in fact, oxygen requires energy to dislodge from the hydrocarbon backbone in fuel precursors. In addition, biobased feedstocks are typically porous solids with high water content. Porosity adds volume while water adds volume and mass to the feedstock material. Taken together, biobased feedstocks are lower in volumetric and mass energy density than petroleum or other fossil feedstocks. The logistics of transporting solid, low-energy density, biomass feedstocks reduce the economically-viable distance between feedstock collection and product processing. Therefore, the bioeconomy lends itself to a distributed supply chain where domestic feedstocks support local manufacturing and likely serve domestic markets. A distributed supply chain creates jobs from the farm to the factory to the customer.

With a distributed biobased supply chain, parallel production markets can develop in nations or regions without competing with each other.

Thus parallel biobased supply chains can grow, using similar technologies, but serving different markets. The distributed supply chain can foster knowledge collaboration between producers who would normally compete.

16.3.1 Feedstocks for the Bioeconomy

A significant opportunity to achieve cost competitive biobased products is to utilize "advantaged" feedstocks. We define advantaged feedstocks as available at commercial scale as part of an existing agriculture supply chain, lacking significant demand from markets that support significantly higher prices, and can be readily converted to targeted products. The advantaged feedstock supply chain is impacted by growing and harvesting, flexibility of transport and storage, the stability of the materials to degradation, and their volumetric availability based on region and time frame. Opportunities for advantaged feedstocks are impacted by costs along this supply chain. Advantaged feedstocks include both purpose-grown biomass such as a grass or short-rotation woody energy crop and residues of a feedstock produced for other markets such as the corn stover component of corn. In general, residues from large-scale agricultural products have significant potential as advantaged feedstocks. Competition from other markets, such as between food, feed, fiber and fuel, can significantly impact the price and availability of feedstocks, but this competition is diminished with residues. Food commands a higher price either than fuel or chemical product and will capture the market for biobased feedstocks in virtually all cases. In the US, government mandates or subsidies (*e.g.* RFS2) provide an advantage to fuels over production, which don't have policy incentives, and can capture feedstocks in immature markets. The impacts of mandates and policies have been discussed in Chapters 13 and 14.

16.3.2 Biobased Feedstocks *vs.* Fossil Feedstocks

In the chemical industry, biobased products compete with fossil products, and therefore biobased feedstocks compete with a range of fossil feedstocks including natural gas, natural gas liquids, petroleum, and in some regions, coal and petroleum coke. Fossil feedstocks typically have distinct advantages in terms of concentrated supply (*e.g.* oil wells) as well as volumetric and mass energy density. Fossil feedstocks are easier to transport (liquids, gas, and dense solids) and store. Wet biomass feedstocks are susceptible to microbial degradation and must be handled and stored appropriately.

Fossil feedstocks have little to no oxygen making them more amenable for producing reduced species such as hydrocarbon fuels. The oxygen content of biomass feedstocks gives it a distinct competitive advantage in producing oxygenated species such as alcohols, ethers, ketones, organic acids, and sugar derivatives.

Fossil feedstocks are subject to enormous price variability as has been seen with dramatic rises and crashes in both the oil and natural gas prices over the past decade. Price is impacted by both societal events (*e.g.* war) and natural events (*e.g.* hurricane). With increasing larger tankers, oil markets are extremely interconnected at a global scale. With extensive pipelines in the US and Europe, natural gas markets are similarly interconnected at a continental scale. China doesn't have as mature an infrastructure in natural gas at this time. With the growth of liquefied natural gas terminals and tankers, it is only a matter of time until natural gas markets become globalized. These fossil feedstocks respond to external events very rapidly and sometimes unpredictably.

Biobased feedstock production and supply chains are widely distributed. Wide distribution reduces correlation in response to external events diminishing price uncertainty. For example, in the summer of 2012 the US Midwest experienced the most severe drought in generations. Some regions experienced near total loss of biomass production, but on a national scale, total production was not significantly impacted and price spikes often experienced in fossil markets were avoided in the biomass market. Biomass has a floor price set by the value in food and feed. In general food and feed provide higher revenue to the farmer or feedstock generator and are the primary competitors for utilization of biomass. In addition, food and feed markets have an inelastic component, *i.e.* we will always produce food, and will capture feedstock over a wide price range. From these considerations it is apparent that the bioeconomy competes with both fossil feedstocks and food and feed products.

16.3.3 Processing

Chapters 4–10 discuss a range of conversions and separations technologies. There has been significant public and private investment in processing technology to manufacture biobased products. The research spans biological, chemical, and engineering. These investments have increased throughput, yield, energy efficiency, chemical (atom) efficiency, productivity selectivity, and system automation. Performance improvements have led to more efficient use of manufacturing systems and reduced capital requirements. Investments have also improved energy integration, reduced waste generation, and enabled use of diverse and lower cost feedstocks, resulting in reduced operating costs.

Separations technologies to recover targeted products and recycle intermediates and wastes have improved substantially. Separations trains can be designed to manufacture products of sufficient purity for high-quality applications such as polymerization. These process enhancements enable the bioeconomy to serve high value product markets and increase revenue.

Processing technologies used in the bioeconomy are borrowed from other industries. Some processes, such as pyrolysis or catalytic conversion, have been adapted from the fossil energy and petrochemical industries. Other processes have been adapted from the food and beverage industries, including fermentation and purification of productions in water. In some supply chains, wood is the advantaged feedstock, and the bioeconomy borrows from the pulp and paper industry.

16.3.4 The Advantages of Small

As a new market grows, there are economic advantages to starting production at smaller scales. Small production scales place less capital at risk. Smaller feedstock volumes are required and simpler supply chains can be readily managed. A smaller customer base is required for sale of the manufactured biobased product. The nature of processing in the bioeconomy enables it to downscale more readily than the fossil industry.

Fossil-based processes are typically catalytic and require high temperatures and high pressures. High pressures require reactors fabricated from stainless steel or more exotic metals, increasing costs. In a processing reaction, the amount of product that is produced is dependent on the volume of the reactor. From basic geometry, the volume of the reactor scales with the cube of the radius of the reactor. The amount of metal used in the reactor is dependent on the surface area of the reactor. The surface area of the reactor scales with the square of the radius of the reactor. Therefore, in fossil processes, economies of scale drive you to larger and larger reactors to improve capital efficiency. Historically, fossil refineries have become more centralized manufacturing facilities with larger total volumes as markets grew and technology risk decreased.

The bioeconomy uses bioprocesses such as fermentation. Fermentation occurs at body temperatures and ambient pressure. Therefore biobased products do not require exotic metals for manufacturing. Capital requirements are much less sensitive to production scale for biobased products rather than fossil products. Therefore, costs for producing biobased products scale more linearly and can be competitive at smaller production volumes. Smaller production volumes are more suitable for distributed feedstock. As stated, smaller production scales reduce capital at risk.

16.4 Public Policy and the Bioeconomy

Many trade groups have highlighted positive consumer sentiment when biobased products are described. Typically end-use consumers indicate that they would use biobased products if they were offered at price parity. Some consumers would pay a limited price premium if there was a strong social benefit. Most large companies have corporate sustainability programs that consider biobased products as part of an overall approach to corporate and social responsibility. While the US has formal mandates for conventional, advanced, and cellulosic biofuels, the uncertainty in the cellulosic mandate has hindered investment (Chapter 13). The uncertainty arises from the clause requiring the US Environmental Protection Agency (EPA) to review expected capacity and adjust the mandate volume to the expected production capacity. As discussed later in the essay, this policy makes it difficult to secure capital financing. The overall process from concept to full commissioning of a biorefinery is a multi-year process. As of June 2015, the EPA just issued mandate volumes for 2014, which were due in November 2013. The EPA announced mandates for 2014, 2015, and guidance for 2016. The proposed 2016 mandate appears to be based on a very conservative projection of market capacity and will not reduce investment risk. In general, the EPA does not provide any guidance regarding the mandated volumes for the future. The capital markets don't have the capacity to predict EPA actions. Therefore cellulosic market volume mandates are discounted to essentially zero. Capital markets assume there is no mandate and therefore new production capacity is evaluated with the consideration that they must compete on cost alone. Without a mandate, both market and technology risk are high so projects cannot secure capital financing and the cycle is reinforced.

The US has voluntary policies (*e.g.* biopreferred.gov) to promote purchase and use of biobased products but no formal mandates and no national incentives. Europe has more formal mandate programs and they are growing. In addition, Europe's Emissions Trading System (EU ETS) functions as an indirect incentive. From a high level, we would expect to see rapid worldwide deployment of biobased products in specific and a bioeconomy in general.

16.4.1 Aviation Biofuels as an Example of the Challenge

A detailed analysis of aviation biofuels was presented in Chapter 15. From a market perspective, aviation biofuels behave similar to other biobased products including renewable diesel and ethanol. The market size for aviation fuels is significantly smaller than ground-based fossil and renewable transportation fuels. It is small enough that aviation biofuels do not drive refinery operations, but instead are treated as a co-product. The customer base is highly concentrated. For example, in the Midwest of the US, 20 airports account for approximately 90% of fuel purchases (MASBI 2013). We use the aviation biofuel analysis as a reasonable proxy for biobased products.

In the market, either clear product differentiation or shutdown economics are required. The highly integrated incumbent fossil industry has significant economic advantages including depreciated capital, economies of scale, heat and resource integration, efficient supply chains, and technology maturity. These advantages create an environment in which a novel biobased process using radically different feedstocks and an immature supply chain pose significant risks.

16.4.2 Biogas is a Bright Spot

One bright spot is that in July 2014, the EPA (EPA-420-F-14-045) issued the rule that qualifies biogas as a cellulosic biofuel. Credits for cellulosic biofuels are more valuable than advanced biofuels, the previous designation for biogas. The biogas must be used for transportation as compressed natural gas (CNG), liquefied natural gas (LNG) or electricity to power electric vehicles. Biogas is produced from waste carbon sources such as wastewater sludge, manure, or agricultural and food residuals, and represents an opportunity to expand the local bioeconomy. Cellulosic biofuel credits generated by biogas grew significantly immediately after the EPA ruling.

16.5 Factoring in Risk

While biobased products incorporate novel feedstocks, processes and technology, they compete with highly commoditized fossil fuel products. A new or modified product creates market risk. A new or modified process creates technology risk. Therefore the bioeconomy must overcome both market and technology risk. A study by McKinsey and the Council for Chemical Research (http://bit.ly/McKinsey-CCR) reveals that commercialization with both market and technology risk is extremely challenging in the commodity chemical sector. The chemical sector is much better equipped to use mature technology (low technical risk) and seek competitive advantages from market risk (novel functionality).

The bioeconomy incorporates new feedstocks, new processes, and new product markets, which creates both market and technology risk. The combination of market and technology risk poses a unique challenges to securing financing. Capital markets have difficulty assessing the likelihood of success of the project, and therefore the risk of default. Capital markets have difficulty assessing profitability of the products in commodity markets, and therefore, the rate of return on capital. Taken together, the bioeconomy typically requires projects to use high yield debt, reducing the commercial viability of the technology. With a multi-year timescale between land acquisition, design, construction, and plant startup, the cost for debt service can prevent a project from even starting.

16.6 The Largest Challenge is Project Finance

As discussed, the combination of technology and market risk factors is a large barrier to securing project financing. To absorb the risk, the lender must increase financial returns to cover the uncertainty, which increases overall project costs. At an early stage, biobased products have little to any cost advantage and may actually cost more than conventional products. Therefore, initial deployment of new technologies has cost disadvantages. In a commodity market such as biobased products, there is no strategic advantage to being the first mover. There are competitive advantages to letting others establish proof of the technology, supply chain, and market acceptance of the products before investing. So the expectation is that business will take a wait and see approach until other producers create a supply chain and demonstrate it is viable.

The first adopter absorbs significant risk, which raises cost for debt and increases total capital costs. Therefore, creation of new markets in the bioeconomy requires an "externality" to modify some of the risk, reduce capital costs and overcome this competitive disadvantage. The most common externalities are government mandates, regulations, incentives, or credits. Simply speaking, the public sector uses financial incentives to create markets for products that are expected to benefit society. As discussed in Chapter 13, the US market has mandates and incentives for biofuels. Biobased product mandates and regulations in the US are largely voluntary. Social acceptance of "green products" works if, and only if, the product is near cost parity.

Therefore we have a chicken and egg dilemma. Market and technology maturation requires deployment, and without incentives, deployment risk makes costs uncompetitive. A path forward requires a source of capital with a significantly different risk profile than currently exists in US private capital markets.

16.6.1 Project Finance So Far

The US has made substantial public and private investments in technology for the bioeconomy. Research from universities and research institutes has been spun out into companies focused on advanced biofuels and biobased products. These companies successfully attracted seed and venture financing and several had initial public offerings (IPOs). As these companies moved to commercial deployment, most projects have suffered delays and setbacks due to lack of project finance. Transformational opportunities developed in the public and private laboratories are sitting on the shelf.

After experiencing rapid growth from about 2000 through 2011, capital investments in the bioeconomy have declined. This behavior is uncorrelated with general capital markets, which declined rapidly in 2008 and were starting to recover by 2011. In fact, as of 2014–2015, the bioeconomy market is starting to consolidate with companies being acquired for targeted assets. Some of the more high profile companies such as OPX Biotechnologies (3-hydroxy propionic acid) were acquired outright, and others such as Mascoma have sold some of their most promising technologies (yeast business).

There have been a few launches of cellulosic biofuel production (Ineos BIO, Poet DSM, Abengoa, *etc.*). A notable factor is that a significant fraction of these projects has been dependent on capital financing from major European industrial companies. With more limited fossil resources and stricter environmental regulations, including greenhouse gas (GHG) emissions, European industry is more driven to invest in the bioeconomy. Successful technologies will have a more accepting market (lower market), and therefore, European capital may be able to absorb more technology risk. A downside is that the European energy and products markets are mature and exhibit little-to-no growth. Therefore rapid deployment in Europe will be hindered by market saturation and will require replacement of existing fossil markets. While these projects are very important to broaden the knowledge base, they offer less opportunity for widespread and rapid commercial growth. European-funded projects point the way to a further expansion of the bioeconomy.

16.7 A Path Forward

As with Europe, China has a different risk profile than the US. There is substantial private capital seeking to reduce risk by diversifying investments out of China and into mature international markets. In comparison to domestic markets, projects with technology and market uncertainties in the US may offer similar or better risk profiles than domestic projects in China. While there are projects where Chinese investments may not be acceptable to the US society (*e.g.* projects related to security or critical infrastructure), there is little reason to restrict projects in the bioeconomy. Could Chinese capital be the path forward for building the US bioeconomy?

16.7.1 Capital from Chinese Investors

There are direct benefits to China in investing in the US bioeconomy. Energy and materials consumption are growing rapidly in China. China is passing the US as the largest importer of fossil energy feedstocks. In comparison to the US, China imports a larger fraction of its total fossil energy consumption. Fossil imports represent a much larger fraction of China's overall economy than the US. Therefore relative costs and availability of feedstocks for the products and energy industries are under significantly greater stress in China than the US. The biobased industry will have a larger impact in China than the US. There is significant public and private interest in developing a bioeconomy in China. In comparison to the US and Europe, China has a less developed infrastructure. In China, there is reduced competitive advantage of the mature fossil industries' centralized manufacturing and high-integrated supply chain, especially when delivering energy to rural areas. A less developed infrastructure levels the playing field between biobased and fossil industries. China has significantly lower per capita energy consumption than the US, enabling a distributed biobased supply chain to have a larger overall net impact on the Chinese consumer.

An ancillary incentive is the EB-5 Immigrant Investor program from US Citizenship and Immigration Services (http://www.uscis.gov/eb-5). The EB-5 program was created in 1992 to stimulate the US economy through job creation and capital investment by foreigners. Investing in capital projects in the US could enable the investors to qualify for EB-5 permanent resident "green card" status. Investors can apply for EB-5 visas if they invest at least \$1 million (\$500 000 in a rural or high-unemployment area) and create at least 10 jobs. The EB-5 program could attract investors from China in the bioeconomy, especially considering the additional incentive for rural investment projects.

16.8 The Bioeconomy in China

With economic growth and urbanization, caloric intake and protein consumption are rising rapidly in China. In order to secure a reliable food supply, China is strongly opposed to diverting valuable agricultural food products to biobased products or bioenergy. Therefore the bioeconomy must utilize biomass feedstocks and residues. Technology in biomass conversion is much more mature in the US than China. One way to avoid the long learning curves is for China to invest directly in advanced technology development. China will benefit from learning technology from the US. To be successful, both countries must find value. Mutual benefit will be created if partnerships with China accelerate financing of US bioeconomy projects.

16.8.1 The Needs in China Are Even Greater

Beside dependence on imported fossil feedstocks, China is experiencing significant concerns regarding air and water quality. There is a strong national interest in developing and deploying clean technologies. For example, major food producer COFCO Limited has made significant investments in the Chinese biofuel industry. Biofuels production in China is ~5% of US production but is growing more rapidly. A bioeconomy could enable continued growth of the Chinese economy while reducing the environmental impact of fossil sources for energy and products.

16.8.2 Progress in China

There have been high-level delegation visits to explore collaboration. Secretary Henry Paulson and the Henry Paulson Institute are fostering dialog and partnerships with the goal of building long-term relationships between the US and China on a range of sustainable energy markets (http://www.paulsoninstitute.org/). We present a few examples where partnerships are growing or could prove beneficial. In 2013, Shuanghui Group, China's largest meat producer, acquired Smithfield Foods (US), the world's largest pork producer. This acquisition demonstrates that agricultural markets are becoming more open to partnerships. More directly aligned with the current topic, LanzaTech is a rapidly growing US biofuels technology company. LanzaTech has an ongoing demonstration with Bao Steel in China that utilizes LanzaTech's syngas fermentation platform. LanzaTech has received Asian investments and is more towards commercialization in both the US and China. Sapphire Energy, a leading US algae company is collaborating with Sinopec through the US-China EcoPartnerships program. Cathay Biotechnologies is a Chinese company that produces dibasic carboxylic acids from fossil feedstocks. Cathay has a well-established business producing chemicals from dibasic carboxylic acids. As discussed in Chapters 2, 6, 9, and 11, the US and Europe have made significant R&D investments in the production of dibasic carboxylic acids from biobased feedstocks. Cathay could benefit from establishing collaborations with the US.

16.9 Deployment in the US and China

Success could come from joint development and near-simultaneous build-out of the bioeconomy in the US and China. We propose that the US hosts the first deployment to enable access to technical expertise in the US to debug the technology. After pioneer demonstration, the US and China work together to build production capacity in both countries. Each country uses domestic feedstocks with domestic manufacturing to serve domestic markets. If a joint venture is used, the US and Chinese companies would benefit from each other's success.

The critical driver is that the US provides technical insight and China provides high-risk capital. Once production is growing in both countries, technology risk will be reduced for the US investment community and US companies can expand with domestic capital. The bioeconomy is still serving relatively immature markets and competes with fossil production, so there will still be market risk.

16.9.1 Benefits of Cooperation in the Bioeconomy

Cooperation between the US and China in bioenergy could enhance agriculture and energy security. Deployment of biobased products and bioenergy can enhance environmental performance and provide sustainable economic growth. Joint investment and development of bioenergy projects by the US and China could increase domestic energy security, paving the way for binational cooperation. These benefits have been highlighted throughout the book.

16.9.2 Impacts on the Environment and Greenhouse Gas Emissions

The US and China are the world's dominant greenhouse gas (GHG) emitters. Over the past 40 years, the US has made substantial investments in the environment, which have resulted in improved air and water quality and a decrease in health issues attributed to the environment. With a rapidly urbanizing society, a huge increase in energy consumption and industrial output, China suffers from some of the worst air and water quality in both urban and rural regions. Air quality is most impacted by coal utilization. If the bioeconomy results in a decrease in coal combustion, it will provide enormous air quality benefits. The bioeconomy could address water quality. For example, costs for treating wastewater could be managed by generating biobased products from algae or advanced anaerobic digestion. The US is the leader in algal technology and Europe is the leader in anaerobic digestion.

16.9.3 Agriculture

The US is the most food secure nation in the world and the largest producer and exporter of agricultural products. The rapid growth rate in per capita disposable income in China, coupled with a continued migration of hundreds of millions of new consumers to urban areas, has created challenges for the Chinese crop and livestock sectors. There has been a phenomenal increase in demand for animal protein while the agricultural labor force has decreased. This is one of the drivers for the acquisition of Smithfield Foods. China is a significant importer of grain products including corn, distillers dry grains (DDGS), and soy. Grains are imported to support meat production. Coproduction of food with biobased products from the agricultural sector is a path to growth.

16.9.4 Challenges to Cooperation in the Bioeconomy

Beside financing and technology, there are other challenges to cooperation in the bioeconomy. The US and China have different regulatory structures. Projects that proceed rapidly in one country may be delayed in the other. It is beyond the scope of this essay to identify ways to coordinate regulations.

Perception and trust are also a challenge. The US and China tend to distrust each other in new endeavors. For example, China distrusts the US use of genetically modified organisms (GMOs) in agriculture. The US distrusts the quality of Chinese agricultural products in the food market. Perhaps biofuels and biobased products could serve as the bridge to overcome distrust.

There is still significant distrust of intellectual property (IP) rights between the US and China. Intellectual property protection needs to be addressed before US and Chinese companies are receptive to detailed sharing of technologies. As highlighted throughout this book, IP can cover feedstock production, catalysis, fermentation, separations, and software tools. Designing a system to block reverse engineering will prevent cooperative improvement of technologies.

16.9.5 Making the Case for Cooperation on the Bioeconomy

There have been a number of visits and delegation between the US and China. On the US side, the federal government has included both the US Department of Energy (DOE) and the US Department of Agriculture (USDA). There has been general information sharing but not significant joint technology development. DOE investments have been mostly for monitoring activities and evaluating sustainability criteria.

In precompetitive research, there can be significant collaboration on best practices, sustainability, *etc.* without loss of IP. This would improve environmental performance, reduce GHG emissions, and address water consumption and quality. Sharing IP will be necessary for commercial deployment. This barrier must be addressed if binational cooperation is expected to foster development of the bioeconomy.

There is still a limited market for oceanic trade in biofuels and biobased products. China does not consider bioenergy imports as addressing the primary risks associated with energy dependence and security. It does not leverage local resources.

Through formal and informal mechanisms, there are already many active collaborations on pre-competitive fundamental research. A significant number of Chinese scientists train in the US and return to China with US skills. In parallel, Chinese universities have substantially increased their research performance, resulting in a significant growth in the number and quality of their scientific publications and international patents.

16.9.6 Programs to Foster Cooperation Must Address Intellectual Property Rights

Since 2009, the US–China Clean Energy Research Centers (CERC) program has built relationships in advanced coal technologies, clean vehicles, building energy efficiency and recently, the energy–water nexus (http://www.us-china-cerc.org/). The centers offer a pathway to cooperation and better understanding of IP rights (http://www.us-china-cerc.org/pdfs/June_2015_Steering_Committee_Meeting/IP_June_1_2015_English.pdf). While three of the four centers (coal, vehicles, and water) are related to the bioeconomy, a US–China bioeconomy CERC does not exist; this would be very valuable.

While the governments are creating programs to foster collaboration and consider IP rights, we suspect that US companies will only do cooperative technology development and pioneer projects in joint ventures (JVs) where they retain some control of their IP. IP rights will remain an ongoing challenge.

16.10 Potential Size of the Bioeconomy

The potential size of both the US and Chinese markets are enormous—think crude oil scale. The US government has highlighted the enormous potential of the bioeconomy in the 2012 National Bioeconomy Blueprint (https://www.whitehouse.gov/sites/default/files/microsites/ostp/national_bioeconomy_blueprint_april_2012.pdf). DOE's target is to displace 30% of the transportation fuel supply and create a $200 billion bioeconomy (http://www.biomassboard.gov/pdfs/male_campbell_february_2014.pdf). China is developing similarly ambitious goals.

With technology maturation and financing, both countries can produce significant fractions of their biobased products, liquid fuels, and power needs with sustainable, and largely domestic, feedstocks. These investments will enable both countries to provide some stability to fuel prices and commodities, avoid supply and price disruptions from unstable markets, and decrease GHG emissions. Agriculture will be a critical factor in the success of the bioeconomy. Biomass feedstocks will need to be supplied from less productive land (Chapter 3). Biomass feedstock systems are an opportunity for cooperation to improve environmental performance and build a large, sustainable bioeconomy.

16.11 Conclusions

Biobased products and the bioeconomy offer economic and environmental benefits. The US has made substantial investments in technology but has seen only limited commercial production. China has a tremendous need to develop a bioeconomy. Cooperation between the US and China could lead to accelerated commercialization in both countries. The risk profiles in the US and China for capital investment are quite different. Projects that are considered too high risk for US capital could be considered very favorable to Chinese investors. Cooperation must overcome some traditional distrust between the US and China. The bioeconomy is a strong contender to build trust. By the nature of agricultural feedstock production, the supply chain is distributed and favors local deployment. It is quite possible for the US and China to develop parallel supply chains without directly competing with each other. As the markets grow and mature, competition will grow. Without early market competition, it will be easier to cooperate on commercialization. This begs the author to ask the question: could the bioeconomy foster economic development, environmental sustainability and international cooperation?

Acknowledgements

The submitted manuscript has been created by UChicago Argonne, LLC, Operator of Argonne National Laboratory ("Argonne"). Argonne, a US Department of Energy Office of Science laboratory, is operated under contract no. DE-AC02-06CH11357. The US Government retains for itself, and others acting on its behalf, a paid-up nonexclusive, irrevocable worldwide license in said article to reproduce, prepare derivative works, distribute copies to the public, and perform publicly and display publicly, by or on behalf of the government. The author would like to thank the Henry Paulson Institute (Chicago IL) for suggesting the topic area. The author acknowledges the guidance provided by Evan Feigenbaum and Damien Ma from the Paulson Institute and Benjamin Saydah (Carbon Strategies). The essay represents the opinion of the author.

Subject Index

References to figures are given in *italic* type. References to tables are given in **bold** type.

acetic acid, **9**
acetone–ethanol–butanol (ABE) process, 48–9
acetyl coenzyme A, 180
acrolein, 83
acrylic acid, 263
  applications, **266**
  fossil fuel consumption, **275**
  greenhouse gas emissions, **275**
  water consumption, **279**
adhesives, 148–51
adsorption, 55, 94–5, 119
  by ion exchange, 97–8
advantaged feedstocks, 354
agar, **112**
agricultural-ecological zone (AEZ), 268
Alberta, 220
algal biomass, 265, 278
  carbohydrates, 115–16
  carotenoids, 112–13, **112**, 113–14
  lipids, 116
  production, 110
  proteins, **112**, 114–15
  recovery, 110–11
  separation, 108–16
  water consumption, 278
algal oils, **19**, **21**
Amberlite™ LA-2, 118
American Process, Inc., **22**
anion exchange, 106
aquaculture, 325
arachidonic acid, 167
Archer Daniels Midland, 83
arthropods, 35
aspartic acid, **75**, 85–6
Aspen Plus®, 269
Aspergillus niger, 191
astaxanthin, **112**
ASTM D6866, 320
Audi, 222
Avantium, 81
aviation fuels, 336–7, 357
  capital costs, 343–5
  DCFROR models, 337–41
  Monte Carlo model, 341–6
  policy implications, 347–8
2,2-azinobis-(3-ethylbenzthiazoline-6-sulfonate) (ABTS), 145

bagasse, 173
Battelle Memorial Institute, 14
bimetallic catalysts, 78
Bio-Based Industries Consortium, 331
BioAmber, **262**
biodiesel, 171
Biodiesel Blender Tax, 347
biodiversity, 34–5

bioenergy crops
biodiversity and, 34–5
ecosystem service maximisation, 36–8
flooding and drought resistance, 33–4
nutrient management, 31–2
pest suppression, 36
soil carbon, 29
water quality, 31
water use, 34
yields, 36–8
see also feedstocks
Biofine Renewables, 84
biofuels, 1–2
Biofuels Digest, 15, 20
biogas, 357–8
biomass oils, **21**
BioPreferred program, 322
biorefineries, 17
Biorefinery Assistance, 306–7
bipolar membrane electrodialysis, 120
bipolar membrane electrodialysis (BPED), 108–9
birds, 35
bisphenol A (BPA), 10
Borlaug, Norman, 13
Botryococcus braunii, 114
Braskem, **262**
1,4-butanediol, **9**, **21**, 86, **262**, **266**, **275**, 286
conversion efficiency, **296**
greenhouse gas emissions, **275**
water consumption, **279**
butanol, **21**, 48–50
batch fermentation, 50–1, **52**
performance parameters, **52**
Chinese production, 49–50
continuous systems
cell recycle, 59–61
free cell, 51–3, **54**
immobilized cell, 53–8
product recovery, 61–4
continuous systems, *57*
performance parameters, **63**
feedstocks, 50
butyl-3-methylimidazolium hexafluorophosphate, 228

California Air Resources Board, 261
Canada, 327
carbohydrates, 10, **112**, 115–16
carbon dioxide, 315–16
electrochemical reduction, 221–32
cell configuration, 225–6
formate production, 226–7
higher-order products, 231–2
methane reactions, 219–21
polymer production from, 217–18
salicylic acid production, 217
solar energy and, 232
syngas and, 227–8
urea from, *216*
carbon fibers, 153–4, 228
carbon monoxide, 223, 231
carbon storage, 272–4
carotenoids, 112–13, **112**, 113–14
catalysts, 72–3, 143–5
ethanol synthesis, 80–1
glucaric acid, **75**, 85–6
3-hydroxypropanoic acid, 82–3
3-hydroxybutyrolactone, **75**, 85
itaconic acid, 84–5
levulinic acid, 84–5
lignin depolymerisation, 143–4
sorbitol, 83
succinic acid synthesis, 79–80
cation exchange, 106–7
cation exchange resin, 97
cell immobilization, 55
cell removal, 117
Cellufine™ A-200, 119
cellulose, 86, 173
cellulose-acetate-lignin, 154
CEN, 319

Ceramatec, Inc., 230
chemurgy, **11**, 12
Chevron, 308
China, 352–3, 360
 butanol production, 49–50
Chlorella vulgaris, 113
Chondrus crispus, 115
CLIB2021, 327
climate change, 25
Climate Change and Emissions Management Corporation (CCEMC), 220
Clostridium acetobutylicum, 49, 168
Clostridium beijerinckii, 50
Clostridium spp., 50–1, **52**, 180
Coca-Cola, 258
combined heat and power (CHP), 222
competition, *328*
Congressional Research Service (CRS), 310
ConocoPhillips, 308–9
consumer perception, 321
contour buffers, 37
copper, 225
corn, 28
corn steep liquor (CSL), 176
corn stover, 265, 278, **279**, 294–6, 354
Coskata, **22**
Counter-Rotating Ring Receiver Reactor Recuperator, 232
covalent bonding immobilization, 55
crops *see* bioenergy crops
crude oil, 12
Cryptococcus spp., **169**, **169–70**, **174**, 176
crystallization, 98–9, 124, 192
Cystocseira abies-marina, 114

Debaryomyces hansenii, 123
Delaware, 327
Department of Energy (DOE), 266, 305
DHA, **112**
diacylglycerol (DAG), 183
1,4-dicarboxylic acid, **21**
dimethyl carbonate (DMC), 146
DinCertco, 320
diolefins, 114
Dioxide Materials, 229
discounted cash-flow rate-of-return (DCFROR) models, 337–8
 limitations, 338–41
distillation, 192
docosahexanoic acid, 167
Dokshukino, 53
Dong Energy, 222
Dow Chemical, 168
downstream processing (DSP), 116–17
dried distiller's grains with solubles (DDGS), 270
drop-in replacements, 9, 10
drought, 33–4
Dunaliella salina, 113
DuPont, **262**

Edison, Thomas, 12
eicosapentanoic acid (EPA), 167
Eisenia bicyclis, 114
electrodeionization (EDI), 106–7, 125, 190–1, 195–7, 204–8, 210–12
 solid bed, 208–12, *211*
electrodialysis, 118, 120, 125, 191–2
employment, **309**
Energy Independence and Security Act, 92, 310
Energy Security Act (1980), 14
Enerkem, Inc., 220
entrapment, 55
Environmental Protection Agency, 16, 172, 261, 265, 356–7
Escherichia coli, 204
ethanol, 80–1
 fuel mixtures, 312
 US policy, 310–11
ethylene, 17, **21**
 conversion efficiency, **295**
ethylene glycol, 86
Eucalyptus globulus, 141
European Commission, 332

European Committee for Standardization *see* CEN
European Union, 261
  Bioeconomy Strategy, 331–2
  Emissions Trading System, 357–8
  production capacity, 316–17
  public procurement, 321–2
Exxon Valdez disaster, 13–14
ExxonMobil, 308

farnesene, **21**
fatty acid synthesis, 178
feedstocks, 354–5
  butanol, 50
  see also bioenergy crops
fermentation, 201–10
  aerobic to gluconic acid, 208–10
  anaerobic to succinic acid, 204–8
fertilizers, 268
fibers, 19
Fischer–Tropf process, 13, 229
Flexible Solutions, 85–6
flooding, 33
fluidized bed catalytic cracking, 12, *57*, 58
food *vs* fuel debate, 16
Ford, Henry, 12
formate, 223–5, 226–7
fossil feedstocks, 354–5
fracking, 219
France, 321
free fatty acids (FFA), 178–9
fucoxanthin, 114
functional replacements, 10
funding sources, 318–19
furan-2,5-dicarboxylic acid (FDCA), 93
furanaldehyde, 121
2,5-furandicarboxylic acid, **74**, 81, 122
furans, **21**, **75**, 121–2
furfural, 121

Gas Technology Institute (GTI), 221
gasification, **21**
gasoline, 15–16
genetically modified organisms (GMO), 324
Genomatica, **262**
glucaric acid, **75**, 85, 85–6
gluconic acid, 197–201, *199*, *200*
  separative bioreactors, 208–10, *209*
glucose fructose oxidoreductase (GFOR), 124, *209*
glutamic acid, **75**, 85
glycerol, **74**, 82–3, 123, 125–6, 168
  derivatives, 83
greenhouse gas (GHG) emissions, 25, **267**, **275**
  supply chain analysis, 293–4
guayule, 13

Haematococus spp., 114
hemicellulose, 173
hevea, 13
hexane, 123
high-density polyethylene (HDPE), **262**
highpressure liquid extraction (HPLE), 113–14
Horizon2020, 332
hydrogels, 152–3
hydrogen, 221–2, **344**, **345**
hydrotreated esters and fatty acids synthetic paraffinic kerosene (HEFA-SPK), 337
1-hydroxybenzotriazole (HBT), 145
3-hydroxypropoanoic acid, **74**, 81–2
3-hydroxybutyrolactone, **75**, 85
5-hydroxymethylfurans, **9**
hydroxymethylfurfural, 81, 122

Idaho National Laboratory, 230
Imperial Chemical Industries (ICI), 231
indigo, 11
Ineos, 19, **22**
infrastructure, 321–2, 332–3

insects, 35
investment, 317–19, 358
drivers for, 353–4
investor education, 320–2
project finance, 358–60
Iogen, **22**
ion exchange adsorption (IEX), 97–8, 119, 124
Iron Triangles, 307, *308*
isobutanol, **266**
conversion efficiency, **295**
fossil fuel consumption, **275**
water consumption, **279**
isoprenoids, 13
isosorbide, **9**, 10
itaconic acid, **75**, 84–5

Kennedy pathway, 180
Kraft pulping, 139–40, 151, 154

lactic acid, 20, 118
land-use change, 260
landscape design, 27, 39
Langmuir adsorption isotherm, 96
Lanzatech, 19, **22**
levulinic acid, **75**, 84–5
life cycle assessment, 15, 259–61, *259*
carbon flows, 272, **273**
feedstock choice, 263–4
inherent energy, 274
methodology, 267–74
polyethylene, 276–8
water consumption, 278–9
lignin, 77, 132–3
autohydrolysis, 141–2
carbon fibers, 153–4
copolymers, 151–2
depolymerization, 141–3
biological, 145–8
catalytic, 143–5
foams and gels, 152–3
functional products, 148–51
Kraft pulping, 139–40
organosolv pulping, 140–1
pretreatment, 133–4
structure, 133–8
lignin peroxidase (LiP), 145, *146*
lignin-based phenols, **19**
lignin-phenol-formaldehyde (LPF) resins, 149
Lignoboost, 139
lignosulfonates, 138–9
lipids, **112**, 116
Lipomyces starkeyi, **174**
Liquid Light Corporation, 231
Liquid–liquid extraction, 100–3, 117–18
batch mode, 100–1
continuous, 101–2
Luria broth, 204
lutein, **112**

malic enzyme, 180
manganese peroxidase (MnP), 145, *147*
mass transfer zone (MTZ), 96
medium density fibreboard (MDF), 148–9
membrane electrode assembly (MEA), 229
membrane separation, 103–8, 120
methane, 80, 274
carbon dioxide and, 219–21
methanol, **74**, 77, 230
methoxymethylfurfural (MMF), 81
methyl-*tert*-butyl ether, 15–16
methyl-*tert*-butyl ether (MTBE), 310
microbial oils, 166–7
carbon sources, 167
agricultural waste, 172–5
challenges, 184
cultivation modes, 175
productivity maximisation
biochemical, 175–8
genetic engineering, 180–2
molecular biology, 178–83
microfiltration, 104–5
microwave-assisted extraction, 114
Miscanthus, 28, 30
nutrient management, 32
pest control, 36

water quality and, 31
water use, 32–3
Mitchell, John, 304
modified ammonium lignosulfonate (MAL), 151
monoacylglycerol (MAG), 183
Monte Carlo models, 341
assumptions, 343–5
correlation matrix, 343–6
results, 346–7
uncertainty profiles, 341–3
Mucor circinelloides, 181
multi-objective multi-period mixed-integer nonlinear programming (MINLP), 288–9
case study, 294–6
solution algorithm, 294
municipal waste water (MWW), 278
My 2030s, 326
Myriant, **22**, **262**

NADPH, 180–1
nanofiltration, 118
National Renewable Energy Laboratory (NREL), 77
natural gas, 264, 311, **344**
NatureWorks, 14
Neal Smith National Wildlife Refuge, 37
Netherlands, 315–16, 324–6
economic strengths, 325–6
economic weaknesses, 326–8
nicotinamide adenine dinucleotide phosphate *see* NADPH
nitrous oxide, 260
North American Industry Classification System (NAICS), 309
nucleation, 99–100

Obama, Barack, 306
oil-field chemicals, 232
olefins, 151–2
one-pot reactions, 86
Open-Bio, 322
OPXBio, **262**
organic acids, 116–21
product recovery, 16–21, 195–7
purification, 191–2
separative bioreactors, 210–12
oxygen, 353–4

Pachysolen tannophilus, 123
packed-bed reactor, *57*, 58
partition coefficient, 102
pest control, 36
petrochemicals, 285, 328
petroleum industry, 310
phenol-formaldehyde resins, 149
phenolic resins, **9**
photobioreactors, 110
phycocyanin, **112**
phycoerythrin, **112**
pine, 144
Pioneer Energy, 220
PlantBottle technology, 258
plastic bags, 333
platform chemicals, 266
PNNL, 85
policy, 356–7
poly-(*N*-isopropylacrylamide) (PNIPAAm), 153
poly(2-hydroxyethyl methacrylate) (PHEMA), 152
polyacrylamide (PAAm), 152
poly(acrylic acid) (PAA), 152
polyacrylonitrile (PAN), 153
polyelectrolyte complex (PEC) fibers, 154
poly(ethyl methacrylate) (PEMA), 152
polyethylene, **275**, 276–8
polyethylene glycol (PEG), **9**
poly(ethylene oxide) (PEO), 154
polyethylene terephthalate (PET), 93, 258
polyethyleneimine (PEI), 151
poly(glucosylethyl methacrylate) (PGEMA), 152
polyhydroxyalkanoic acids, 78
poly(hydroxypropylmethacrylamide) (PHPMA), 152
polylactic acid (PLA), 10, 10–11, **20**

poly(methacrylic acid) (PMA), 152
polysaccharides, 115–16
polytrimethylene terephthalate (PTT), 266
polyvinyl chloride (PVC), 14, 152
poplars, 26
processing, 355–6
Proctor & Gamble, 168
product recovery
  algal biomass, 110–12
  2,5-furandicarboxylic acid, 122
  furans, 121–2
  glycerol, 125–6
  hydroxymethylfurfural, 122
  organic acids, 16–21, 195–7
  xylitol, 123–4
  see also separation technologies; separations technologies
1,3-propanediol, **20**, 266
  applications, **266**
  feedstocks, 264
  fossil fuel consumption, **275**
propylene glycol, **20**
  fossil fuel consumption, **275**
  greenhouse gas emissions, **275**
proteins, **112**, 114–15
Pseudomonas putida, 77
public funding structures, 322–4
public procurement, 332
  European Union, 321–2
pyrolysis oil, 144

Quantiam Technologies, 220

rail transport, **296**
refined lignin, 154
regulation, 324
Renewable Energy Directive (RED), 261
Renewable Fuel Standard, 310
resin-wafer electrodeionization (RW-EDI), 204–10
reverse osmosis, 105–6, 118, 192
Rhizopus orizae, 191
Rhodosporidium toruloides, **170**, 171, 177
Rhodotorula spp., **174**
risk, 358
Robert Gordon University, 220–1
Rotterdam, 325
Roundup, 14

salens, *218*
salicylic acid, 217–18
SDTC, 327
security, 324
Senergy, 83
separation technologies
  adsorption, 94–5, 94–8
  applications, 108–26
  crystallization, 98–100
  electrodeionization, 195–6
  furans, 121–2
  liquid–liquid extraction, 100–3
  membrane
    bipolar membrane electrodialysis, 108–9
    microfiltration, 104–5
  organic acids, 116–21
  sugar alcohols, 122–3
  supercritical fluid extraction, 103
  see also product recovery
separative bioreactors, 190–1
  design, 192–7
  gluconic acid and sorbitol production, 197–201
  integrated fermentation, 201–10
  organic acids, 210–12
  process blocks, *193*
  succinic acid production, 204–8
short rotation coppice (SRC), 34–5
size exclusion chromatography (SEC), 141
SK Innovations, 218
sodium hydroxide, 139
soil
  carbon content, 29
  physical properties, 28–9
  rooting habits, 29–30

solar energy, 232
solid oxide electrolysis cells (SOEC), 230
sorbitol, **74**, 123, 124–5, 197–201
sorghum, 264
soybeans, 12
Spirulina platensis, 168
standards, 319–21
steam reforming, 80
stirred-tank reactor, *57*, 58
succinic acid, **20**, **74**, 79–80, 86, 118, **262**
    applications, **266**
    conversion efficiency, **295**
    fossil fuel consumption, **275**
    greenhouse gas emissions, **275**
    separative bioreactors, anaerobic fermentation, 204–8
    water consumption, **279**
sulfite pulping, 138–9
Sunshine to Petrol, 232
supercritical fluid extraction, 103, 114
supply chains, 285–6, 353–4
    case study, 294–8
    costs, 291–2
    greenhouse gas emissions, 293–4
    model constraints, 289–91
    optimization, 297–9
    process models, *287*
    processing, 289–91
    revenue, 292–3
    transport chains, **396**
sustainability, 329–31
switchgrass, 26, 28
    nutrient management, 32
    soil carbon, 29
    water use, 34
syngas, 77, 220, 222, 227–31
    via high-temperature electrolysis, 230

Taber Alberta, 85
tax policy, 323–4, 333–4
tetrahydrofuran, 86
triacylglyceride (TAG), 116
triacylglycerol (TAG), 180, 182–3
Trichosporon spp., **174**
truck transport, **296**

ultrasound assisted extraction (UAE), 113
Ulva pertusa, 116
uncertainty profiles, 341–3
United States
    bioproducts policy, 304–7
        development strategies, 313
        Farm Security and Rural Investment Act of 2002, 305
        reasons for lack of action, 307–10
    Clean Air Act, 310
United States Navy, 17
University of British Columbia, 232
urea, 216

Vilsack, Tom, 306
Vincotte, 320
vinyl-chloride-vinyl-acetic acid (VC-VAc), 152
Virent, 258

wastewater, 171–2
water consumption, 267, **279**
    life cycle assessment, 278–9
water quality, 31
Watergate scandal, 304
Weizmann, Chaim, 49
willow, 28, 30, 31–2
    water use, 33
World Resources Institute, 25
World War II, 12

xylitol, **75**, 123–4

Yarrowia lipolytica, 168, 177, 181–3
yttria-stabilized zirconia (YSZ), 230

zeaxanthin, **112**
Zymomonas mobilis, 124, 125